SHEEP AND WOOL

SHEEP AND WOOL:
Science, Production, and Management

M. P. Botkin   Ray A. Field   C. LeRoy Johnson
Department of Animal Science
University of Wyoming

PRENTICE HALL, Englewood Cliffs, New Jersey 07632

Library of Congress Cataloging-in-Publication Data

Botkin, M. P., 1922-
  Sheep and wool : science, production, and management / M.P.
Botkin, Ray A. Field, C. LeRoy Johnson.
     p.   cm.
  Bibliography: p.
  Includes index.
  ISBN 0-13-808494-7
  1. Sheep. 2. Wool.   I. Field, Ray A., 1953-   . II. Johnson, C.
LeRoy.   III. Title.
SF375.B64 1988
636.3--dc19   87-30443
                                                                  CIP

Editorial/production supervision and
  interior design: Tom Aloisi
Cover design: Diane Saxe
Manufacturing buyer: Peter Havens

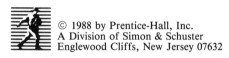

© 1988 by Prentice-Hall, Inc.
A Division of Simon & Schuster
Englewood Cliffs, New Jersey 07632

All rights reserved. No part of this book may be
reproduced, in any form or by any means,
without permission in writing from the publisher.

Printed in the United States of America

10 9 8 7 6 5 4 3 2 1

ISBN 0-13-808494-7 025

Prentice-Hall International (UK) Limited, *London*
Prentice-Hall of Australia Pty. Limited, *Sydney*
Prentice-Hall Canada Inc., *Toronto*
Prentice-Hall Hispanoamericana, S.A., *Mexico*
Prentice-Hall of India Private Limited, *New Delhi*
Prentice-Hall of Japan, Inc., *Tokyo*
Simon & Schuster Asia Pte. Ltd., *Singapore*
Editora Prentice-Hall do Brasil, Ltda., *Rio de Janeiro*

Contents

PREFACE xi

PART I   BASIC CONCEPTS

1   INTRODUCTION   1

Origin and Characteristics of Domestic Sheep   *1*
Sheep Compared to Other Ruminants   *3*
Form and Function   *5*
Life and Growth of Sheep   *7*
Importance and Distribution of Sheep   *9*
Changes and Projected Future   *10*

2   OPPORTUNITIES AND MANAGEMENT SYSTEMS   17

Advantages   *17*
Disadvantages   *21*
Management Systems   *22*

3 PRODUCTION REQUIREMENTS 31

Capital *31*
Feeds and Land *31*
Markets *32*
Labor *33*
Animals *34*
Records *35*
Information Sources *35*
Buildings and Equipment *37*

4 POLITICAL CONSIDERATIONS 53

Public Lands *53*
Support Prices *55*
Foreign Influences *59*
Predator Control *60*
Growers' Organizations *61*

5 BREEDS OF SHEEP 62

Breed Classification *62*
Fine-wooled Breeds *63*
Medium-wooled Breeds *66*
Coarse- or Long-wooled Breeds *81*
Other Breeds *83*

PART II BIOLOGICAL FUNCTIONS

6 REPRODUCTIVE PHYSIOLOGY 85

Anatomy and Physiology of the Reproductive System *85*
Breeding Behavior and Reproduction *94*
Factors Influencing Productivity *97*

Contents vii

7 SHEEP BREEDING AND APPLIED GENETICS 108

Mechanism of Inheritance *108*
Types of Gene Action and Types of Traits *110*
Traits Influencing Productivity and Their Inheritance *114*
Correlations among Traits *122*
Selection *123*
Systems of Breeding *133*

8 SHEEP NUTRITION 144

Nutrients *144*
Requirements *147*
Balancing Rations and Formulating Diets *156*
Composition of Feeds *164*
Providing Feed and Controlling Costs *171*

PART III MANAGEMENT

9 PREBREEDING AND BREEDING SEASON 180

Realizing Present Potential *180*
Increasing Production *194*

10 MANAGEMENT DURING GESTATION 199

Feeding during Gestation *199*
Routine Management Chores *201*
Shearing *205*

11 LAMBING TIME 206

Getting Ready for Lambing *206*
Causes of Death in Lambs *210*

Identification *224*
Feeding Ewes at Lambing Time *224*
Feeding Newborn Lambs *225*

12 MANAGEMENT DURING LACTATION 227

Management after Lambing *227*
Milk Production of Ewes *230*
Feeding Ewes during Lactation *230*
Artificial Rearing *232*
Creep Feeding Lambs *233*
Docking and Castrating Lambs *234*

13 WEANING AND POSTWEANING MANAGEMENT 237

Weaning Time *237*
Early Weaning *238*
Feeding Lambs for Replacements *239*
Finishing Lambs *240*

14 MARKETING 245

Feeder Lambs *247*
Slaughter Lambs *252*
Electronic Marketing *260*
Slaughter Ewes and Rams *262*
Breeding Sheep *263*
Purebred Sheep *264*

PART IV   PRODUCTS

15 MEAT FROM SHEEP 266

Carcass Value *266*
Inspection *269*

Contents

    Grading *270*
    Cuts of Lamb *276*
    Lamb and Mutton Processing *276*
    By-Products *281*
    Lamb—A Healthful Food *284*
    Lamb Consumption *288*

16   LAMB GROWTH AND CARCASS COMPOSITION   289

    Introduction *289*
    Influence of Age and Weight *291*
    Influence of Sex *295*
    Influence of Diet *298*
    Influence of Breed *300*

17   PALATABILITY OF LAMB AND MUTTON   307

    Influence of Age and Weight *308*
    Influence of Sex *310*
    Influence of Diet *312*
    Influence of Breed *316*
    Influence of Frozen Storage *317*
    Vacuum-packaged Fresh Lamb *318*

18   WOOL EVALUATION AND MARKETING   321

    The Wool Situation *321*
    The Language of Wool *323*
    Description of Raw Wool *323*
    Pricing Raw Wool *336*
    Marketing Systems *338*
    Government Influences *340*
    Harvesting Wool *344*

19 WOOL GROWTH AND STRUCTURE 354

Wool Growth *354*
Wool Fiber Structure and Properties *363*
Applications *370*

20 WOOL MEASUREMENT 376

Choice of Method *376*
Subjective Measures *377*
Objective Measures *377*
Quantitative Measures *378*
Qualitative Measures *386*

21 INHERITANCE AND ENVIRONMENT IN WOOL PRODUCTION 395

Breeding for Improved Wool Production *396*
Environmental Effects on Wool Production *412*

22 WOOL SHOWS AND WOOL JUDGING 419

Wool Shows *419*
Wool Judging *428*

GLOSSARY OF WOOL TERMS 438

INDEX 445

Preface

The manuscript for this book was prepared by three authors at the University of Wyoming. The chapters prepared by each were as follows:

Chapters 1 and 4 through 13	M. P. Botkin
Chapters 14 through 17	Ray A. Field
Chapters 2, 3, and 18 through 22	C. LeRoy Johnson

There is no specific section devoted to a discussion of sheep diseases and their control. Diseases are mentioned only as a part of routine management. Recommended books that cover the subject of sheep diseases in detail are as follows:

Gates, N. L. 1984. *A Practical Guide to Sheep Disease Management.* Washington State University, Pullman, Washington,

Jensen, Rue, and Brinton L. Swift. 1982. *Diseases of Sheep.* Lea and Febiger, Philadelphia, Pa.

In addition to the authors, many people have contributed to the preparation of this book by lending encouragement, by helping to make materials such as illustrations available for the manuscript or by reviewing and making helpful suggestions for improvement of the manuscript. To all those who have helped in any way, the authors wish to extend their sincere appreciation.

1 Introduction

ORIGIN AND CHARACTERISTICS OF DOMESTIC SHEEP

Sheep are thought to be the oldest species of domesticated animals. According to Wentworth,[9] the most ancient relics suggest that sheep were the first food animals to arrive when Neolithic man drove back the Cro-Magnons in western Europe about 10,000 B.C. Apparently, sheep were domesticated by that time. The most common assumption as to origin is that domestic sheep are descendants of the Urial of western Asia (*Ovis vignei*) and of the Mouflon of southeastern Europe (*Ovis musimomn*). These species of wild sheep both possess characteristics similar to domestic sheep. A Mouflon ram and ewe are pictured in Figure 1-1, and a Mouflon cross lamb in Figure 1-2.

From the Urial and Mouflon foundation, four main types of domestic sheep have been developed:

1. Fine-wooled sheep in the Mediterranean area
2. Medium-wooled sheep in Europe and England
3. Long-wooled sheep in Europe and England
4. Carpet-wool type of sheep in Asia

The position of domestic sheep among vertebrates of the animal kingdom is shown in the following from Lydekker[4] and Ryder and Stephenson.[6]

Figure 1-1 Mouflon ram and ewe. (Courtesy LeBar Ranch, Douglas, Wyoming)

Class *Mammalia*. mammals, those which suckle young
  Order *Ungulatea*. Hoofed mammals
    Suborder *Artiodactyla*. Even-toed ungulates
      Section *Pecora*. Typical ruminants
        Family *Bovidae*. Hollow-horned ruminants
          Subfamily *Caprinae*. Sheep and goats
            Genus *Ovis*. Sheep
              Species *Ovis aries,* domesticated sheep
                  Ovis ammon, Argali
                  Ovis canadensis, North American Bighorn
                  Ovis orientalis, Urial
                  Ovis laristanica, Urial
                  Ovis musimom, Mouflon
                  Ovis tragelaphus, North African Aoudad
                  Ovis vignei, Asiatic Urial

Figure 1-2 Mouflon × Rambouillet crossbred lamb. (Courtesy LeBar Ranch, Douglas, Wyoming)

Changes that have accompanied domestication are development of tails, development of polled sheep, and presence of fleeces free of the outer hairy coat found in wild sheep. Selection has been the main force in development of polled sheep and in eliminating the outer hairy coat, but development of tails likely resulted from domestication itself since there was no longer a need for adaptation to the natural environment. Horns are still common in some breeds of domestic sheep, and it is fairly common to see traces of the hairy coat in newborn lambs of many breeds.

Within the species *Ovis aries* there are large variations in almost every trait. Color variations range from black to white, with many degrees of variation in between. Black with white face, black or brown with white spots, white with black or brown faces and legs, and white with one or more dark spots are common. Even changes of color within the same fiber are sometimes seen. Horns of varied size and shape are present in some sheep. Wide spreading horns or those growing close to the head are common variations. In early development of polled strains, partial horn development is often seen, and scurs or vestigial horns are common in fine-wooled ewes. In some cases these scurs or partial horns turn inward toward the face and become a problem.

Size of frame and shape of body are variable both among breeds (Figure 1-3) and between individuals of the same breed. Fat-tailed and fat-rumped sheep from other countries present an entirely different appearance than sheep found in the United States. Breed differences still are a major part of the variations in size and shape. Wool is probably the most variable of all, ranging from fairly uniform, fine white wool to long coarse wool, with every possible degree of variation in between with regard to fiber diameter and length. Also, wool is variable in color, crimp, luster, and yield of clean wool.

Undesirable traits or defects are still a common problem in some sheep and represent variations among individuals as well as among breeds. For example, wool blindness and body folds or wrinkles continue to plague breeders of fine-wooled sheep in particular, and dark-colored wool fibers or fleeces show up in many flocks.

SHEEP COMPARED TO OTHER RUMINANTS

Ruminants are of the section *Pecora* and are characterized by their digestive system. Cattle, sheep, and goats are all in this classification. They do not have any incisor teeth in the upper jaw, but rather a dental pad, and they eat by manipulating the incisor teeth of the lower jaw against the pad with help from the lips and tongue. Their stomach is made up of four compartments: rumen, reticulum, omasum, and abomasum or true stomach. This allows consumption of large amounts of roughages, which are digested with the aid of rumen microorganisms.

Compared to goats. Sheep resemble goats in many respects, but are distinct and do not cross. The number of chromosomes is different. Male sheep do not have beard hairs growing from the chin as do male goats, and generally wool and horn

Figure 1-3 Lambs prepared for show, contrasting in size, shape, and color. (University of Wyoming photo)

characteristics of sheep differ from hair and horns of goats. However, some sheep have hairlike wool similar to mohair from Angora goats, and others tend to have short hair similar to that of dairy goats. Tails of goats are short and usually upright, while sheep's[2] tails are down. Moreover, goats are more susceptible to exposure or stress than are sheep.

Compared to cattle. Sheep have many similarities to cattle in form and function, yet differ markedly in many respects. Sheep are smaller than cattle, and they differ in breeding habits (sheep are seasonal breeders) and grazing habits. Sheep have a narrow muzzle and the shape of the external mouth parts, in particular the cleft upper lip, allows them to graze more selectively and closer to the ground.

The skin of sheep is much thinner than the hide of cattle, ranging from 1 to 3 mm in thickness. It ranges from a pink color in white-faced sheep to gray or black in those with black faces and legs. Sheep skin is easily torn. Skin provides an external protective covering, is the base for production of wool fibers and is embedded with glands (sweat glands and sebaceous glands) that help in regulation of body temperature. Sebaceous glands secrete the yolk or oily material that gives raw wool its oily feel.

FORM AND FUNCTION

General shape or conformation is determined largely by the shape of the skeleton or bone framework (Figure 1-4). Outward appearance is modified by muscle growth and fat deposits. The shape of the head is almost completely determined by bones since muscles and fat in this area are not very bulky. The neck is comprised of seven cervical vertebrae that are irregular in shape and very flexible. Differences in length of the neck are probably due to different bone lengths. Outward appearance of the neck is influenced by muscle development. In rams, muscle development along the neck is much more extensive than in ewes or wethers. The thoracic vertebrae normally number 13, but some variations occur. To these vertebrae are attached 13 pairs of ribs, with some variations, since 12 or 14 have been reported. Eight pairs of ribs are attached by individual cartilages to the sternum (breastbone). Four pairs (asternal ribs) are attached to each other by cartilage, but not directly to the sternum. The last pair of ribs is not attached to the sternum and is often called the floating pair of ribs.

Spinal processes also are attached to the thoracic vertebrae and extend upward. Variations in length of these spinal processes influence to some extent the shape of the back. Width over the back is influenced more by the covering of muscles and fat than by spread of the ribs, but depth of body is more influenced by the bone framework. Lumbar vertebrae extend over the loin area, and generally there are six or seven. This area too is largely influenced by muscle and fat in determining apparent shape. Four sacral vertebrae are fused into one bone to make up the sacrum, the lower side of which forms the upper wall of the pelvis. The ilium is attached to the sacrum by a joint with little movement and is the point of the hip. Coccygeal vertebrae form the skeletal base of the tail, and their number varies. In

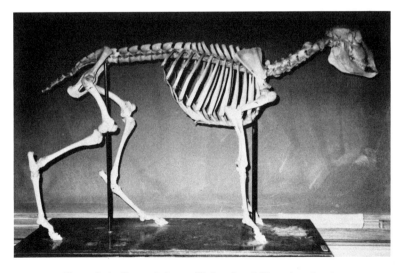

Figure 1-4 Sheep skeleton. (University of Wyoming photo)

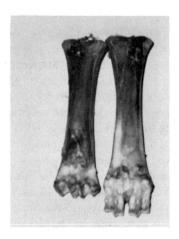

Figure 1-5 Break joint (left) versus spool or round joint (right). This is one way to estimate age in the meat trade. The forelegs of lambs are severed at the break joint. (University of Wyoming photo)

some tailless or short-tailed sheep there may be four to six, while in long-tailed sheep there are twenty or more.

Bones of the legs make up the remainder of the skeleton. Forelegs do not have a rigid connection to other skeletal parts, but the bones of the hind legs do. Leg bones vary in length, thickness, and placement. Length of leg bones along with depth of body is responsible for differences in height. Just above the hooves in the

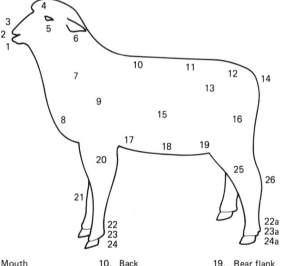

1. Mouth
2. Nose
3. Face
4. Forehead (Poll)
5. Eye
6. Ear
7. Neck
8. Breast
9. Shoulder
10. Back
11. Loin
12. Rump
13. Hip
14. Dock
15. Side
16. Thigh
17. Fore flank
18. Belly
19. Rear flank
20. Foreleg
21. Knee
22, 22a. Dew claw
23, 23a. Pastern
24, 24a. Hoof
25. Hind leg
26. Hock

Figure 1-6 Parts of the sheep. (Drawn from University of Wyoming photo)

leg bones are the epiphyseal cartilages, which are known as break joints in young lambs and become ossified as lambs mature (Figure 1-5). External parts of the sheep are shown and labeled in Figure 1-6.

A knowledge of skeletal shape and external shape is helpful in understanding the differences in the value of wholesale cuts from the different parts of the carcass.

Body temperature of sheep normally ranges from 101° to 103.8°F, with the average being 102.3°F.

Sheep are gregarious, tending to stay together in groups and to behave as a group with regard to grazing, resting, and seeking protection from sun, wind, or storm. This trait is much more pronounced in sheep of the fine-wooled breeds than in other breeds. They band together more closely and show less individuality of behavior. It is this gregarious nature that allows one person to herd or watch over large flocks.

LIFE AND GROWTH OF SHEEP

At birth lambs range in weight from 3 to 20 lb. with an average of 10 to 12. This varies with age of ewe, type of birth, breed, and level of nutrition. Skeletal structure is more nearly developed at birth than is trunk depth or width. Growth of sheep, as of all species, follows a typical S-shaped curve, as seen in Figure 1-7.

Early growth is normally faster than growth later in the lamb's life. Likewise, this early growth is the most efficient. In some situations, lambs are fed so as to reach market weight by the time they are weaned, but in many cases lambs are still considered feeder lambs at this time. Weaning age is extremely variable in today's sheep business, ranging from one day to 4 or 5 months. Early weaning is essential in accelerated programs, and in most operations there are advantages to weaning

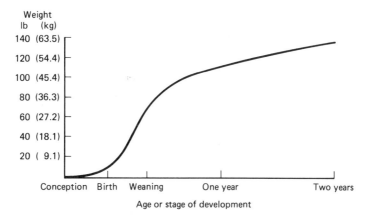

Figure 1-7 Pattern of growth in sheep.

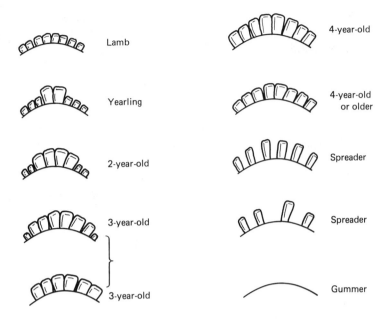

Figure 1-8 Changes in incisor teeth by advancing age.

lambs early. However, in traditional types of sheep operations, all advantages are more than offset by reduced utilization of native feeds.

Postweaning growth is also variable. Many lambs are placed in feedlots and finished as quickly as possible, whereas others are on deferred programs that use cheap feeds (crop aftermath) and are then moved to feedlots. Feeding and management of replacement lambs is just as variable. Many are fed to reach maturity as fast as possible, and many are allowed to graze. In range areas particularly, both age at puberty and age at mature size are delayed. Thus feed conditions and methods of management often have a greater influence on growth than does the inherent ability to grow. In spite of all variations, mature size is usually reached at 2 years. Once mature size has been reached, any fluctuations in size or weight are the result of differences in body fat or condition as influenced by feed or by stage of the productive cycle.

There are several indications of age in sheep. In most cases young lambs can be distinguished from older lambs, and definitely lambs can be distinguished from mature sheep simply by observations as to size, activity, and characteristics of the wool tip (whether or not wool has been shorn). As sheep become older, beyond 6 or 7 years of age, their general condition and thrift tend to decline, although many cases have been noted of ewes producing and maintaining good condition and health for as long as 11 or 12 years in desirable environments.

Determining approximate age of sheep by examining their teeth is common among sheep producers. Sheep have 20 temporary teeth, but have 32 permanent teeth. Eight of these are incisor teeth in the front of the lower jaw, and it is these that can reveal the sheep's age. By parting the sheep's lips, one can observe the number of permanent incisor teeth and of temporary or lamb teeth, as well as correctness of teeth placement and soundness of mouth in older sheep. Figure 1-8 shows examples of differences in incisor teeth as sheep advance in age.

IMPORTANCE AND DISTRIBUTION OF SHEEP

The sheep industry is worldwide and is of great importance in some countries (Tables 1-1 and 1-2). Australia and the Soviet Union are the leading countries, and both are much ahead of China, which ranks third, in sheep numbers. These three countries, along with New Zealand, Turkey, and India, have about two-thirds of the world's sheep population. The United States ranks nineteenth among world countries, with 10.4 million head as of January 1985.

In the United States, sheep were brought into the early Colonies almost as soon as they were settled.[9] Sheep were brought to Jamestown, Virginia, in 1609 and were taken into Maryland at its first settlement. Sheep were brought to New York from Holland in 1725 and Swedish sheep were brought to New Jersey and Delaware. The first Spanish Merino flock was established in 1801. An interesting detailed account of the early history of sheep in the United States has been prepared by Wentworth.[9]

The earliest estimate of total numbers of sheep in the United States was about 7 million in 1810. By 1840 there were over 19 million, with about 60% in New England and the mid-Atlantic states. After this the westward movement began. Numbers increased to over 53 million head in 1844. By 1908 there were about 48 million, of which over half were in the western states and Texas. Sheep numbers declined for 15 years and then quickly increased to 50 million head in 1930.[7] Numbers remained high and the peak was reached in 1942 with 56 million head. Numbers

TABLE 1-1 WORLD SHEEP POPULATION

Region	Number of Sheep, 1985 (Millions)
Africa	192
North and Central America	19
South America	102
Asia	311
Europe	133
Oceania	220
USSR	142
World	1,121

From Devendra and Coop.[2] and *FAO Production Yearbook*[3]

TABLE 1-2 LEADING COUNTRIES IN SHEEP NUMBERS

Country	Sheep Numbers, 1985 (1000 head)
Australia	149,747
USSR	142,876
China	95,191
New Zealand	70,600
India	41,300
Turkey	40,391
South Africa	30,256
Argentina	29,000
United Kingdom	23,946
Ethiopia	23,500
Uruguay	20,600
Sudan	19,000
Romania	18,637
Algeria	18,000
Spain	17,485
Nigeria	12,800
Morocco	12,000
France	10,824
Bulgaria	10,500
United States	10,443
Italy	9,500

[a]From *FAO Production Yearbook*[3] and USDA *Agricultural Statistics*.[8]

rapidly declined for the next 8 years, stayed fairly constant for another 13 years, and since 1963 sheep numbers have declined steadily (Table 1-3).

Regardless of total numbers of sheep for the last 30 years, about two-thirds of them were in the western states and Texas, as can be seen in the 1984 census numbers (Table 1-4). Sheep are of relatively minor importance in the United States compared to cattle and swine, but in some states sheep remain an important part of the economy. Numbers of sheep in a particular state do not reflect the number of sheep producers (Table 1-4), as there are more producers in some states with relatively low sheep numbers than in western states with high sheep numbers. As is true on a world basis, sheep are more numerous in states of vast land areas and relatively sparse human population.

CHANGES AND PROJECTED FUTURE

The most significant change in the U.S. sheep industry during past years has been the decrease in sheep numbers, as shown in Table 1-3. This change is shown graphi-

TABLE 1-3 SHEEP NUMBERS AND PRODUCTION BY YEAR

Year	All Sheep and Lambs (1000)	Ewes, 1 Year and Older	Lamb Crop	Lambing Percent	Sheep and Lambs Shorn	Fleece Weight (lb)
1985	10,443	7,288				
1984	11,487	7,874	7,772	99	12,015	7.74
1983	11,904	8,165	8,158	98	12,365	8.00
1982	12,966	8,788	8,499	97	13,199	8.04
1981	12,936	8,771	8,825	101	13,498	8.14
1980	12,687	8,524	8,249	97	13,255	7.95
1979	12,365	8,366	7,974	98	13,068	8.02
1978	12,421	8,588	7,927	94	12,719	8.09
1977	12,722	8,850	8,573	97	13,217	8.12
1976	13,311	9,314	8,888	95	13,536	8.21
1975	14,515	10,083	9,587	98	14,403	8.30
1974	16,310	11,058	10,509	95	15,956	8.23
1973	17,641	12,049	11,500	95	17,425	8.25
1972	18,789	12,909	12,599	97	18,770	8.44
1971	19,731	13,609	12,998	95	19,036	8.41
1970	20,423	13,923	13,465	96	19,163	8.43
1969	21,350	14,707	13,723	93	19,584	8.46
1968	22,223	15,290	14,443	94	20,759	8.55
1967	23,898	16,218	15,040	93	21,982	8.56
1966	24,734	16,850	15,881	94	22,923	8.51
1965	25,127	17,502	16,312	94	23,756	8.48
1964	29,176	18,723	16,994	92	25,455	8.34
1963	30,969	20,028	18,516	94	27,264	8.53
1962	32,725	21,252	19,712	94	29,193	8.45
1961	33,170	22,199	20,782	95	30,454	8.51

From USDA *Agricultural Statistics*.[8]

cally in Figure 1-9. Currently, meat from sheep represents less that 1% of annual per capita consumption of red meat, and wool makes up less than 2% of annual per capita fiber consumption in the United States.

Causes of the decline have been studied by many groups through surveys and in conferences, and programs have been devised to reverse this trend, but as yet none of the programs has been successful. Sheep numbers remain at a low level and sheep industry leaders are still striving to find new solutions to their problems.

Other changes tend to indicate improved production and to lead to some optimism for the future. Average size of sheep has increased and average weight of slaughter lambs has continually increased. See Figures 1-10 and 1-11. Also in very recent years there seems to be a decided upward trend in production per head (see Figure 1-12). Introduction of Finnish Landrace or Finn sheep and testing of other

TABLE 1-4 ALL SHEEP AND LAMBS AND OPERATIONS WITH SHEEP BY STATE, 1985

State	All Sheep and Lambs		Number of Operations with Sheep
	1000 Head	Rank (T = tie)	
Alaska	2.9	37	30
Arizona	266.0	11	500
California	1,065.0	2	6,000
Colorado	675.0	4	2,500
Connecticut	9.2	34T	440
Idaho	313.0	10	2,500
Illinois	136.0	17	6,200
Indiana	110.0	20T	5,200
Iowa	360.0	9	10,500
Kansas	245.0	14	2,600
Kentucky	27.0	28	900
Louisiana	9.2	34T	700
Maine	19.0	29	680
Maryland	16.0	30	1,000
Massachusetts	11.0	33	600
Michigan	110.0	20T	2,700
Minnesota	255.0	13	6,000
Missouri	123.0	19	3,900
Montana	515.0	7T	2,900
Nebraska	165.0	16	3,400
Nevada	100.0	21	270
New Hampshire	12.1	32T	500
New Jersey	12.0	32T	700
New Mexico	538.0	6	1,500
New York	57.0	26	2,200
North Carolina	8.4	36	500
North Dakota	215.0	15	2,000
Ohio	265.0	12	7,400
Oklahoma	85.0	23	2,300
Oregon	445.0	8	5,400
Pennsylvania	88.0	22	4,900
South Dakota	639.0	5	5,400
Tennessee	9.0	35	470
Texas	1,810.0	1	8,800
Utah	515.0	7T	2,500
Vermont	15.0	31	630
Virginia	125.0	18	2,800
Washington	50.0	27	2,200
West Virginia	76.0	25	2,700
Wisconsin	81.0	24	3,500
Wyoming	860.0	3	1,300
United States	10,442.7		117,220

The ten high states in numbers have 71.1% of the sheep and 39.9% of the operations.

From American Sheep Producers Council.[1]

Changes and Projected Future

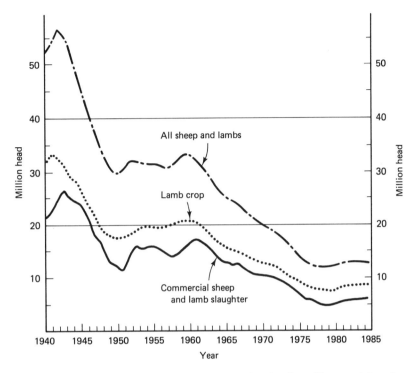

Figure 1-9 Sheep inventory, lamb crop, and number slaughtered by year. Adapted with permission from Parker and Pope.[5]

breeds as sources of new genetic material lead to the belief that this upward trend in production per head will continue.

Increased use of the technology already available, as well as further developments through research, is a requirement for any projected improvements in productivity or for significant increases in numbers.

Regardless of the past history and the continual decline in sheep numbers in the United States, the sheep remains an important agricultural resource. Sheep are unsurpassed in their ability to convert large quantities of forages into food and fiber for human consumption. Therefore, it is logical that sheep can be productive while using feeds that are not required for human use.

Most predictions for the future of the sheep industry in the United States foresee an increase in the proportion of farm flock production, with a corresponding decrease in the proportion of strictly range operations. These predictions are based largely on past performance and political considerations. A discussion of politics and the sheep business is included in Chapter 4.

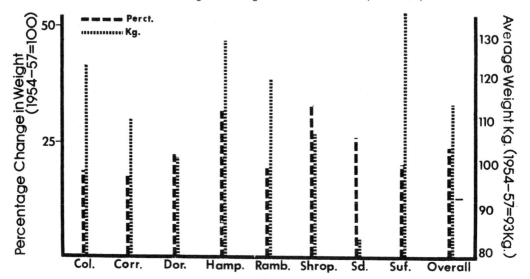

Figure 1-10 Percentage and actual weight averages for top yearling rams of eight purebred breeds at the Ohio State Fair, 1954 to 1982. Reproduced with permission from Parker and Pope.[5]

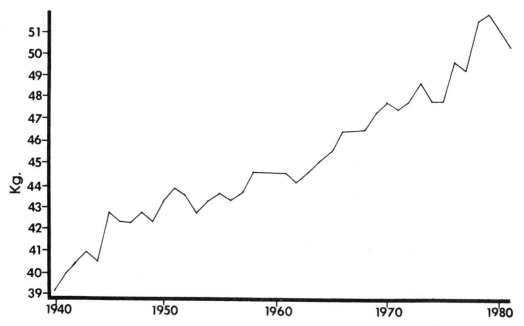

Figure 1-11 Average live weight of sheep and lambs slaughtered under federal inspection. Adapted with permission from Parker and Pope.[5]

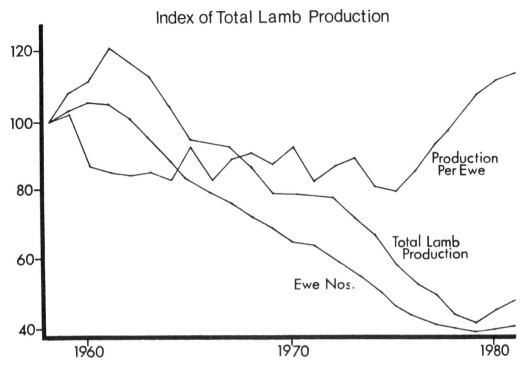

Figure 1-12 Total lamb meat production index, numbers (1 year +) and production per ewe, 1958 to 1981. Reproduced with permission from Parker and Pope.[5]

REFERENCES CITED

1. American Sheep Producers Council. 1985–86. *Situation Outlook Report.* Sheep Industry Development Program. Denver, Colo.
2. Devendra, C., and I. E. Coop. 1982. *World Animal Science C-1.* Elsevier Science Publishing Co., New York.
3. Food and Agriculture Organization of the United Nations. 1985. *FAO Production Yearbook,* Vol. 39, Rome.
4. Lydekker, R. 1913. *The Sheep and Its Cousins.* E. P. Dutton, New York.
5. Parker, C. F., and A. L. Pope. 1983. The U.S. sheep industry: Changes and challenges. *J. Animal Sci.* 57, Supplement 2, p. 75.
6. Ryder, M. L., and S. K. Stephenson. 1968. *Wool Growth.* Academic Press, New York.
7. Terrill, C. E. 1958. Fifty years of progress in sheep breeding. *J. Animal Sci.* 17:944.
8. U.S. Department of Agriculture. *Agricultural Statistics.* Agricultural Marketing Service, Washington, D.C.
9. Wentworth, Edward N. 1948. *America's Sheep Trails.* Iowa State College Press, Ames.

SUGGESTED READING

Ensminger, M. E. 1970. *Sheep and Wool Science,* 4th ed. Interstate Printers and Publishers, Danville, Ill.

Kammlade, William G., Sr., and William G. Kammlade, Jr. 1955. *Sheep Science.* J. B. Lippincott Co., Philadelphia.

2 Opportunities and Management Systems

Sheep production is frequently described in terms of its "opportunities" or "limitations" with major emphasis on either. Both negative and positive experience, or lack of experience contribute to these apparently biased analyses. However, people who have raised sheep recognize that elements of both approaches are always present. The more capable managers are those who minimize the disadvantages while exploiting the inherent advantages of sheep production.

ADVANTAGES

Evaluation of opportunities and problems associated with sheep production is usually made in comparison to other red-meat-producing species and particularly to cattle, because both depend on forages as a primary feed source.

Two Crops per Year

Except for lamb feeding operations, sheep production provides a crop of lambs and a crop of wool each year (Figure 2-1). This is a particular advantage in terms of annual cash flow. The relative importance of each is directly related to the efficiency of lamb production in most commercial operations (Table 18-1). However, certain specialized systems may emphasize either crop, and range operators consider wool as a protective factor along with its monetary value.

Figure 2-1 Sheep production offers a crop of lambs (A) and wool (B) each year. (University of Wyoming photos)

Feed Sources

Because of a cleft upper lip, which lets sheep graze or eat harvested feeds very selectively (Figure 2-2), and their natural preference for broad-leaved plants, a wide variety of forages can be utilized for sheep production. Aftermath, such as cereal grain kernels lost during harvesting along with stubble and straw, sugar beet tops, and corn fodder represent feed sources that are often wasted but can be harvested

Figure 2-2 Sheep can eat and actually prefer the broad leaves of browse and forbes. (University of Wyoming photo)

by sheep (Figure 2-3). Many farms have nontillable areas, such as steep hills, under trees, or along drainage systems, that can be grazed by sheep.

Browse and forbes on native ranges are more palatable and/or more readily eaten by sheep than by cattle, which has led to the suggestion that deer and antelope compete more directly with sheep than cattle for a large portion of the forage on these areas. A sheep's preference for broad-leaved plants is probably largely responsible for the observation that the carrying capacity of mixed-plant pastures or ranges is significantly greater when cattle and sheep are grazed together or in sequence than when either species is grazed singly. Also, sheep can graze in rougher terrain than cattle, reaching plants normally not utilized except by game animals.

When considering feed sources, the biggest advantage for sheep as compared to other domestic animals is the small proportion of concentrated feeds required. Grazed or harvested forages can provide all the nutritional requirements of ewes during most of a typical production year.[5] Cereal grains and/or protein supplements are normally offered to ewes in relatively small amounts for short periods at breeding, late gestation, and early lactation (Chapter 8). Many lambs reach market weight and quality without ever receiving concentrated feeds.

Figure 2-3 Aftermath of cash crops can provide a good source of feed for sheep. (University of Wyoming photo)

Reproductive Rate

In terms of reproduction efficiency, the natural potential for multiple birth (Figures 2-4 and 7-23) combined with a relatively short gestation (Chapter 6) places sheep well ahead of cattle. Although weaning rates reported across the United States are disappointing when compared to realistic potentials, many producers are reaching goals of 1.4 to over 1.8 lambs weaned per ewe mated on an annual lambing basis. Also, some commercial operations have developed systems for accelerating production by lambing twice per year or three times every two years (Chapter 9), resulting in weaning rates of two or more lambs per ewe per year.

Investment Return Rate

The short gestation length of ewes combined with the rapid growth rate of lambs allows for relatively quick return on investments in commercial ewe flocks. It is possible to buy open ewes in the fall and have a wool check plus a lamb check in less than one year. Turnaround time for feeding lambs is also relatively short, ranging from 60 to 90 days, for the majority of lambs fed in traditional feedlots. Longer periods are required in systems where aftermath is utilized at the beginning of the feeding period.

Management Flexibility

There are a wide variety of systems for managing sheep, ranging from very intensive to very extensive and involving a dozen or less to several thousand head. This suggests that sheep production might well be considered as either a primary or secondary enterprise for most farm or ranch units.

Figure 2-4 The natural potential for multiple birth in sheep offers excellent reproductive efficiency. (University of Wyoming photo)

DISADVANTAGES

As for many agricultural commodities, the production of lamb and wool requires dealing with some serious challenges, usually described as disadvantages.

Predation

Probably due largely to their fearful nature and their relative size, sheep are vulnerable to destruction by predators (Chapter 11). Among the wild species, coyotes are especially troublesome, along with fox, lions, bobcats, eagles, and bears. Attacks by uncontrolled dogs (Figure 2-5) are a familiar problem to sheep raisers in all areas of the United States. Predation is the most frequent reason cited for discontinuing sheep production and is probably a primary factor associated with the decrease in sheep numbers in the United States (Chapter 1).

Weather and Climate

Extremes in weather conditions pose special problems for sheep producers, particularly in range systems where housing is normally unavailable. Freshly shorn ewes and newborn lambs are especially vulnerable to temperature extremes. Reports of major losses due to snow storms (Figure 2-6) are an annual occurrence, and losses due to extremely hot temperatures or heavy rains occur frequently.

Humid, wet climates are challenging to sheep producers because of associated skin problems, highly succulent forages, and internal parasite infestation. On the other hand, feed and water shortages due to droughts and excessive accumulations of sand in fleeces are reported frequently in the drier areas of the United States. Excessive snow cover on ranges saved for winter grazing are reported almost annually somewhere in the Intermountain West.

Figure 2-5 Sheep killed by dogs. (University of Wyoming photo)

Figure 2-6 Snow storms can cause serious losses of young lambs in flocks lambing with only natural protection. (University of Wyoming photo)

Labor

It is generally agreed that the annual labor requirement per animal unit is somewhat higher for sheep than cattle. In areas where flocks are managed without fences on open ranges, a lack of qualified herders is a constant problem. Lambing is seasonal in most operations, and the availability of skilled workers to help with lambing is extremely limited. On the other hand, off-season labor can be utilized for lambing, particularly in farm flock systems, in which case the availability of labor should be described as an advantage for sheep production.

MANAGEMENT SYSTEMS

The adaptability of sheep, particularly from the standpoint of breeds (Chapter 5), to a wide variety of environments and marketing and management systems suggests that sheep production can be included as a rewarding effort in many different situations.

Purpose

Before beginning a new sheep production enterprise, expanding an existing one, or changing a management system, producers are encouraged to establish their goals. In most cases, *profit* is the primary purpose. However, many small flocks are established for personal satisfaction.

From a management standpoint, purposes for specific units should be more clearly defined than merely noting the importance of profitability. Examples of specific purposes are as follows:

Management Systems

1. To utilize crop aftermath that would normally be wasted.
2. To provide gainful employment for labor during periods when other farm work is reduced.
3. To activate existing farm buildings that are currently not used.
4. To market low-quality cereal grains and by-products of human food manufacturing systems such as sunflower hulls or sugar-beet pulp.
5. To market forages required in crop rotations.
6. To market forages that cannot be harvested mechanically.

Many other purposes for raising sheep have been identified. Whatever they are, it is essential they be clearly defined and used as guides for constructive management decisions.

Range

Flocks classified as "range" imply relatively large numbers. However, attempts to describe management systems in terms of flock size may be misleading. A more correct approach is to define range flocks as those for which the majority of the annual feed supply is from standing forage[3], which normally cannot be harvested mechanically or is more efficiently harvested by grazing (Figures 2-7 and 12-5). Range flocks are further classified as fenced or migratory. Both systems typically supplement grazing with either hay or concentrated feeds for short periods during flushing (breeding), pre- and postlambing, and when required because of drought or heavy snow cover.

Range flocks usually represent a primary rather than supplemental source of income for the units involved and are located mostly in the western states and Texas.

Figure 2-7 Sparse mixed-plant ranges are harvested efficiently by sheep. (University of Wyoming photo)

They represent a minority of operations with sheep but produce a distinct majority of the U.S. lamb and wool crop[1,2] (Table 1-4).

Although some range sheep production systems include sheds for lambing, most of them lamb without constructed shelter and facilities such as corrals and jugs. In these situations, the only protection from weather extremes is natural (brush, trees, and topographical features).

The white-faced breeds (usually referred to as *wool* breeds) predominate in range flocks. They are hardy and excellent foragers and their gregarious nature aids in herding and handling. Their fleeces are essentially bright, free of colored fibers, fine to medium in grade, and relatively free of vegetable matter. The bulk of the lambs are sold at weaning as "feeders" weighing 70 to 90 lb (32 to 41 kg), but there may be a proportion that are properly finished and heavy enough to go directly to slaughter in years when range forages are abundant.

Where range operations have meadows or irrigated units, weaned lambs may be held for short periods to utilize regrowth on areas previously harvested for hay or aftermath of cash crops before moving to feed lots. Some ranchers prefer to retain ownership of their lambs until slaughtered rather than sell them as feeders (Chapter 14); they do so by contract arrangements with commercial feedlots or by establishing their own feedlots.

Farm Flock

Ewe flocks that are managed in relatively small numbers in the "farming areas" of the United States are described as "farm flocks" (Figure 2-8). Such flocks average less than 40 head and produce only about one-third of the American sheep and wool crop, but represent a large majority of producers (Table 1-4). Traditionally, farm flocks are used to provide a supplemental source of income to other agricultural or

Figure 2-8 A typical small farm flock. (University of Wyoming photo)

even nonagricultural enterprises. They may be grazed during all or part of the growing season or confined year-around in corrals adjacent to shelter and fed harvested feeds.

In areas where forage production is high, per acre production of lamb and wool can be increased by harvesting forages mechanically and feeding sheep in confinement. This approach avoids waste by tramping, permits harvesting when plant growth is optimum, and allows for control of feed intake to meet recommended levels for the various production stages.

Accelerated lambing systems are increasing in the United States and may represent a primary income source, but they are usually classified as farm flocks because of the intensity of management required. To be successful, they require a great deal of management ability.

A wide variety of "management systems within a system" are represented by farm flocks. It is this flexibility that helps make sheep production attractive to many farmers, part-time farmers, and urban workers with acreages. Breeds used vary as much as methods of handling them, but the meat breeds and their crosses tend to predominate.

Except in accelerated systems, lambing occurs anytime between early December and mid-June, depending on the availability of housing, labor, and feed. Although wool-market options are often inadequate (Chapter 18), there are many market opportunities for lambs produced (Chapter 14).

Attempts have been made to establish guidelines for farm flock owners regarding number of ewes required for an "adequate economic production unit." [2] A minimum of 30 to 50 ewes has been proposed to justify the purchase of a good quality ram. However, a minimum of 200 to 250 ewes has also been suggested, because even modest equipment costs required for up to 50 head might well be used for 200 or more. Because available land, labor, housing, feed sources, and other factors are so extremely variable in farm flock management systems, it is recommended that *commercial farm flocks should be large enough to be of economic importance to the unit and therefore not neglected.*

Purebred

Purebred flocks represent the primary sources of seedstock for American sheep production at all levels. Mating must be carefully controlled, and breeding records require identification of all animals so they are usually managed similar to farm flocks. However, purebred flocks differ from commercial farm flocks in terms of their marketing options. Rather than utilizing the standard channels available (Chapter 14), they must, in a sense, create their own markets by calling the attention of prospective buyers to their breed and the special merits of each flock, line of breeding, and so on.

The biggest profit potential for purebred flocks is from the sale of rams. Well-grown ram lambs are quite acceptable for farm flocks, but range operators usually prefer to buy rams as "yearlings" or older. This requires carrying them up to 18

Figure 2-9 Participation in sheep shows is one way of advertising breeding animals. (University of Wyoming photo)

months, and an important consideration is when and how to evaluate important traits and estimate relative breeding merit (Chapters 7 and 21).

The purebred sheep business is very competitive, and creating a demand for animals from individual flocks requires a great deal of effort. Paid advertising in popular sheep publications, show ring exposure (Figure 2-9) and winnings, farm field days, performance testing, participation in local, state, and national auctions, and personal contact are some of the methods that have been shown to be effective.

Many beginners in the purebred sheep business have been disappointed because their rams have not sold as well, privately or at public auction, as those from the more established flocks. These breeders need to recognize that, during the years they are establishing their reputations, they may show more profit by selling the bulk of their rams as market lambs than by entering them in the "ram market." As their breeding programs develop and their visibility increases, they can expect to sell greater proportions of the rams they produce for breeding purposes at competitive prices.

Lamb Feeding

Lamb feeding (Figures 13-1 and 13-2) is an excellent way of marketing harvested forages, cereal grains (often of qualities unacceptable for humans), and by-products of human food and beverage manufacturing systems. The *traditional* commercial lamb feedlot is designed to grow lambs weighing 65 to 90 lb (30 to 40 kg) from light and/or underfinished to "not-too-fat" but more acceptable slaughter weights of 100 to 120 lb (45 to 55 kg) (Chapter 14 and 15). Because it is more economical to haul lambs to feed than feed to lambs, they are usually located in areas where feed supplies are abundant. Locations within a few hundred miles of slaughtering establishments are preferable.

Feed lots range in size from a few hundred head to provide a market for home-grown feeds, which are often handfed, to highly mechanized operations involving upward from 100,000 head annually. To maximize use of facilities, labor, and equipment, commercial operations attempt to keep their lots filled to capacity all year long. However, lambs suitable for feeding are seldom available in adequate numbers during late spring and early summer.

Cereal grains usually represent the primary feed source, combined with relatively small amounts of harvested forage and protein supplements. However, lambs have been shown to gain well on balanced diets ranging from nearly all roughage to all concentrates, so the types of feeds used vary widely between and within years based on cost–quality analysis of those most readily available. A typical procedure is to *start* lambs on rations that are relatively high in forages (and usually cheaper) and then increase the proportion of concentrated feeds as they become adjusted to the lot and approach slaughter weight.

Lambs in the larger commercial lots are either owned by the operator or are fed on a *custom* basis, with those who raised the lambs retaining ownership until slaughter. The most frequent feeding arrangements are based on a price per unit of gain basis or a charge for amounts of selected rations actually fed. Both include yardage and handling fees, with specific agreements concerning death loss. Owned or custom fed, break-even price estimates (Table 2-1) are useful for evaluating purchase price, sale price, and feed-cost relationships.

Many aspects of lamb feeding are considered "art" rather than "science" because managers must be quick to recognize how well their lambs are performing (doing). If they are not gaining weight as rapidly as expected or appear unthrifty, the cause must be identified and corrected promptly. Also, lambs entering feedlots may come from widely different backgrounds and thus may require different feeding, health, and general management programs. For example, they may be light, unshorn, newly weaned and strictly grass fed from either range or farm flocks. On the other hand, they may be much older, shorn, vaccinated, drenched, and nearly ready for slaughter from grazing aftermath such as beet tops, regrowth of meadows

TABLE 2-1 BREAK-EVEN PRICES FOR SLAUGHTER LAMBS IN LAMB FEEDING ENTERPRISES[a,b,c]

Net Cost/Cwt. Delivered at Lot	Approximate Feed Costs/Lb. (454 g) of Gain			
	$0.40	$0.50	$0.60	$0.70
	Sale price/Cwt.			
$70.00	$60.90	$64.55	$68.20	$71.80
65.00	57.75	61.35	65.00	68.00
60.00	54.55	58.20	61.80	65.45
55.00	51.35	55.00	58.65	62.25
50.00	48.20	51.80	55.45	59.10
45.00	45.00	48.65	52.25	55.90
40.00	41.80	45.45	49.10	52.70
35.00	38.65	42.25	45.90	49.55

[a]Calculated for lambs delivered at 70 lb (32 kg) and sold at 110 lb (50 kg).

[b]Includes feed costs plus $2.00 per head for veterinary, building, equipment, and miscellaneous expenses, which may vary greatly among lots.

[c]Does not include labor and management costs.

that have been harvested for hay, or waste kernels and forage left in corn fields harvested for grain.

Specialty

Specialized sheep management systems are seldom large enough to be classified as primary sources of income and are usually managed to meet the demands of very selective markets. A few examples are:

1. The production of *hot-house lambs* weighing 30 to 60 pounds (13.5 to 27 kg) and sold primarily in the Boston and New York areas between Christmas and Easter is a typical "specialized sheep production management system." It is a seasonal market but can be very profitable for producers in those areas. Out-of-season lambing is required, so Dorset, Rambouillet, and Merino ewes are preferred and usually mated to "meat" breed rams.
2. There is a growing interest among handcrafters in using natural-colored fleeces for spinning, weaving, and knitting. Although the market is not large, prices paid for these fleeces are very high compared to those sold through normal channels (Chapter 18). Small flocks producing fleeces ranging from brown to black to gray to spotted can provide good supplemental income with small investments in sheep, equipment, feed, and labor.
3. *Club lamb* producing for sale to youth involved in 4-H and FFA projects is becoming an increasingly popular sheep management system. It is much more widespread in the United States than other systems designed for selective markets. Club lambs are usually produced in small flocks or small parts of larger purebred or commercial flocks and bred specifically for this purpose. Many breeds and breed combinations are used, but long-bodied, upstanding, and muscular lambs are preferred in anticipation of winning or at least placing well in Junior Lamb Shows.
4. Total confinement systems for raising ewes or feeding lambs have frequently been classified as "specialized" because of the dense concentration of animals and specialized management required. However, the lambs and wool produced from them are marketed through typical channels, so they are more appropriately classified as farm flock or lamb feeding systems.
5. In areas where sheep numbers are high and orphan (bum) lambs are readily available during lambing, artificial rearing systems can be developed to provide profitable part-time employment where acreage is limited (Chapter 12).

Many small flocks are managed without concern for profit potentials. Some examples are the following:

1. Small flocks of ewes are often raised in confinement or semiconfinement on "acreages" near suburban areas to provide enjoyable "busy-work" for off-farm workers or retired persons, leading to the term "pastime sheep far-

ming."[4] *Hobby* flocks of purebreds are usually raised specifically for recreation and participation in shows is a primary purpose. Besides recreation, these flocks can provide meat and/or wool for family and friends. Personal preference rather than economic potential determines the breeds used, and harvested feeds required are normally purchased from area farmers to avoid the need for extensive equipment.

2. Sheep (particularly newly weaned lambs) are often the species of choice for introducing young people to the production of domestic animals (Figure 2-10). Because of their size, they can be handled easily by 8 to 12 year olds and

Figure 2-10 Sheep projects offer young people learning experiences (A) with a potential for recognition (B and C). (Photos courtesy of Wyoming State 4-H Club Office)

provide an opportunity for learning the responsibilities of daily care. Small pens and shelters can be provided cheaply and feed costs are minimal. Although prices paid for *club lambs* are often extremely high, they need not be. Many youngsters have had rewarding experiences with orphaned (bum) lambs that are either given to them or that cost only a few dollars.

REFERENCES CITED

1. Gee, C. Kerry, and Albert G. Madsen. 1983. *Sheep Production in the Seventeen Western States.* Colorado State University Agriculture Experiment Station, Fort Collins, and U.S.D.A. Economic Research Service. Special Series 24.
2. Hall, James T. 1973. Earning a living with sheep, ESC-576. U.S. Department of Agriculture, Washington D.C.
3. Hultz, Fred S., and John A. Hill. 1931. *Range Sheep and Wool,* p. 2. John Wiley & Sons, New York.
4. Muller, Karl. 1985. Pastime sheep farmers predominate. *Sheep!* May, p. 14.
5. National Research Council. 1985. *Nutrient Requirements of Sheep,* 6th rev. ed. The Council, National Academy of Sciences. Washington, D.C.

SUGGESTED READING

CAST, 1982. *The United States Sheep and Goat Industry: Products, Opportunities and Limitations.* Report No. 94. CAST—The Council for Agricultural Science and Technology. Carl S. Menzies, Task Force Chairman. Ames, Iowa.

Collins, Spelman, B. 1956. *Profitable Sheep.* Macmillan, New York.

Ensminger, M. E. 1970. *Sheep and Wool Science,* 4th ed. Interstate Printers and Publishers, Danville, Ill.

Hultz, Fred S., and John A. Hill. 1931. *Range Sheep and Wool.* John Wiley & Sons, New York.

Kammlade, William G., Sr., and William G. Kammlade, Jr. 1955. *Sheep Science,* rev. ed. J. P. Lippincott Co., Philadelphia.

McKinney, John. 1959. *The Sheep Book.* John Wiley & Sons, New York.

Parker, Ron. 1983. *The Sheep Book.* Charles Scribner's Sons, New York.

Scott, George. 1975. *The Sheepman's Production Handbook,* 2nd ed. Sheep Industry Development Program, American Sheep Producers Council, Inc., Denver, Colo.

3 Production Requirements

Sources of capital and the availability of different types of land, feeds, labor, and animals should be evaluated carefully *before* selecting a management system. Management information sources, record-keeping procedures, and especially buildings and equipment needs are also important.

CAPITAL

A source of money and estimates of cash flow for the establishment and operation of each unit are basic to profitable sheep production. However, money management is often omitted from discussions of sheep production, probably because its importance is so readily apparent it is assumed understood. Producers are reminded that their operations can fail from lack of financial planning as readily as from inadequate production management, such as poor nutrition or ill-planned breeding or marketing programs.

FEEDS AND LAND

Because they represent the major cost of producing sheep, careful evaluation of feed potentials must be made when establishing, enlarging, or changing sheep management systems. Sheep feeds are classified broadly as harvested (mechanically) or grazed. Harvested feeds are described as concentrates (cereal grains and protein supplements) or as many forms of forages.

The amounts, types, and qualities of feeds available for sheep production are usually directly related to the soil types, topography, and climatic influences prevalent in the areas where the feeds are grown. Because of their density, concentrates can be moved greater distances at a lower cost per unit of energy or protein than roughages. Therefore, the availability of forages is often the first limiting factor in determining which management system has the greatest profit potential in a given area. In countries where sheep production is based primarily on grazed forages, relative efficiency is expressed in terms of production per unit of land rather than per head or animal unit, which is the usual approach in the United States. Real differences in forage cost per animal unit in the United States are often surprisingly small when all costs are included. For example, when comparing quantities of harvested forages grown on a unit of land, wide variations are observed and purchase or lease prices tend to vary accordingly. Likewise, where forages are normally grazed rather than harvested mechanically, costs of moving sheep, fencing, water supply, and/or herding are considered harvesting expenses and included as a cost of forage.

Factors other than feed-producing potential are important when considering land for special aspects of sheep production, such as sites for building corrals. In these situations, soil type, drainage, and water supply become more important than the potential for growing feed.

MARKETS

Ideal feeder lambs, correctly finished slaughter lambs, and top-quality breeding animals or wool cannot help producers show a profit if badly marketed (Chapters 14 and 18). An evaluation of the relative demand for these products, which often varies by areas where grown, should be conducted when establishing a management system (Chapter 2). Some disappointing marketing situations that have been reported are as follows:

1. Feedlots established long distances from packing houses or sources of feeder lambs. Although feeds and climate indicate a potential for efficient lamb feeding, the cost of moving lambs into the lot or to slaughter represents a major limiting factor in many areas of the United States.
2. Lack of competition among buyers. Buyers tend to concentrate their efforts in areas where the types of products they need are most plentiful, resulting in differential pricing of similar products by area. Wool and feeder lambs are good examples.
3. Poor choice of breed. Breed adaptability is important and influences the demand for breeding animals in different areas and for different management systems. The fine-wooled breeds and their crosses are typical examples because they are well adapted to the more arid areas of the United States, but less so

where humidity and rainfall are high. Similarly, the meat breeds tend to perform better in farm flock than in range systems.

LABOR

A lack of qualified labor, particularly for herding and lambing, is often cited as a key limiting factor in commercial sheep production. A system is needed for training potential workers because as sheep numbers and the number of farms and ranches involved with sheep decreases (Chapter 1), the number of people with a sheep-oriented *experience base* also decreases. Attempts to accomplish this by formal training have been initiated by a few junior colleges, universities, and vocational schools. Student participation in these programs and identification of employment opportunities need expansion.

It has been suggested that the sheep industry is dealing with a "labor management" rather than a "labor" problem. Where family labor is inadequate, individual producers may need to become "teachers" (Figure 3-1) by hiring unskilled but interested people and combining training with financial incentives that will motivate them to remain involved for relatively long terms.

When considering labor requirements for handling sheep, dogs should not be overlooked. A well-trained dog can accomplish more when moving sheep than several people. One problem expressed by producers contemplating the use of dogs is that they have little or no dog-handling experience. It is a valid concern because reports of dogs with a great deal of natural ability or even well-trained dogs that are ineffective when handled by inexperienced people are numerous.

People who have not worked sheep with a dog, but recognize that one or more might be useful in their operations, are advised to contact a reputable breeder of one of the several breeds of sheep dogs. Many breeders offer trained dogs for sale, and some will include instruction and practice in handling them as part of the pur-

Figure 3-1 A rancher directing the activity of his herder. (University of Wyoming photo, courtesy of Carl A. Larson, Lyman, Wyoming)

chase price. Whether purchased by inexperienced shepherds or raised and trained by those with experience in handling dogs, sheep dogs are considered sound investments for most sheep management systems.

ANIMALS

For new operations, or those expanding or changing their management systems or feedlots, obtaining suitable animals can be a frustrating experience. In some operations such as those where lambs are marketed for slaughter at four to five months of age, it has been shown that purchasing replacement ewes is economically superior to raising replacements from within the flock. However, ewes of the ages and breeds or breed combinations desired are not always available for purchases at reasonable prices; such operations have had to develop their own systems for raising replacement ewes.

Cull ewes from range flocks have been suggested as a good source of farm flock ewes because they are typically only 5 or 6 years old. Their teeth are worn from grazing sandy forages but most are reproductively sound, with a small proportion of spoiled udders and other health problems. With ample feed, they can be expected to produce at least one and often two or more lamb crops before being sold for slaughter.[4] This procedure should be approached with caution for beginning producers. These "old ewes" often require a great deal of special care because health, udder, and barrenness problems that may not be readily apparent at purchase can lead to disappointing lambing seasons.

Ewe lambs selected from groups sold as feeder lambs (feeders) are usually too small for breeding the year they are purchased so must be fed a year before they can be bred. Good quality yearling ewes are not a dependable source of breeding animals because in years when lamb prices are strong ewe lambs are sold for slaughter rather than held for the yearling ewe market. The seasonal availability of feeders is a serious problem for commercial lots.

Although public auctions are considered a competitive method for selling sheep (Chapter 14), it is a "buyer beware" procedure for the inexperienced. *Newcomers* and many *old timers* might well be advised to get help from more experienced people when buying sheep of all classes. Order buyers, some central market and sale barn managers, and wool warehouse people charge small fees for their services. They are often familiar with the breeding and management practices of the producers offering sheep for sale. If not, they know the general management practices followed in their areas and are very much aware of the availability and current market prices of the different classes. The commissions they charge are considered a good investment for most operations requiring sheep purchases. A great deal of *shopping time* and travel can be saved, and the balance between quality and prices paid will probably be as close to optimum as possible.

Ram buyers are encouraged to insist on as many objective measures of production potential as possible (Chapters 7 and 21).

RECORDS

Purebred flock managers are required to keep breeding and lambing records by individual animal. Many of them also keep performance records obtained at central test stations or within their own operations. Most breed associations offer suggestions as to the information they prefer included and forms for accumulating these data easily and accurately.

Although the large numbers and lambing practices used prohibit individual animal records in range flocks, group records of production can be useful as guides to more efficient management. Farm flock managers are encouraged to collect and use records of as many production factors as possible (Chapters 7 and 9).

Good financial records are basic to the operation of all systems where profit is considered a goal. Important costs such as interest, depreciation, and repairs are frequently overlooked. Taxes require special consideration by producers. Some items, such as feed, medication, insurance, and marketing aids, can be deducted from gross incomes. Others, such as sheep, machinery, buildings, and corrals, can be deducted on a depreciation basis as tax credits.

Regardless of type of operation, a good budget needs to be prepared and updated throughout the year. Figure 3-2 shows a typical list of expected expenses. It is presented as a guide for preparing an outline specific for a given unit, rather than a form for use by everyone. For example, Figure 3-2 does not itemize herder salaries and their personal needs. These may be included in the cost of forage, which is appropriate in range operations, or identified separately. On the other hand, it does include a ram charge, which is inappropriate for feedlots, and building charges, which are normally not considered in range operations.

As the accessibility to computers (personal or through university extension services, financial institutions, or breed associations) increases, their importance to the sheep industry is expected to become more apparent.[1] Some universities[6,14] have offered computerized accumulation and summarization of performance data for several years; others are expected to follow suit and also to develop useful software permitting more widespread use of computer systems for handling both performance and financial records.

INFORMATION SOURCES

Many good research projects designed to identify methods for increasing production efficiency have been and are currently generated by university experiment stations and private industry. However, most producers recognize a serious information gap

Item	Estimated cost per ewe
Pasture	_____
Concentrate feed	_____
Roughage	_____
Shearing	_____
Personal property tax on livestock	_____
General overhead (supplies, fuel, etc.)	_____
Interest on above costs	_____
Ram cost	_____
Replacement ewe cost	_____
Interest on ewe	_____
Equipment (DIRTI)*	_____
Buildings (DIRTI)*	_____
Total	_____

Estimated income required per ewe to break even with no return to land, labor, or management:

_____ lb wool @ _____ = _____

_____ lb lamb @ _____ = _____

Total = _____

*Depreciation, interest, repairs, taxes and insurance

Figure 3-2 A simplified budget format. (Adapted with permission from the *Sheepman's Production Handbook*[12])

and are searching for ways to keep updated. Recognizing that the best sources of pertinent information for some operators may be inadequate for others, the following sources are offered as options:

1. Field days, seminars, and workshops may be unacceptable if held during seasons of heavy work responsibility such as lambing or for producers holding off-farm jobs. On the other hand, if they occur during relatively slow work periods, they can serve as a combination of information gathering and vacation. They normally permit excellent opportunities for one-to-one contact with specialists.

2. Scientific journals are good sources of the most current research information. Also, the American Sheep Producers Council publishes reports of pertinent sheep production research,[2] which are prepared especially for producers three times each year.
3. Popular publications (i.e., general farm magazines and newspapers, breed magazines, and other sheep-oriented publications) provide a broad spectrum of information, ranging from market reports to promotional efforts to all aspects of production management. They provide good thought stimulation, are inexpensive, and can be studied as time permits.
4. Veterinarians should be consulted regarding flock health, particularly when developing disease prevention programs.
5. Representatives of agricultural lending businesses can provide help in financial planning.
6. The Agricultural Extension Service is readily available in virtually every county in the United States. Personnel in county offices and area and state specialists represent an excellent source of updated information regarding all aspects of production, including money management and marketing. They may be reached on a personal basis for specific problems and provide regular informational meetings, fact sheets, and radio and TV programs.
7. The *Sheep Housing and Equipment Handbook,*[7] is a low-cost but excellent reference for ideas as well as specific plans for lot layouts, housing, handling facilities, and equipment items.
8. The *Sheepman's Production Handbook*[12] is one of the most complete sources of general sheep production information available. It is considered an ideal desk reference for all management systems.

BUILDINGS AND EQUIPMENT

Next to feed and animals, fixed costs of buildings and handling facilities can be the most expensive aspect of sheep production. Inspection of facilities used by producers indicates that they vary from elaborate to inadequate.

Investment in buildings, handling facilities, and equipment must be considered simultaneously with labor costs. In many cases, utilization of off-season or family labor may be preferable to equipment purchases. For example, daily handfeeding may be more cost effective than a powerwagon for fence line feeders. On the other hand, if labor is a limiting factor, investment in mechanized systems of feeding may be justified.

Fencing

Sheep fencing has two purposes: to keep sheep in designated areas and to keep out other species such as coyotes, dogs and, in some cases, deer and other game animals.

Except for migratory range flocks, it represents a major investment for sheep producers, so cost of materials and ease of construction along with effectiveness of purpose are basic considerations. Materials vary from wood to tubular steel to many types of wire. Wood and metal are preferred for those areas that are used most frequently or where sheep are crowded when handled, as in lanes and corrals.

Traditionally, perimeter fences have been built of a combination of standard woven wire and barbed wire with wood and steel posts. One barbed wire may be located near the ground to discourage predators from crawling under. Woven wire measuring 24 in. (60 cm) or more high is placed above it, followed by one or two strands of barbed wire, with the top one 40 to 50 in. (100 to 125 cm) above the ground (Figure 3-3). In hilly areas, five barbed wires are often used rather than a woven and barbed wire combination because it is easier to follow the terrain during construction. Also, some producers prefer five barbed wires to woven wire because sheep are less likely to get their heads *stuck* when reaching through, as they often do with the woven wire.

Normal post spacing for these permanent fences is 16 ft (4.8 m). However, the *suspension* concept in fencing is becoming more popular in many areas, particularly on fairly level land. The reduction in costs for posts is significant because they are placed 80 to 120 ft (24 to 36 m) apart and twisted wire *stays* (Figure 3-3) are threaded over the barbed wires every 16 ft (4.8 m) as a substitute for posts.

Multi-strand (four to nine wire) electric fences made from smooth, light-

Figure 3-3 A standard woven wire fence (A) and stays used to substitute for posts in a barbed wire fence (B). (University of Wyoming photos)

weight, high-tension wires with alternating energized strands have been used successfully for perimeter fences and/or to aid in predator control. A wide variety of plans and materials for building permanent, semipermanent, or temporary fences are available commercially from lumberyards and livestock supply companies. Some are described in advertisements in sheep-oriented magazines. Producers need to consider costs of materials, labor for construction, and maintenance, as well as the relative permanence required when choosing between temporary fencing and the more traditional (permanent) types.

There are many options for building temporary fences used to control grazing on highly productive pastures or contain sheep on crop aftermath. Woven wire can be hand stretched between two sturdy posts and then tightened by driving steel posts in a zigzag fashion. Two electrically energized smooth or barbed wires placed 8 to 12 and 18 to 24 in. (20 to 30 and 45 to 60 cm) above ground level have worked for many producers. Lightweight *nets* made from polyethylene or poly–steel combinations with fiber-glass posts are easy to install and remove.

Housing

Housing for sheep need not be elaborate. It should provide easy access for feeding and handling sheep and manure removal, along with its primary purpose of *protection* from weather that can reduce production efficiency. Sheep barns should be located on well-drained sites, and adjacent yards with east and south slopes are preferable.

In some climates, open-faced shelters of pole-type construction (Figure 3-4) are adequate. In others, sheds must be built so they can be completely closed. In very cold and humid climates, insulation and ventilation are basic concerns. Fiberglass panels or glass windows can be located to utilize solar energy during winter months without reducing the shaded area during summer. For sheep with an inch or more of wool growth, traditional sheep barns are used only when weather is extremely severe. However, when freshly shorn and during lambing, adequate housing is essential.

Figure 3-4 Open faced, pole-type shelter with fence-line feeder, portable feeder for expansion of permanent feeding space, and lamb creep with self-feeder in protected area. (University of Wyoming photo)

Some producers have expressed interest in buildings that permit lambs to be fed or ewe flocks housed on a year-around basis. A few universities[8,9,13] have been leaders in conducting research involving total confinement of sheep. Rations, breeds, and disease control, along with construction features associated with feeding, manure handling, floor type, and ventilation, have been studied. Economic aspects have been carefully documented.

Detailed plans are readily available for anyone interested in building new barns and lots. The extension service in most states has good sources of plans to choose from, as do many lumberyards. The Midwest Plan Service[7] describes many good options for housing and lot layouts that can be applied directly or adapted to fit the needs of most management systems and building sites. Producers are reminded to always include opportunities for expansion when building new sheep barns.

The most challenging housing problems occur when producers want to change or expand management systems where buildings are already in place. Building costs may be reduced significantly if existing structures that are basically sound can be remodeled. An example of remodeling a dairy barn for use in sheep production is shown in Figure 3-5. Extension agricultural engineers can provide excellent guidance in resolving specific construction problems.

Whether building new or remodeling, producers are encouraged to visit other operations with similar management systems for ideas that may be helpful.

Handling Facilities and Equipment

The purposes for constructing sheep handling facilities are many and the options for accomplishing them are almost unlimited. In the broad sense, their purpose is to aid in performing the necessary production practices as required with a *minimum of effort by handlers and a minimum of trauma for the sheep.* Handling of sheep is required for sorting, loading, vaccinating, weighing, drenching, lambing, shearing, and so on. Depending on the overall management system used, each of these activities varies greatly among production units, but *the cost-benefit ratio must be considered in each case.* When planning facilities to handle sheep and feed them, the aids and procedures discussed for *housing* are appropriate.

Feeds. Methods for handling feeds vary from hand feeding to highly automated augers or belts leading from central storage and electronically controlled processing and mixing facilities. For migratory range operations, where flocks are moved frequently, feeding on the ground is considered the best method for providing supplemental forages, corn, or pelleted concentrates (Figure 8-6). Bed-ground areas are relatively clean, and the savings that could be realized from reducing waste do not justify the labor for frequent moving and the investment in the large numbers of portable feeders required.

In farm flocks, feeders are considered a sound financial investment. Fence-line types (Figure 3-4) permit access without competition from crowding sheep and, as their description indicates, serve as both fence and feeder. Mixed or single feeds

Buildings and Equipment

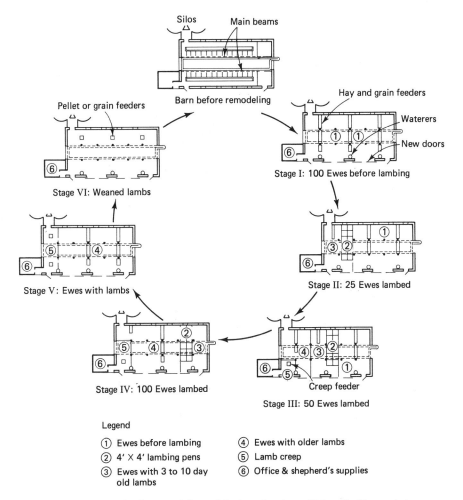

Figure 3-5 A plan for remodeling a dairy barn for sheep. (Adapted with permission from Brevik[3] and Midwest Plan Service[7])

may be distributed from power wagons or grinder-mixers or hand fed. Covers add to the cost but may be justified in areas where snows and rains are frequent and/or heavy because they reduce feed waste and labor for cleaning. Stacks of hay can be used as temporary windbreaks until fed (Figure 3-4).

Portable feeders designed for either forages or concentrates are useful in farm flock management systems. Although they need to be sturdy, they should not be heavy or cumbersome because frequent moving for manure removal or for changes in pen arrangements is required. When built in a rectangular shape, they provide feeding from both sides, which reduces the cost per unit of feeder space to less than half that of fence-line feeders. Portable feeders can be used to provide adequate

Figure 3-6 Portable feeders[5] for forages and/or grain. (University of Wyoming photo)

Figure 3-7 Portable feeders made from salvage lumber and metal. (University of Wyoming photo)

Buildings and Equipment

feeder space in small or interior pens where space for fence-line feeders is inadequate (Figure 3-4). They may also be used as *dividers* for groups of sheep receiving the same ration (Figure 3-6).

Round or five- or six-sided feeders are lightweight and excellent for feeding small numbers in pens. Metal feeders are more expensive but more durable than those made from wood. On many farms, lumber salvaged from remodeling or removing old buildings and metal from discarded machinery can be used to build portable feeders. (Figure 3-7).

Self-feeders of several sizes and shapes (Figure 3-8) have been used successfully in many farm flock operations. They are especially good for creep feeding lambs and are available commercially in many sizes and shapes. For ewes, complete diets offered in self-feeders must be chopped, ground, or pelleted and balanced with intake to meet production goals (Chapter 8). Bridging of chopped or ground feeds, particularly with a high proportion of forage, can be a problem, but it is usually minimized in feeders with perpendicular sides. Accessible space required is reduced to approximately 25% or less of that required for limit feeding.

In commercial feedlots, all types of rations are handled mechanically. Self-feeding is typical and accomplished by use of augers or self-feeders or simply by

Figure 3-8 Large (A) and small (B) self-feeders. (Photo A courtesy North Dakota State University[8] and photo B University of Wyoming photo)

keeping fence-line feeders filled from power-unloading trucks, trailers or grinder-mixers.

Sheep. *Working* or *handling facilities* for sheep are often inadequate, contributing to ill-timing or avoidance of many important management practices. For range sheep management systems, they may be either permanent and located near where sheep are usually handled or portable for quick setup whenever needed. A corral or preferably a series of progressively smaller corrals leading to a sorting or loading chute is adequate for most functions involving range sheep. By serving as a moving, solid panel, a 3 ft (90 cm) wide piece of canvas (called a "shoofly"), long enough to reach across a large corral and held by two or more handlers, can be used as an effective and low-cost way of crowding sheep into smaller areas.

As management systems become more intensive, the need for handling sheep increases and proper facilities are essential. *Pens, corrals, lots,* and *yards* are words used interchangeably to describe areas where sheep are confined. Sheep that are crowded into small areas exert a great deal of pressure on fences and gates, so they should be made of wood or steel materials. Wire fences are inadequate. Sorting and loading chutes should be solid-sided and smooth (Figure 3-9) so sheep will not be distracted by side influences and can move freely. They should be wide enough so that the type of sheep to be handled can move through them freely without permitting turnaround. Typical chute widths for mature sheep vary between 16 and 24 in. (40 and 60 cm) depending on breed, sex, and fleece length. In units where sheep and lambs need to be handled separately, one adjustable side or appropriate inserts help reduce the turnaround problem when working lambs.

Permanent handling facilities should be located in well-drained areas, easily accessible by way of lanes from pastures or larger lots. Shade and protection from wind, rain, and snow can increase the usefulness of sorting chutes, scales, and areas where individual handling such as foot trimming is required. Gates with strong but easily operated latches are best located at corners, and systems for *funneling* groups of sheep into progressively smaller areas are helpful. Round corrals with swinging gates mounted in the center offer maximum flexibility for size of groups to be handled (Figure 3-10). Perimeter corral fences should be approximately 4 ft (120 cm)

Figure 3-9 Smooth-sided funneling corral and loading chute. (University of Wyoming photo, courtesy of Harper Livestock, Eaton, CO.)

high, but 3-ft (90 cm) fences and panels are adequate for sorting chutes and interior fences. They are also more convenient for handlers.

Adequate facilities for handling sheep in feedlots are especially important because unloading, loading, and sorting of both incoming and outgoing lambs are required on a regular basis. Provisions must also be made for isolating incoming groups and treating health problems on a daily basis. Housing for feedlot lambs is not considered cost effective, but good drainage is essential. Mounds or ridges built parallel to the prevailing slope help keep lambs dry and reduce accumulation of mud in fleeces during wet periods. They can be combined with windbreaks.[7] All feedlots and some farm flock operations need to consider Environmental Protection Agency (EPA) regulations for runoff where livestock production is concentrated. An outline of limitations is available from the EPA, Washington, D.C.

Regardless of management system, it is important to consider the behavior of sheep when constructing facilities to handle them. Some basics are as follows:

1. Sheep move toward freedom and other sheep and away from movement or noises that frighten them.
2. Sheep prefer lighted areas, so dark barns, chutes, and the like are to be avoided.
3. Sheep prefer to move uphill and around slight corners or curves in structures.
4. Sheep move better away from rather than toward bright sunlight in early mornings.

Figure 3-10 Round corral with swinging, center-mounted gate. (Photo courtesy of Ohio agricultural Research and Development Center[9])

Lambing facilities used by producers during this labor-intensive season vary from virtually none in range systems, where ewes are lambed without herders in fenced areas, to environmentally controlled barns with slotted or mesh floors for manure removal and automated feeding systems. All have been shown successful in meeting specific goals. In each case, trade-offs between labor, facilities, and lamb survival during different lambing seasons must be considered carefully. The relative profitability of ewe flocks is closely related to lamb survival during their first week of life (Chapter 11) and facilities can be an important factor.

The primary purpose of lambing facilities is to increase lamb survival by providing protection from extremes in weather, to provide an opportunity for newborn lambs to nurse and be "claimed" by their dams, and to aid shepherds in correcting health and parturition problems. To accomplish this, arrangements need to be made for handling feeds, bedding, manure, and sheep as efficiently as possible in areas protected from extremes in weather. A small tractor-loader can be one of the most useful items for handling all of these lambing-related tasks.

Plans are available for building a variety of feeders for either bulky or dense feeds offered to drop bands and ewes with lambs (see "Feeds," this chapter) and for smaller, portable feeders used in lambing pens, grouping pens, and lamb creeps.[7] Many of these can be purchased ready-made from livestock supply companies.

Lambing pens (called *jugs*) are usually built about 3 ft (90 cm) high and measure 4 by 4 ft (120 by 120 cm) to 4 by 5 ft (120 by 150 cm) or 6 by 6 (180 by 180 cm) in size. They may be open (Figure 11-3) or solid sided to reduce drafts or help concentrate supplemental heat provided to each pen (Figure 3-11). They are usually portable and two sides may be hinged. A minimum of 10 per 100 ewes lambing is suggested for most operations, but more may be required if ewes are sychronized at breeding or for flocks with a high rate of multiple births.

Because the thermoregulatory systems of newborn lambs is relatively ineffi-

Figure 3-11 Solid-siding lambing jugs hold heat supplied from ducts (against wall) leading from furnace room in back. (Photo courtesy of Lloyd Snider, Powell, Wyoming)

Buildings and Equipment

Figure 3-12 An open-sided, low-cost cover for lambing pens. (Photo from University of Wyoming wool library)

cient, they require protection from extremes in temperature for their first two to three days of life. Simple structures (Figures 3-12 and 11-2) covering lambing pens are adequate in the warmer climates of the United States where wind, rain, and direct sunlight are primary concerns.

In the colder climates, heat sources may be important in addition to a closed building that allows for 16 to 24 ft^2 (5 to 7 m^2) per ewe, depending on their size and whether they are shorn. There are two approaches for providing heat for lambs while they are in jugs. The entire area may be heated (Figure 3-13) or heat may be directed to the jug area only. Devices such as heat lamps (Figure 11-19) or controlled central heating units work well for concentrating supplemental heat in jugs. Figure 3-11 shows a method where a standard home furnace provides heat to individual pens by way of a duct system. Also, the heat of fermentation from a manure pack can be an "in the pen" source of warmth for jugged lambs. There is some disagreement among managers of successful lambing operations as to whether jugs should be cleaned and disinfected after each ewe and lamb are removed. Proponents of

Figure 3-13 Overhead gas furnace for heating lambing room. (University of Wyoming photo)

Figure 3-14 Low-cost shelter for small groups of ewes and lambs moved from jugs. (University of Wyoming photo)

the "manure buildup" approach cite heat from this source and reduced labor as advantages. Others cite sanitation as justification for frequent cleaning. Either way, a porous sand-gravel floor is superior to cement floors because pens remain drier and the buildup of ammonia in the air of the jug area is reduced.

Although feeders, heat sources, and jug designs are important concerns in shed lambing operations, the *layout* of the entire system (Chapter 11) is the aspect that requires the most careful planning. Good diagrams of layouts that permit good flow of sheep and supplies are available[7] and visits to successful operations can be helpful.

Most shed lambing operations are adequate in terms of size and location of drop band and jug areas. However, their systems for "bunching" groups of ewes and lambs as they leave the jugs is often limited, resulting in excess miss-mothering (bumming) of lambs. Ideally, arrangements (which need not be expensive) should be made for small groups of three or four ewes and their lambs to remain together for one day after they leave jugs to make sure the ewes are caring for their lambs, particularly in the case of twins (Figure 3-14 and 12-2). These groups should then be combined on subsequent days to form progressively larger groups of 6, 12, 24 or 8, 16, 32, and so on. *Hospital pens* for problem cases are essential.

Figure 3-15 A permanent facility for shearing large numbers of sheep. (Photo from University of Wyoming wool library)

In special situations where the incidence of multiple births is high, facilities such as grafting stanchions and adequate jugs (15 to 20 per 100 ewes) to allow extra time for observation and care of grafted, twin, and triplet born lambs are helpful. An arrangement for artificial rearing of lambs may be justified (Chapter 12).

Adequate *shearing facilities* are important for all management systems. They should provide ease of handling by shearers without injury to the sheep and proper care of wool (Chapter 18). Because shearing is a once-a-year function in ewe flocks, facilities are too often neglected or the investment required for good ones is not considered profitable. A well-constructed permanent shearing shed (Figure 3-15) requires that relatively large numbers be handled each year to justify the cost of constructing and maintaining it. These sheds, which combine chutes leading to small catch pens for unshorn sheep and away from shearer stations for those that have been shorn, with ample space for shearers and handling fleeces, are ideal. However, the relative density of sheep in most areas and the cost of driving or hauling them prevents shearing large numbers at central locations.

Shearing trailers (Figure 3-16) have been substituted for permanent sheds as a method of moving facilities to sheep rather than sheep to facilities, and they are gaining popularity in the United States. Shearer stations inside these trailers are essentially permanent and the corrals and chutes that they carry are set up quickly

Figure 3-16 Front (A) and rear (B) views of typical shearing trailer. (Photo by Kurt Piel, courtesy of Dale Aagard, Worland, Wyoming)

for easy movement of sheep. They usually do not include areas for skirting or sorting fleeces (Chapter 18), so these must be provided by the producer.

For most farm flocks, existing buildings can be arranged for shearing with lambing or other portable panels used elsewhere during the rest of the year. Several plans for doing this have been prepared.[7,11] Figure 3-17 offers an approach that can be adapted as needed. Essential parts of the design shown are the wrangling alley, holding pens, shearing floor, and sacking area. Useful additions can include a chute for handling sheep after shearing for branding, spraying, vaccinating, and the like, before they are released.

If shearing is done after lambing, a sorting chute for cutting off lambs on the way into the shed and a "mothering-up" pen for after shearing are helpful. A V-shaped chute leading into the wrangling alley is necessary. Suggested sizes and materials for essential parts of the diagram shown in Figure 3-17 are as follows:

- *Wrangling alley:* Solid board sides about 3 ft (90 cm) high and 2 to 3 ft (60 to 90 cm) wide. Floors of wrangling alley and holding pens are best when made of slats to allow urine and feces to pass through and prevent the sheep's feet from dragging in contaminants such as bedding. These may be built in sections for easy installation and removal.
- *Holding pens:* One per shearer, filled from back, usually 5 by 6 ft (150 by 180 cm) with curtains (wool sacks or canvas pieces work well) in front to serve as access gates. Fill gates may be hinged or sliding.
- *Shearing floor:* Areas that provide about 5 by 6 ft (150 by 180 cm) spaces for each shearer should be made of wood and the wool-handling area should permit frequent sweeping to reduce fleece contamination.

Portable sacking stands for tramping fleeces into burlap bags are useful for small clips,[7] but hydraulic sackers reduce labor and permit more fleeces per bag (Figure 3-18).

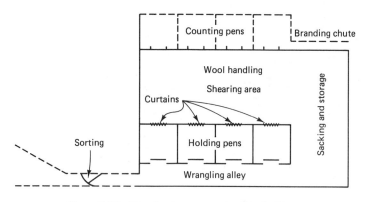

Figure 3-17 Plan for temporary shearing facility.

Figure 3-18 Hydraulic sacker.[10] (University of Wyoming photo)

Miscellaneous. Equipment items such as tilting squeeze chutes, sheep chairs, foot baths, and trimming stands can be helpful in some operations and are considered essential in others. They may be built from available plans[7] or purchased from sheep equipment supply companies. Permanent or portable scales for weighing individual sheep are recommended for purebred and some farm flocks. On the other hand, scales for weighing groups of sheep are more appropriate for range operations and feedlots.

A cabinet or drawers for providing a systematic way of storing small tools and supplies required throughout the year should be located at the "handiest" place available. Useful items include those required during lambing (Chapter 11), hoof trimmers, bolling guns or forceps, syringes, drenching guns, branding irons, ear tags, spare light bulbs, and so on. A refrigerator should be located close by for keeping supplies that require temperature control, such as frozen colostrum and several medications and vaccines.

REFERENCES CITED

1. Anonymous. 1985. Computers can help sheep producers. *National Wool Growers.* June, p. 22.
2. Anonymous, n.d. *Research Digest.* Sheep Industry Development Program. American Sheep Producers Council, Denver, Colo.
3. Brevik, T. J. 1977. Housing your flock. Sheep production and management No. A2830. University of Wisconsin—Extension, Madison.
4. Faller, Timothy C. 1984. Productivity of aged western whitefaced ewes under North Dakota conditions. 25th Western Dakota Sheep Day Bulletin. North Dakota State University Research Extension Center, Hettinger.
5. Gold Circle Inc., Centerville, Iowa.
6. Haugen, Roger G. 1981. The North Dakota Sheep Production Testing Program. North Dakota State University Extension Service, Fargo.

7. Midwest Plan Service. 1982. MWPS-3, *Sheep Housing and Equipment Handbook*. The Service, Ames, Iowa.
8. North Dakota State University. Research Extension Center, Hettinger.
9. Ohio State University. Ohio Agricultural Research and Development Center, Wooster.
10. Rapid City Tank Supply Co., Rapid City, S.D.
11. Sachse, James M. 1985. Shearing Facilities and Wool Preparation at Shearing. Circular 518. New Mexico State University Extension Service, Las Cruces.
12. Scott, George. 1975. *The Sheepman's Production Handbook*. SID—Sheep Industry Development Program, American Sheep Producers Council, Denver, Colo.
13. University of Illinois. Dixon Springs Agricultural Center, Simpson.
14. Wharton, W. W. 1970. *Ohio Sheep Production Testing Program*. Ohio State University Extension Service, Columbus.

SUGGESTED READING

Ricketts, G. E., R. D. Scoggins, and D. L. Thomas. 1983. Recommendations for a Sheep Management Program. Circular 1221. University of Illinois Extension Service, Urbana–Champaign.

Midwest Plans Service. 1982. MWPS-3, Sheep Housing and Equipment Handbook, The Service, Ames, Iowa.

Sachse, James M. 1978. Sheep Production and Management Circular 480. New Mexico State University Extension Service, Las Cruces.

SID. 1968. Symposium on Production and Business Management. SID-Sheep Industry Development Program. The Council, American Sheep Producers Council. Denver, Colo.

SID. 1981. National Lamb Marketing Symposium. National Blueprint and Sheep Industry Development Program. SID-Sheep Industry Development Program. The Council, American Sheep Producers Council. Denver, Colo.

Also see Suggested Reading for Chapter 2.

4 Political Considerations

The sheep business is involved politically in several ways and is influenced by decisions or rulings that are political in nature: grazing on public lands, tariffs, import or export regulations, support prices, predator control, and use of foreign laborers.

These issues are constantly being aired by news media and are the subject of constant public concern as reported in all publications from grower organizations. In many cases, this removes decision making from the owners or managers, which often results in frustration. These types of involvement affect the range sheep business to a greater extent than they do farm flock operations, but no one in the sheep business is immune from political influence. Ensminger[1] has cited the political influence on wool as a factor unfavorable to sheep production due to the difficulty of predicting prices. More and more of these issues have become sociological in nature and also highly emotional issues. Sheep producers and all agricultural producers are a minority, and it is a real disadvantage to have far-reaching decisions made for them, particularly when decisions are sometimes highly influenced by emotion.

PUBLIC LANDS

Vast areas of land, particularly in the western states, are available and used for livestock grazing. These include land owned by the federal government, the states, Native Americans, railroads, and individuals. In the early days of grazing on public lands, there was little control and as a result extensive areas were overgrazed, but that is no longer the case. In most sheep operations, good grazing management is practiced, whether sheep are grazed on arid to semiarid public lands during fall and

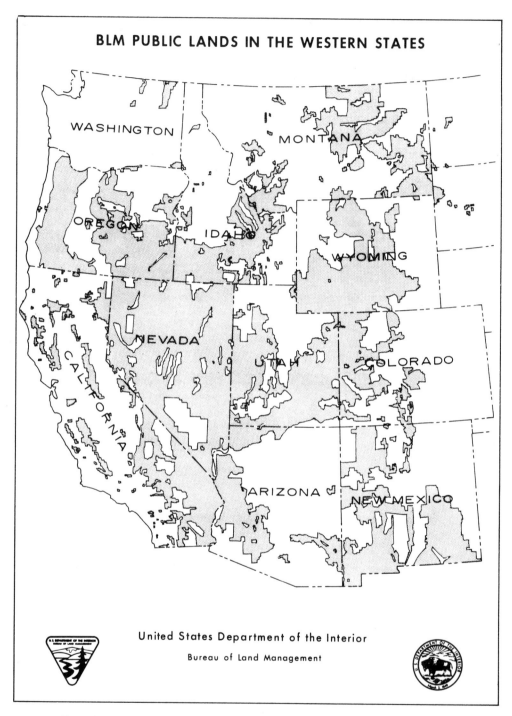

Figure 4-1 BLM public lands in the western states. (Reprinted with permission from USDA and USDI, *Grazing Fee Review and Evaluation*[3])

winter, or national forests during summer, or on improved pastures on private lands. Drought in some years and in some areas does result in more grazing pressures, but this is only temporary. Livestock producers should constantly strive to become familiar with forage plants and management systems so as to ensure adequate nutrition for their grazing animals and also preserve the quality of vegetation in grazing areas.

The Bureau of Land Management (BLM) of the U.S. Department of the Interior and the Forest Service of the U.S. Department of Agriculture administer livestock grazing on approximately 307 million acres of public rangeland located within the 16 western states, covered by the Public Rangelands Improvement Act of 1978.[3] About 56% of this is administered by the BLM and 43% by the Forest Service. Maps showing the location and distribution of these areas are given in Figures 4-1 and 4-2.

The western states have about 70% of the forests and rangelands in the 48 contiguous states, but this provides 91% of the total public and private land grazed. Nearly half the sheep producers with 2500 or more sheep use federal rangelands, which provide about 42% of their annual forage requirements.

Permittees and lessees are charged for public rangeland grazing use according to the number of animal unit mouths (AUM) they are authorized to use (see Table 4-1). Grazing fee receipts are distributed according to legislative requirements to range betterment funds, to the state or county of origin, and to the U.S. Treasury. Grazing fees for the BLM and Forest Service were established on different bases, but since 1969 the goal has been to bring their fees to the same level. The fee system under the Public Rangelands Improvement Act in 1978 was the first to be legislated by Congress. Funds for range betterment are used for improving public rangelands through such practices as brush control, fencing, and water development in cooperation with permittees.

Many individuals and groups are concerned with present government policy, and there is constant pressure on Congress to increase grazing fees, resulting in the development of what is called an omnibus range bill. The future of the sheep industry in terms of numbers and distribution as well as type of operation is bound to be influenced by political decisions regarding public land use and the grazing fee structure.

SUPPORT PRICES

Since the Wool Act was passed in 1954, incentive payments have been made to wool producers. At present the Wool Act is again foremost in the political picture as a part of an omnibus farm bill. The incentive payment level has been increased over the years, as shown in Figure 4-3. Income from import duties is used as the source of funding. Out of the incentive payment, a deduction has been made to provide funding for the American Sheep Producers Council (ASPC). This organization is responsible for advertising and promoting the sheep industry and its products. In

TABLE 4-1 NUMBER OF PUBLIC LAND AUMS BY STATE, 1983.

State	Forest Service		Bureau of Land Management			Total BLM/FS
	AUMs	AUMs[a]	Sec. 15	Sec. 3	BLM Total	
1983						
Arizona	1,488,319	1,240,266	154,638	462,345	616,983	1,857,249
California	630,747	525,623	142,050	200,598	342,648	868,271
Colorado	946,167	788,473	48,152	490,210	538,362	1,326,835
Idaho	943,870	786,558	34,795	1,063,926	1,098,721	1,885,279
Kansas			244		244	244
Montana	664,548	553,790	243,803	959,130	1,202,933	1,756,723
Nebraska	99,960	83,300	1,257		1,257	84,557
Nevada	351,723	293,103	20,856	1,431,710	1,452,566	1,745,669
New Mexico	954,539	795,449	232,809	1,259,952	1,492,761	2,288,210
North Dakota			9,718		9,718	9,718
Oklahoma	6,663	5,553	475		475	6,028
Oregon	568,801	474,001	93,317	943,532	1,036,849	1,510,850
South Dakota	133,095	110,913	74,563		74,563	185,476
Utah	750,990	625,825		919,227	191,227	1,545,052
Washington	135,442	112,868	30,155		30,155	143,023
Wyoming	620,188	516,823	466,735	1,638,267	1,635,002	2,151,825
Total	8,295,052	6,912,543	1,553,567	8,898,897	10,452,467	17,365,007

1982

Arizona	1,396,981	1,164,151	155,592	440,326	595,918	1,760,069
California	601,806	501,505	243,502	243,097	486,599	988,104
Colorado	960,300	800,250	53,694	456,732	510,426	1,310,676
Idaho	933,274	777,728	40,317	1,101,472	1,141,789	1,919,517
Kansas			244		244	244
Montana	666,241	555,201	168,157	969,010	1,137,167	1,692,368
Nebraska	102,665	85,554	1,257		1,257	86,811
Nevada	356,501	297,084	49,787	1,558,334	1,608,121	1,905,205
New Mexico	950,010	791,675	283,370	1,354,404	1,637,774	2,429,449
North Dakota			9,718		9,718	9,718
Oklahoma	6,035	5,029	475		475	5,504
Oregon	574,137	478,448	79,924	916,181	996,105	1,474,553
South Dakota	128,989	107,491	74,563		74,563	182,054
Utah	719,967	599,973		945,538	945,538	1,545,511
Washington	129,415	107,846	30,155		30,155	138,001
Wyoming	621,022	517,518	463,593	1,166,934	1,630,527	2,148,045
Total	8,147,343	6,789,452	1,654,348	9,152,028	10,806,376	17,595,828

[a]Forest service AUMs are converted to a unit that is similar to BLM AUMs by dividing by 1.2. Reprinted with permission from USDA and USDI, *Grazing fee review and evaluation*.[3]

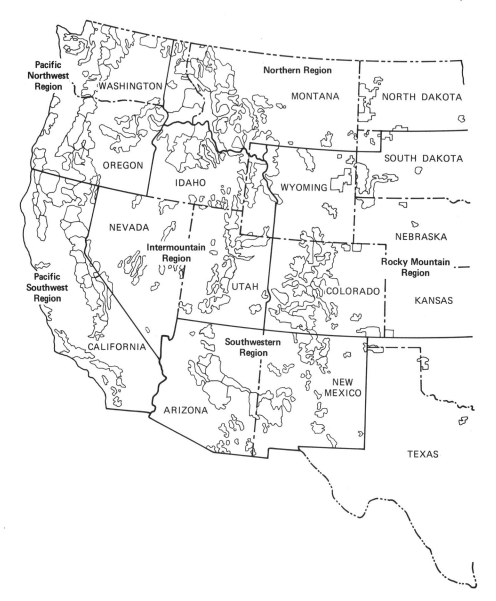

Figure 4–2 Public rangelands managed by the Forest Service. (Reprinted with permission from USDA and USDI, *Grazing Fee Review and Evaluation*[3])

addition, the ASPC has been helpful in providing industry services, producer education programs, lamb and wool merchandising, and new product development.

Both the imposition of import duties and level of imports either of lamb or wool are controversial subjects. Not only is controversy evident among nations, but also among interested groups and individuals within this country. Sheep producers,

Foreign Influences

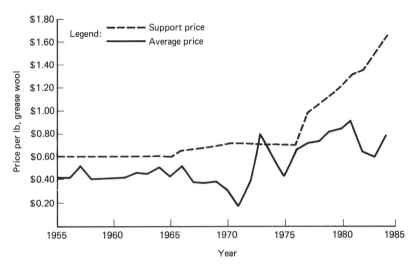

Figure 4-3 Trend of actual wool prices and of support level.

too, tend to have varying opinions on the matter of support prices. The ASPC is not free of controversy, either. Opponents criticize the idea of promoting a product that is in short supply. Others mention the lack of incentive, citing the decrease in sheep numbers. Still others would have the ASPC extend its influence beyond promotion or advertising and increase its support of research.

FOREIGN INFLUENCES

Foreign trade has been a traditional issue for sheep producers. Since the United States is in the position of an importing nation, this becomes a political issue with regard to both import duties and in some cases import restrictions on lamb or wool or both. Restrictions on the importation of sheep from other countries by disease-control agencies are also a problem for the sheep industry. Such restrictions limit access to new genetic material that has the potential of increasing efficiency of production. In addition, foreign influences plague the sheep industry due to the use of foreign herders and other laborers by many sheep producers. In some cases, these producers are highly dependent on aliens for their work force, since these are the ones most willing to perform menial tasks around the ranch. Along with other agricultural enterprises, as well as several other industries, the use of illegal aliens becomes the crux of the controversy. Legislators are continually trying to find solutions and develop workable programs to control the problem of illegal aliens.

All these foreign influences combine to create a complicated situation for producers. Furthermore, this issue is influenced by emotion, which makes decisions difficult and often results in frustration to those who are involved.

PREDATOR CONTROL

Predator control probably causes the greatest disagreement between the sheep industry and nonagricultural groups. This controversy is truly charged with emotion to such an extent that decision making by any agency or political entity is extremely difficult.[4] The hazards of losses to predation limit the stocking of sheep or goats, with consequent reduction in efficiency of range forage utilization. Many former sheep producers who have gone out of the sheep business cite losses to predation and their inability to use effective control methods as their primary reasons for terminating their operations. Public attitudes toward the control of predators by producers or by any established program can also cause discouragement and thus be a reason for sheep producers to go out of business.

Animal damage control is supported by a wide range of private efforts, county and state programs, federal programs, and those who cooperate with county, state, and federal agencies. Since 1939, the federal program has been the responsibility of the U.S. Department of the Interior. Operational programs are mostly carried out in the western states, but information and advice are available on request to producers in any state.

Opposition to animal damage control programs has increased and as a result the control methods used by the Department of Interior and other agencies have been reduced. Losses to predation have become more serious (see Figure 4-4). Re-

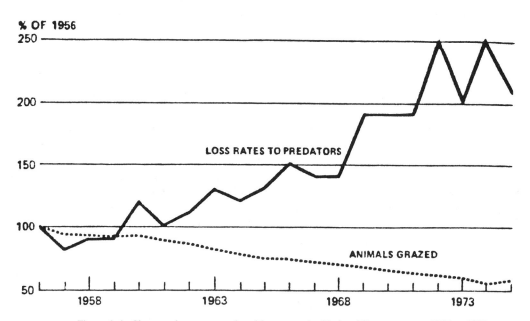

Figure 4-4 Sheep and goats grazed and losses on the National Forest ranges, 1956 to 1975. (Adapted with permission from Parker and Pope[2])

search efforts have been intensified in recent years, but the problems are far from being solved.

Transfer of the animal damage control program from the U.S. Department of the Interior to the U.S. Department of Agriculture has recently been completed.

GROWERS' ORGANIZATIONS

Many organizations have been formed to provide a source of support for various programs and for the mutual solutions of various problems. County pest control boards have made an attempt to unify efforts for controlling losses due to predation. In many cases, producers within counties form groups to discuss and determine mutual needs. State growers' organizations are numerous and most are associated with the National Wool Growers Association. Both state and national organizations are involved in a variety of efforts, such as holding sales, sponsoring tours, arranging educational programs, providing guidance for promotional programs, and promoting social activities. The biggest portion of their efforts, though, is aimed at representing growers in lobbying efforts. Exerting political pressure on legislative bodies with regard to all the previously mentioned political aspects of the sheep business is an almost full-time effort, although the social aspect of group meetings and entertainment is also important.

The limitations imposed on the sheep industry by political implications are not easily solved and likely will continue. Being aware of potential problems is necessary, but avoiding all political entanglements is impossible. A new sheep producer or one contemplating starting a sheep enterprise would be well advised to be familiar with the political aspects of the business. If one or more problems can be avoided, there is a greater probability of success. In spite of all the potential difficulties, sheep production is among the noblest of all professions. Providing food and fiber for peoples' use is a satisfying occupation.

REFERENCES CITED

1. Ensminger, M. E. 1970. *Sheep and Wool Science,* 4th ed. Interstate Printers and Publishers, Danville, Ill.
2. Parker, C. F., and A. L. Pope. 1983. The U.S. sheep industry: Changes and challenges. *J. Animal Sci.* 57, Supplement 2, p. 75.
3. USDA Forest Service and USDI. Bureau of Land Management. 1985. Grazing Fee Review and Evaluation. Draft Report.
4. Wade, Dale A. 1982. *Impacts, Incidence and Control of Predation on Livestock in the United States, with Particular Reference to Predation by Coyotes.* Council for Agricultural Science and Technology, Special Publication No. 10.

5 Breeds of Sheep

During the last 25 years, annual registrations of purebred sheep have increased despite a continued decrease in sheep numbers.[3] Likewise there has been a steady increase in auction sale prices of purebred sheep. The greatest part of this has been in Suffolks, which account for about half of all sheep registrations. Primary selection emphasis has favored body size both within and among breeds, and thus small-sized breeds of sheep have tended to decrease in popularity. Even though purebred sheep make up a small portion of the total sheep population, purebred breeders are generally more involved in setting trends than are commercial producers raising grade sheep.

BREED CLASSIFICATION

Sheep breeds have been classified in various ways, based on their conformation, wool and meat type, origin, face color, or their use as a sire breed and dam breed.[4] One system of classification may seem more appropriate than another to some producers, for some production systems, or in some geographic areas. For simplicity, breeds discussed here are separated according to grade of wool, as follows:

- Fine wool: Merino, Rambouillet, Debouillet
- Medium wool: Columbia, Corriedale, Panama, Targhee, Montadale, Suffolk, Hampshire, Shropshire, Dorset, Oxford, Southdown, Cheviot, Finn, Polypay, Clun Forest, Ryeland, Tunis
- Coarse or long wool: Lincoln, Cotswold, Leicester, Romney, Black-faced Highland, Barbados

These classifications are according to an approximate wool grade and do not imply that all within the group are alike in terms of wool-producing ability.

FINE-WOOLED BREEDS

Merino

The earliest of all to become popular in the United States were Merinos, and even though they are not very numerous in this country at present, they are one of the most important of all breeds in world distribution. In the United States, Merino sheep are noted more for their contribution to the early sheep populations than for their present popularity. Also, Merinos were the forerunners of Rambouillets; both modern Merinos and Rambouillets trace their ancestry to the early Spanish Merino. They have gone through different phases, having at one time been designated as A, B, or C type according to extent of body folds or wrinkles. At present, Merinos in this country are smooth bodied and relatively open faced, and there is no longer any separation into types of Merinos.

Delaine Merinos and Texas Delaines, even though having different breed associations, are related and are similar in appearance. They are white sheep, intermediate in size, with rams in good condition weighing from 180 to 250 lb (80 to 115 kg). They tend to be covered with wool on their legs. Rams generally are horned and ewes are polled, although there are strains of polled Merinos and it is not unusual for ewes to have scurs.

Merinos are noted for producing heavy shearing fleeces that grade 64's spinning count or finer; however, there are variations among flocks and among countries as to exact fineness of wool desired by breeders. Their wool is attractive in appearance and unsurpassed in quality for use in the manufacture of clothing materials.

Merinos are hardy in that they can live and produce in areas of climatic extremes and under relatively poor feed conditions. They are gregarious and will breed at nearly any time of the year. At present they are of only minor importance in commercial sheep production in this country.

Rambouillet

These are the most numerous of all breeds in the United States when considered from the standpoint of percentage of blood. Rambouillets and Rambouillet crosses make up the greatest portion of range sheep in the western states, which in turn make up over half of the U.S. sheep numbers. The Rambouillet was developed in France from the Spanish Merino. Following their importation into this country, they became very popular and still maintain that popularity.

Rambouillets are big sheep, and breeders are constantly striving to maintain or increase their size. Mature ewes weigh from 140 to 180 lb (64 to 82 kg) and mature rams weigh from 250 to 300 lb (115 to 137 kg) or even more when in good condition (Figures 5-1, 5-2, and 5-3). They are white sheep, and traditionally rams have horns while ewes are polled; but polled rams are becoming more numerous each year and in some areas polled rams are relatively popular. Figure 5-4 shows a high-performing polled Rambouillet ram. Horned rams should have spreading horns that spiral away from the head. Those with horns growing close to their head or those with short

Figure 5-1 Rambouillet yearling ram. This ram was a high-indexing certified ram, a register of merit ram, and a champion show ram. (University of Wyoming photo)

Figure 5-2 Rambouillet herd sires at Texas Agricultural Experiment Station, Sonora, Texas. (University of Wyoming photo, Courtesy of Texas Agricultural Experiment Station)

Figure 5-3 Group of Rambouillet yearling ewes. (University of Wyoming photo, Courtesy of Richard Strom, Laramie, Wyoming)

Fine-Wool Breeds

Figure 5-4 A high-performing Rambouillet ram. (University of Wyoming photo)

horns that tend to grow inward are discriminated against. Modern Rambouillets should be smooth and free of skin folds and open faced, although various degrees of wool covering on the face and folds occur in many flocks and create difficulty in selection when they do occur; heavy folds are shown in Figure 5-5.

Rambouillets shear heavy fleeces of high quality, mostly in the 62's and 64's spinning count. They are excellent wool producers, and highly adaptable sheep. Due to their hardiness and grazing ability, they can live and produce under extremely hot or cold climatic conditions. They are gregarious and tend to stay together, thus being particularly adaptable to extensive range conditions. They are less seasonal in their breeding habits than are sheep of most other breeds, and thus are adaptable to out-of-season and accelerated lambing programs.

Although Rambouillets have many strong points, they are not as desirable in carcass meatiness as are sheep of the black-faced breeds, and it is common for Rambouillet lambs to be feeder lambs when weaned, thus requiring some time in feedlots (Figure 5-6). The Rambouillet is often characterized as relatively low in

Figure 5-5 Rambouillet ram with heavy body folds. (University of Wyoming photo)

Figure 5-6 Rambouillet lambs at weaning. (University of Wyoming photo)

prolificacy, largely because of its being raised under less than optimum feed and climate conditions; but this is not noticed when they are kept under similar conditions as are sheep of other breeds. In crossbreeding programs the Rambouillet ewe is outstanding in total production of lamb and wool.

The American Rambouillet Association actively sponsors Performance Testing and Register of Merit programs for the benefit of ram producers.

Debouillet

This breed was developed by the late Amos Dee Jones from an original cross of Delaine Merinos and Rambouillets. Selection favored long-stapled fine wool and large, smooth-bodied sheep. In appearance, Debouillet sheep are white, fairly large, and have heavy shearing fleeces that grade 64 in spinning count or finer. Rams with horns or polled rams are both acceptable for registration. They should be open faced below the eyes and free of body folds. Before being registered, Debouillet sheep must meet certain production requirements as specified by their association.

MEDIUM-WOOLED BREEDS

Columbia

This breed was developed by the U.S. Sheep Experiment Station at Dubois, Idaho, from original matings of Lincoln rams to Rambouillet ewes and intermating of the first cross offspring. By selection, large sheep with fleeces intermediate between fine wool and coarse wool were developed. Their popularity has spread through much of the United States, and in numbers of registrations Columbias ranked fifth in 1985.[1]

Medium-Wooled Breeds

Figure 5-7 Columbia ram. (University of Wyoming photo)

Columbias are big-framed sheep that are well muscled and have heavy shearing fleeces in the medium range of fiber diameter. The association accepts sheep for registration with fleeces ranging from 46 to 62 in spinning count. However many breeders prefer those toward the finer extreme, 60's to 62's, while some prefer to keep all their sheep in the coarser range, 48 to 50 in spinning count. These sheep are white, with chalky white open faces. They stand straight and tall and are long bodied (Figures 5-7 and 5-8). All are polled, with even small scurs being discriminated against. They will shear fleeces weighing 12 lb (5.4 kg) for ewes and 18 lb (8.2 kg) for rams when kept on good feed.

Columbia breeders have attempted to ensure breed characteristics and freedom from defects by having sheep inspected before registration. Most purebred Columbias are raised in farm flock situations or small irrigated ranches, but Columbia rams are used extensively for breeding in range flocks. As a result, many range ewes are crosses of Rambouillet and Columbia. Figure 5-9 shows typical Columbia lambs at weaning.

Figure 5-8 Columbia ewes at breeding. (University of Wyoming photo)

Figure 5-9 Columbia lambs at weaning. (University of Wyoming photo)

Corriedale

This breed was developed in New Zealand in an attempt to combine the desired traits of long-wool and fine-wool sheep into one breed. The original crosses were Lincoln and Leicester rams mated to Merino ewes. From this crossbred foundation, an intermediate, true-breeding population was obtained with equal emphasis on lamb and wool production. Since its first importation into the United States in 1914, the Corriedale has become widespread; it is found in many parts of the country and is intermediate in popularity at present.

In appearance, Corriedales are white with dark pigmented muzzles and hooves. Size is variable, with a tendency toward a greater emphasis on size in Corriedales raised in the western states, but generally they are medium sized. Figures 5-10 and 5-11 show examples of Corriedales.

Figure 5-10 Corriedale ram lamb. (Courtesy of American Corriedale Association, Seneca, Illinois)

Medium-Wooled Breeds

Figure 5-11 Corriedale yearling rams, Wyoming State Fair. (University of Wyoming photo)

One strong point of Corriedales is their fleeces, which are medium in grade and highly productive. They are noted for attractive fleeces with a bold crimp and a bright lustrous appearance. The wool provides a complete covering, including the legs, but Corriedales should be open faced.

Panama

The original cross used in the formation of the Panama breed was Rambouillet rams mated to Lincoln ewes, the reciprocal of that used in forming the Columbia. Private breeders, such as James Laidlaw in Idaho, rather than state or federal experiment stations, have been the guiding influence in the breed development. Popularity of Panamas has been limited to Idaho and neighboring states.

In appearance, Panamas are similar to Columbias (Figure 5-12). They are large-framed white sheep with medium wool. Any differences from Columbias

Figure 5-12 Panama yearling ewes. (Courtesy of N.C. 111 Technical Committee)

Figure 5-13 Targhee ram. (Courtesy of U.S. Targhee Association, Jordan, Montana)

would be the result of selection, either in the original sheep used in forming the breed or in selection pressure applied in later generations.

Targhee

The goal in the development of Targhees was to produce a sheep that would breed true for fleeces grading half-blood, and still maintain desirable traits of the fine-wooled sheep with regard to adaptability and longevity. They originated as a "backcross" or "comeback" to fine-wool breeding, and basically their ancestry is one-quarter Lincoln and three-quarter Rambouillet. They were developed at the U.S. Sheep Experiment Station at Dubois, Idaho. Since their origin, the popularity of Targhees has continued to increase and at present they are fairly widespread throughout the country.

The appearance of Targhees is similar to other white-faced, medium-wooled sheep. They are large-framed, white in color, polled, and with fleece grade being

Figure 5-14 Targhee ewe at Wyoming State Fair. (University of Wyoming photo)

their chief distinguishing trait (Figures 5-13 and 5-14). There are many flocks of sheep in the western range area that are similar in appearance, often called western white-faced sheep or often thought of as Targhee-type sheep. Targhees are highly productive, with heavy shearing, high-quality fleeces and a relative high reproductive rate. They are readily adaptive to a wide variety of conditions.

Montadale

This is a relatively new breed developed in Missouri by crossing Columbias and Cheviots, with the goal being to produce dual-purpose sheep. Montadales have continually increased in popularity, particularly in the Mid-west.

They are white sheep, medium in size, polled, and bare faced and bare legged with an alert appearance similar to that of Cheviots. They produce relatively heavy fleeces of medium grade and are muscular and desirable in conformation. Figure 5-15 shows Montadales.

Figure 5-15 Montadale ewes at Top of the Rockies sale. (University of Wyoming photo)

Suffolk

This breed originated in England and was not widespread in this country in the early days of sheep production, but at present this is the most popular of all breeds in the United States, particularly as measured by number of registrations or number of transfers.[1,2] Also, the Suffolk is outstanding in growth rate and in meatiness or carcass cutability. Suffolks are popular in all sheep-growing areas in this country. Suffolk rams are used extensively as sires for crossbreeding programs in all types of conditions. Resulting crossbred ewes also are commonly used in farm flock production, but under range conditions this is not very common.

In appearance, they are tall and long bodied; they have the capability of growing to extremely heavy weights. They are black faced, black legged, and also bare faced and bare legged. Long, heavy, slightly drooping ears that are slick black are typical of the Suffolk. Those with short ears or with brown color on the face and legs or with wool covering over the poll are discriminated against. Although ears

Figure 5-16 Suffolk ram lamb ready for show. (University of Wyoming photo)

Figure 5-17 Suffolk show ewe. (Courtesy of Suffolk Banner, Cuba, Illinois)

and color or shape of face are not related to productivity or are not usable products, they do have a pronounced influence on value in the purebred business.

Suffolk lambs are born with black or dark fleeces and generally turn white except for extremities by the age of 3 to 4 months (Figure 7-5). The skin of Suffolks is often dark gray or black and, as might be expected, black fibers are often present as contaminants in the fleece, even after the color has turned to white. They are polled sheep, but it is not uncommon for rams to have small scurs. Wool production is not their "long suit" as their fleeces usually are light in weight (7 to 8 lb; 3 to 4 kg) and often are contaminated with hair, as well as with black or brown fibers. Although not desirable, they sometimes become bare bellied. Examples of Suffolks are seen in Figures 5-16 through 5-19. Suffolks are quite prolific. Suffolks are not as long-lived as are white-faced fine-wooled or medium-wooled sheep. Neither are they as good grazers under arid to semiarid range conditions.

Figure 5-18 Suffolk ewes with ram during breeding. (University of Wyoming photo)

Medium-Wooled Breeds

Figure 5-19 Suffolk lambs at weaning. (University of Wyoming photo)

In the United States, two breed associations (American and National) register and promote the Suffolk breed. Both are active and there are some sheep registered in both associations. Show ring trends have been toward extremes in height and length, with continued emphasis on color and attractive appearance.

Hampshire

These are similar to Suffolks in many respects. Hampshires also were developed in England, and they have the same desirable traits as Suffolks, such as growth and carcass meatiness, and tend to have the same problems with wool production. They rank second to the Suffolks in numbers of registrations and have been widespread throughout the country for many years.[1,2] In recent years, Hampshires have been changed in their appearance mostly with regard to wool covering over the face. They are black faced and black about their legs, but they have a wool cap extending over their poll and some wool growing on their legs (Figures 5-20, 5-21, and 5-22). Their ears are long and black, and light colors, such as brown or gray on the face and ears are objectionable. Hampshires are big sheep with large frames and are polled. Hampshire rams are commonly used in crossbreeding programs. Both straight Hampshire and Hampshire cross lambs produce excellent carcasses at desired mar-

Figure 5-20 Hampshire yearling ewe. (Courtesy of *Sheep Breeder and Sheepman Magazine*, Columbia, Missouri)

Figure 5-21 Hampshire ewes with ram during breeding. (University of Wyoming photo)

Figure 5-22 Hampshire lambs at weaning. (University of Wyoming photo)

ket weights. Since their skin is usually dark and they are so dark about the extremities, their fleeces have some problems with color contamination, and they shear relatively lightweight fleeces.

Shropshire

These also originated in England. As a breed in the United States, the Shropshire has had some fluctuations in popularity, likely due to their trend toward wool blindness and then selection away from such. Their popularity has been mostly in farm flock production, as Shropshires have not been used much even as sires in the range areas in recent years.

In appearance, Shropshires are medium sized with dark faces and dark markings on their legs (Figure 5-23). They are open faced with wool extending over their poll. Both rams and ewes are polled with no sign of scurs allowed. They are alert-appearing sheep with small semi-erect ears. Breeders attempt to maintain a high level of production for both wool and lamb. Thus their fleeces are relatively heavy, are

Medium-Wooled Breeds

Figure 5-23 Shropshire yearling ewe. (Courtesy of American Shropshire Association, Monticello, Illinois)

of medium grade, and should be white and free of dark fiber contamination. Also, they are prolific sheep that produce lambs excellent in carcass conformation.

Dorset

These originated in southern England, and Dorset horn sheep were brought into the United States in the late 1880s. In the early years of the sheep industry in this country, this breed was the only one in which horns were common on both rams and ewes. The advent of polled Dorsets during recent years has given the breed a boost in popularity, and now the Dorset ranks third among U.S. breeds in registration numbers.[1] Part of the increase in popularity is also a result of their adaptability. Dorsets have a longer breeding season than most other breeds and are similar to fine-wooled sheep in this respect. Their ability to breed out of season makes them useful for early lambing programs such as hot-house lamb production and for accelerated programs.

In appearance Dorsets are medium-sized white sheep and produce fleeces that are of medium grade and fairly lightweight as compared to western white-faced sheep (Figures 5-24 and 5-25). Either horned or polled sheep are acceptable accord-

Figure 5-24 Dorset yearling rams. (Courtesy of Continental Dorset Club, Hudson, Iowa)

Figure 5-25 Dorset yearling ewe. (Courtesy of Continental Dorset Club, Hudson, Iowa)

ing to breed standards. Dorsets are also excellent from the standpoint of carcass conformation.

Oxford

Oxford was one of the later breeds to be developed in England and was the result of crossing Cotswolds and Hampshires with selection based on size and production. Its introduction into this country occurred at about the same time as that of several "Down breeds," but it has never reached the widespread popularity of some others.

Oxfords are large sheep with dark brown or gray faces with fairly complete wool covering. In past years there was a tendency toward wool covering both the ears and faces. Sheep of both sexes are polled and their fleeces reflect the influence of the Cotswold in their ancestry, being longer, coarser, and heavier than those of other Down breeds. Their fleeces are relatively free of dark fiber contamination, but are neither as heavy nor as high in quality as fleeces from the fine-wooled sheep. Their appearance in the show ring is somewhat artificial, with the topknot a result of grooming for the show (Figures 5-26 and 5-27).

Figure 5-26 Oxford yearling ram. (Courtesy of American Oxford Sheep Association, Ottawa, Illinois)

Figure 5-27 A group of Oxford lambs. (Courtesy of American Oxford Sheep Association, Ottawa, Illinois)

Southdown

This is considered as one of the oldest breeds of sheep; it was developed in England and was used in the formation of other English breeds. It achieved fairly widespread popularity in many areas due to its extreme muscularity and carcass conformation. However, some of that popularity has been lost, and Southdowns have never been used in range production due to their lack of size.

In color, Southdowns are white with mouse-brown to gray faces and legs. They are polled and are small sheep in compared to most other domestic breeds (Figures 5-28 and 5-29). Their fleeces often grade 58 to 60 in spinning count, but are light in weight. Southdowns are useful in crossing programs for production of lightweight carcasses, such as hot-house lambs. When fed to current desired weights for market lambs, they produce wasty carcasses with excess fat.

Figure 5-28 Southdown yearling ram. (Courtesy of American Southdown Breeders Association, Bellefonte, Pennsylvania)

Figure 5-29 Southdown ewe lamb. (Courtesy of American Southdown Breeders Association, Bellefonte, Pennsylvania)

Cheviot

This, too, is one of the oldest breeds, originating in the hill country of Scotland, where they were developed for hardiness and vigor. There are different strains of Cheviots, some known as Border Cheviots and some known as North Country Cheviots, which are larger but still have the same hardiness.

These sheep are striking in appearance, having small erect ears that make them look alert (Figures 5-30 and 5-31). They are white and polled, with bare faces and legs, and are small in size; even North Country Cheviots are small framed compared to such breeds as Suffolks and Hampshires. They are meaty sheep with excellent carcass conformation, but tend to fatten at lighter weights than currently desired. They produce lightweight, medium-grade fleeces. They are generally known as active sheep that are good grazers and are still considered as hardy sheep.

Figure 5-30 North Country Cheviot ram. (Courtesy of *Sheep Breeder and Sheepman Magazine,* Columbia, Missouri)

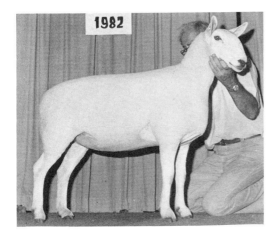

Figure 5-31 North Country Cheviot ewe. (Courtesy of *Sheep Breeder and Sheepman Magazine,* Columbia, Missouri)

Finn

The greatest change in the breed structure of sheep in the United States during recent years has resulted from the introduction and expansion of Finn-sheep breeding. Imported as Finnish Landrace, they have become known simply as Finns. Since more emphasis is being placed on reproductive rate, the Finn has made a large contribution. Finns are prolific, being far ahead of any breed or cross previously used in this country. Not only are Finns more prolific, but they also reach puberty at an earlier age than do sheep of our common domestic breeds. They have proved that they can transmit part of their advantage in this respect to their crossbred offspring. Thus they have been used extensively in crossbreeding programs by many state and federal research stations, as well as by private breeders. Finn and Finn cross rams have been used in many range flocks, but as yet an evaluation of their success is not readily available. Whether these rams can weather severe storms during breeding season and still maintain their aggressive breeding behavior is not known. They have been used in the development of the Polypay, to be discussed later, and in the formation of several composite breeding groups aimed at a high level of production. This popularity of Finns in new combinations is likely to continue as long as they have such an advantage over other breeds in the most important of all traits.

In appearance, Finn sheep are small, white, and polled, with bare faces and bare legs (Figures 5-32, 5-33, and 5-34). Their wool is of medium grade and is light weight. Pure Finns have a small, short, ratlike tail, which is used as an indication that they are pure. They are not very impressive in terms of size and ruggedness, nor in fleece production nor carcass cutability when compared with our traditional standards. Thus their contribution to production may have an influence on what is considered as the ideal in terms of overall production.

Figure 5-32 Finn ewes. (Courtesy of Finnsheep Breeders Association, Inc., Indianapolis, Indiana)

Figure 5-33 Finn ram. (Courtesy of Finnsheep Breeders Association, Inc., Indianapolis, Indiana)

Figure 5-34 Finn ewe with lambs. (Courtesy of J. K. Judy, Columbus, Ohio)

Polypay

This recently developed breed originated at the U.S. Sheep Experiment Station at Dubois, Idaho; it is a combination of four breeds: Dorset, Finn, Rambouillet, and Targhee. The goal was to develop a breed of sheep that is white, polled medium wooled, and medium sized, yet is highly prolific and heavy shearing and will breed at any time of year. Reports from the station at Dubois indicate that these sheep are highly productive.

Figure 5-35 Polypay ewes and lambs. (Courtesy of U.S. Sheep Experiment Station, Dubois, Idaho)

In appearance, they are white, polled, medium-sized sheep and have fleeces in the medium grades (Figure 5-35). Pedigree requirements suggested by the Polypay association are as follows:

1. Each contain or be directly descended from ancestors each containing no less than three-sixteenths blood from each of the four breeds, or
2. Result from at least two top crosses of Polypay rams or ewes descended from one or any combination of the four foundation breeds.

In addition, the certification program is designed to create an elite registry of superior Polypays. Breeders are encouraged to certify all eligible sheep.

COARSE- OR LONG-WOOLED BREEDS

Lincoln

This is a breed developed and used in England that has contributed to the U.S. sheep industry through crossing programs and establishment of other breeds. Columbias, Targhees, Corriedales, and Romeldales have Lincoln breeding in their ancestry. Also, many western white-faced sheep are based on similar breed combinations that are basically from one-half to one-quarter Lincoln breeding. Warhill sheep, developed by the Warren Livestock Co., for example, trace their ancestry in part to Lincolns. For many years Lincoln x Rambouillet crossbred rams were a part of several state or national wool growers sales in western states, and some of these are still advertised by private breeders.

In appearance, Lincolns are big, white rugged sheep with long, coarse wool. They tend to have wool hanging down over their face, but not growing on the face.

Cotswold

This is another breed of English origin and in many respects Cotswolds are similar to Lincolns. They have never been very widespread in the United States and have not been used very extensively in crossbreeding programs. In appearance they are large rugged sheep that are white and polled (Figure 5-36). They have long, coarse fleeces that tend to be in locks.

Figure 5-36 Cotswold yearling ewes. (Courtesy of American Cotswold Record Association, Rochester, New Hampshire)

Leicester

Although there are two breeds (types), the English Leicester and the Border Leicester, they have similar characteristics in regard to size and wool. These originated in different areas in England and have become distinct in appearance. Some Leicesters have been used in early development of Corriedales and Columbias, but Leicesters have never become very widespread in the United States.

They are white sheep, polled, and have course wool, but neither as coarse or as long as that of Lincolns. In size, they are intermediate among the more common breeds in the United States. Border Leicesters appear more bare faced and bare legged than English Leicesters.

Romney

This is considered a long-wooled breed, but is somewhat different from the previously discussed breeds also classed as long-wooled. It originated in southeastern England and has found some popularity in the far western part of the United States. The breed, however, has never become very widespread in the United States.

The Romney is medium sized, white, and polled (Figure 5-37). It is common for Romneys to have dark-pigmented nostrils and hooves. Their fleeces are more dense and somewhat finer than are fleeces of the other long-wooled breeds, but still

Figure 5-37 Romney ewe. (Courtesy of American Romney Breeders Association, Corvallis, Oregon)

they are coarse, with grade averaging low quarter-blood. Romney breeders feel that these sheep are more adaptable than are most breeds to climates with high humidity. Also, Romneys seem to be more resistant to foot rot than are sheep of other breeds.

OTHER BREEDS

It is common to see advertisements indicating that sheep of several breeds or crosses are being produced. Such would include Black-faced Highland, Karakul, Tunis, Ryeland, Barbados, St. Croix, and Clun Forest. In addition, colored sheep are being produced, either as representing a particular breed or merely designated by color. New interest in home spinning of wool has created a demand for colored fleeces, either black, brown or multicolored, and it is not unusual for these to bring a higher price than pure white wool. Further information can be obtained from sheep publications such as *Sheep Breeder Magazine, Shepherd, National Wool Grower,* and *Sheep Magazine.*

REFERENCES CITED

1. Henson, Elizabeth. 1985. *North American Livestock Census.* American Minor Breeds Conservancy. Pittsboro, N.C.
2. National Society of Livestock Record Associations. 1983-1984. Annual Report and Directory.
3. Parker, C. F., and A. L. Pope. 1983. The U.S. sheep industry: Changes and Challenges. *J. Animal Sci.* 57, Supplement 2, p. 75.
4. Scott, George E. 1975. The Sheepman's Production Handbook. Sheep Industry Development Program, Denver, Colo.

SUGGESTED READING

Briggs, Hilton M. 1969. *Modern Breeds of Livestock,* 3rd ed. Macmillan Co., Collier-Macmillan Canada Ltd., Toronto.

Ensminger, M. E. 1970. *Sheep and Wool Science,* 4th ed. Interstate Printers and Publishers, Danville, Ill.

Kammlade, William G., Sr., and William G. Kammlade, Jr. 1955. *Sheep Science.* J. B. Lippincott Co., Philadelphia.

6 Reproductive Physiology

ANATOMY AND PHYSIOLOGY OF THE REPRODUCTIVE SYSTEMS

Reproduction is the first goal in any livestock operation, and rate of reproduction is the single trait having the greatest influence on the efficiency or profitability of a sheep enterprise. To understand the causes of variation in lambing rate and be able to avoid reproductive failures, it is essential to have a knowledge of the anatomy of the reproductive systems.

Female Reproductive System

The reproductive tract of the ewe is illustrated in Figure 6-1. This is a very complex system, as the ewe provides the ova. She also provides the environment for developing ova, for fertilization of these ova, and for development of embryos following fertilization. She nourishes the developing fetus during pregnancy, gives birth to the lambs, and then provides milk for nourishment of her lambs during their early life. A summary of the functions of the female reproductive organs is given in Table 6-1.

Ovaries. Ovaries are the primary organs of the female reproductive tract since here is where the ova are produced. Ovaries are also the site of female hormone production (estrogens and progestins). They are somewhat almond shaped and in ewes are small, about 0.6 in. (1.5 to 2 cm) long at their longest axis and each weighs less than 0.01 lb (2 to 4 grams).

Developing ova are contained in small sacs (Graafian follicles) found in the outer layer of germinal epithelium. These are numerous and are in various stages

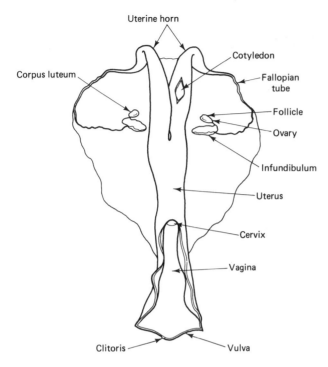

Figure 6-1 Reproductive organs of the ewe. (Drawn under the direction of Dr. C. E. Terrill by artist Denver D. Baker. From *Sheep and Wool Science*, 4th edition, by M. E. Ensminger.[5] © 1970 by the Interstate Printers and Publishers, Inc., Danville, Illinois. Used with permission.)

TABLE 6-1 REPRODUCTIVE ORGANS OF THE FEMALE WITH THEIR MAJOR FUNCTIONS

Organ	Function(s)
Ovary	Production of oocytes Production of estrogens (Graafian follicle) Production of progestins (corpus luteum)
Oviduct	Gamete transport (spermatozoa and oocytes) Site of fertilization
Uterus	Retains and nourishes the embryo and fetus
Cervix	Prevents microbial contamination of uterus Reservoir for semen and transport of spermatozoa Site of semen deposit during natural mating in sows and mares
Vagina	Organ of copulation Site of semen deposit during natural mating in cows and ewes Birth canal
Vulva	External opening to reproductive tract

Reproduced with permission from Bearden, and Fuquay.[1]

Anatomy and Physiology of the Reproductive Systems

of development. In addition to an ovum, each follicle contains fluid that increases in amount as the follicle increases in size. As the ovum matures during this follicular growth, the secretion of estrogen is accelerated until ultimately the follicle ruptures and drops the ovum in a process called ovulation. Multiple ovulation normally is the result of two or more follicles developing and rupturing at or near the same time. This can occur as multiple ovulation on one ovary or as the rupture of one or more follicles on each ovary.

Following ovulation, a corpus luteum is formed at the ovulation site and it begins to secrete progesterone. Should fertilization occur, progesterone production continues to help in maintaining pregnancy and to prevent ripening of other follicles. If fertilization does not occur, the corpus luteum will regress, progesterone production will decrease, and a new follicle or follicles begin to mature as a new cycle begins. The ovary in cross section is seen in Figure 6-2.

Oviducts. Oviducts are small tortuous tubes, often called Fallopian tubes, that lead from the ovaries to the horns of the uterus. The ends near the ovaries flare out to catch ova released from the ovaries in the process of ovulation, but there is no attachment of the oviducts to the ovaries. Within the oviducts a ciliated hairlike lining produces a wavelike motion, and aided by movement of the tubes themselves, the ova are moved through the duct toward the uterus. Sperm from the male pass upward into the oviducts and this is normally the site of fertilization. Oviducts terminate at the horns of the uterus.

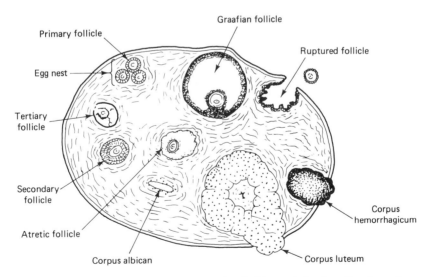

Figure 6-2 Ovary in cross section. (Reproduced with permission from Bearden and Fuquay[1])

Uterus. This is the largest part of the female reproductive tract, consisting of two horns and a body. It is muscular and is supplied with many blood vessels, glands, and lymph spaces. It is elastic and is capable of extreme expansion during pregnancy and subsequent reduction in size following parturition. The muscular character of the uterus aids in expulsion of the fetus during parturition. Uterine horns are about 4 to 5 in. (10 to 12 cm) in length and, after the horns merge, the body of the uterus is only about 1 in. (2 to 3 cm) in length. The entire uterus lies within the body cavity. Its inner lining has many cotyledons (from 75 to 150), or buttons, that attach to the membranes surrounding the fetus.

Cervix. This is often called the neck of the uterus; it is thick walled and muscular in nature with many folds. It is about 1.5 in. (3 to 4 cm) in length and serves as a protective mechanism, being tightly closed except during estrus and parturition. Sperm cells are capable of moving through the cervix. During pregnancy, a mucus is secreted by cervical cells to form a plug, which helps prevent invasion by disease organisms. The cervix begins about 8 in. (20 cm) forward from the exterior opening of the vagina.

Vagina. The female organ of copulation is the vagina, which is thin walled, elastic, and tubular in shape, and about 5 to 6 in. (12 to 15 cm) in length. Semen is deposited near the anterior end where the vagina joins the cervix. The vagina is a passageway for the fetus during parturition and is capable of expansion at this time.

Vulva. This is the external genital organ of the ewe. It is a continuation of the vagina and acts as a passageway for both reproductive and urinary systems. It is comprised of a vestibule from the external urethral opening to the exterior end of the labia, or folds, which are the terminal end of the reproductive tract.

Clitoris. This is the erectile organ of the ewe and lies about 0.5 in (1 cm) forward from the labia at the lower part of the vestibule. It contains erectile tissue and is well supplied with sensory nerves, but is not very prominent in the ewe.

Support. The primary supporting structure for the female reproductive tract is the broad ligament, which suspends the ovaries, oviducts, and uterus from either side of the dorsal wall of the pelvis. Both blood vessels and nerves pass through the broad ligament to the reproductive system.

Mammary system. Although not a part of the reproductive system as such, the mammary system is necessary for the nourishment and survival of newborn lambs and thus is an integral part of the reproductive process. The ewe's mammary system is comprised of two mammary glands with one teat or nipple for each. These are fused together, one on each side of the midline, in the inguinal region (see Figure 6-3). The external part of the mammary system, or udder, is supported by suspensory ligaments with some help from the skin.

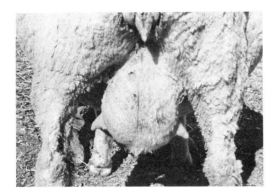

Figure 6-3 Rear view of ewe's udder, 2 weeks prior to lambing. (University of Wyoming photo)

The mammary glands not only produce milk for nourishment of lambs, but also produce a substance called colostrum, which is the first milk produced following parturition. This colostrum contains antibodies that provide the initial resistance to disease in newborn lambs.

In sheep it is common to see variation in the shape and size of the udder. When functioning normally, the udder is relatively soft and pliable in texture. It is not unusual to see two extra supernumerary teats in front of the udder, and attempts have been made to select sheep with four functional teats. However, in most sheep with four teats, only two are functional. The udder and teats are vulnerable to injury, and care is required in handling and particularly in shearing to avoid partial impairment of the mammary system.

Male Reproductive System

The role of the ram is to produce sperm cells capable of fertilizing ova and to introduce these into the female tract at the proper time, when ova are present and capable of fertilization. A diagrammatic sketch of the ram's reproductive system is shown in Figure 6-4, and a summary of the functions of each organ is given in Table 6-2.

Testes (testicles). These are the primary organs of reproduction in the male, as they produce both sperm cells and male hormones responsible for libido or sex drive. The testes do not remain in the body cavity, but descend through the inguinal canals into the scrotum before birth; but it is not uncommon for this to occur shortly after birth. Rams' testes are large in proportion to body size as compared to cattle or swine, weighing approximately 0.5 lb (200 to 300 g). Scrotal circumference in mature rams ranges as high as 16 in. (40 cm), which is nearly as great as that of bulls. Due to their relative size and the length of the scrotum, rams' testes are carried in a vulnerable position, particularly in warm weather, and are subject to injury more easily than is the case in other species.

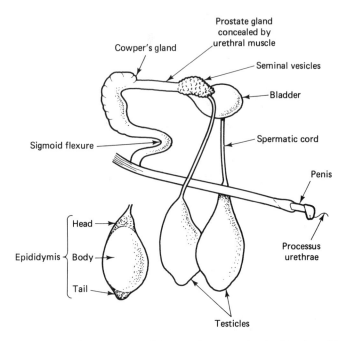

Figure 6-4 Reproductive organs of the ram. (Drawn under the direction of Dr. C. E. Terrill by artist Denver D. Baker. From *Sheep and Wool Science*, 4th edition, by M. E. Ensminger.[5] © 1970 by the Interstate Printers and Publishers, Inc., Danville, Illinois. Used with permission.)

Once the ram has reached sexual maturity, both sperm production and production of hormones (mainly testosterone) are continuous processes. Sperm cells are small, about 5 microns in length, and are highly concentrated in ram semen. Average volume of a single ejaculate is about 1 ml, and sperm cell concentration ranges from 1 to 4 billion/ml. Testosterone is the most important male sex hormone and is essential for normal development of secondary male characteristics and for normal mating behavior. Also, it aids in proper functioning of accessory sex glands and in production of sperm and their transport. Testosterone concentration is highly associated with testicular size and libido.[4]

Scrotum. This is a two-lobed sac that encloses the testes. Its functions are to provide protection and to aid in temperature control. The scrotal sac becomes long in hot weather, during which time surface area is relatively large, and draws up short and nearer the body during cold weather (see Figure 6-5).

Temperature of the testes is maintained at about 3° to 5° F, below that of body temperature, and without this differential, sperm production is reduced or discontinued. There is great difference in the appearance of the scrotum of different rams due to breed and time of year, with some variation within breed. Some have wool covering the scrotum; some have a thin hair covering. Some have a smooth

TABLE 6-2 REPRODUCTIVE ORGANS OF THE MALE WITH MAJOR FUNCTIONS

Organ	Function(s)
Testis	Production of spermatozoa
	Production of androgens
Scrotum	Support of the testes
	Temperature control of the testes
	Protection of the testes
Spermatic cord	Support of the testes
	Temperature control of the testes
Epididymis	Concentration of spermatozoa
	Storage of spermatozoa
	Maturation of spermatozoa
	Transport of spermatozoa
Vas deferens	Transport of spermatozoa
Urethra	Transport of semen
Vesicular glands	Contributes fluid, energy substrates, and buffers to semen
Prostate gland	Contributes fluid and inorganic ions to semen
Bulbourethral glands	Flushes urine residue from urethra
Penis	Male organ of copulation
Prepuce	Encloses free end of penis

Reproduced with permission from Bearden, and Fuquay.[1]

appearance at the lower end as the sacs are fused all the way and some have a cleft appearance. In either case, it should give an even or balanced appearance in normal healthy rams. Abnormalities such as orchitis, epididymitis or single chryptorchidism (only one testicle) often can be detected by observation of the scrotum. This is an area highly susceptible to shearing injuries, particularly in rams with wool covering the scrotum.

Epididymis. The epididymis fused to each testis or testicle is an external duct leading from the testis. The head (caput) of each is flattened at the apex of each testis, and there are 12 to 15 smaller ducts (vas efferentia) that merge, thus forming the single duct. The body (corpus) is a continuation from the head and is a single duct that extends over the long axis of the testis and then becomes the tail (caudus) of the epididymis. The lumen or interior of the tail becomes larger than that of the body. The total length of this entire convoluted duct is nearly 100 ft (30 m). Figure 6-6 shows how the epididymis fuses to the testis.

As a duct leading from the testis, the epididymis becomes a mechanism for sperm transport, with the time involved in transport through the entire epididymis ranging from 12 to 15 days. Ejaculation tends to speed up movement of sperm through the duct. During transport, sperm cells become more concentrated due to absorption of fluids. The change is from about 100 million sperm per milliliter up to from 2 to 4 billion per milliliter. Maturation of sperm cells also occurs during

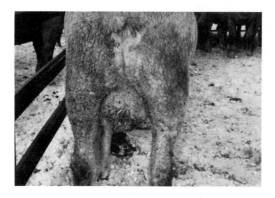

Figure 6-5 Scrotum of Rambouillet ram in winter (upper) and of two Rambouillet rams in summer (lower). (University of Wyoming photos)

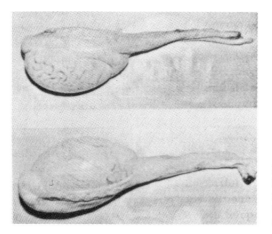

Figure 6-6 View of the epididymis as it is fused to the surface of the testis. Note the tunica albuginea testis that covers the testis. (Reproduced with permission from Bearden and Fuquay[1])

transport. Newly formed sperm cells entering the head of the epididymis from the vas efferentia are neither motile nor capable of fertilizing ova. Storage of sperm cells occurs mostly in the tail of the epididymis. If stored for too long a period, sperm cells can deteriorate.

Vas deferens. This is the duct that leads from the epididymis, one from each, and passes along the spermatic cord through the inguinal canal to the pelvic region. Here both of them merge with the urethra at its origin near its opening to the bladder. The enlarged ends of the vas deferentia near the urethra are called ampullae. The only function of the vas deferens is transport of sperm.

Sterilization (vasectomy) of rams is a surgical procedure in which a portion of each vas deferens is removed and both ends of remaining ducts are tied off. When the emasculatome (burdizzo) is used properly for castrating lambs, each vas deferens is severed without cutting the skin, and the nerves as well as blood vessels continue to function.

Urethra. This is a single duct leading from the ampullae of the vas deferentia to the end of the penis, through which both urine and semen are excreted. Following ejaculation, sperm cells are mixed with fluids from accessory glands along the urethra to form semen.

Accessory glands. *Vesicular glands* (seminal vesicles) open into the urethra about where it originates by merging with the ampullae, and fluid from these glands makes up a large portion of the fluid volume of semen. This fluid contains buffers to protect sperm against changing pH, as well as compounds that provide energy for the sperm cells.

The *prostate* is a single gland completely embedded in urethral muscles. It contributes only a small portion of the fluid volume; its secretion is high in sodium, chlorine, calcium, and magnesium, all in solution.

Bulbourethral glands (Cowpers glands) are a pair of glands located along the urethra where it exits from the pelvis. These contribute very little of the fluid volume of semen.

Penis. This is the organ of copulation. It has a sigmoid flexure (S-shaped bend), which allows it to be retracted completely into the body. A pair of retractor muscles anterior to the flexure relax to permit extension and contraction to draw the penis back into the body. The glans penis (free end) is fibroelastic, contains small amounts of erectile tissue, and is liberally supplied with sensory nerves. A thin twisted projection (urethral process) extends about 1.5 in. (3 to 4 cm) beyond the tip of the penis. The *prepuce,* or sheath, encloses the free end of the penis to provide protection.

BREEDING BEHAVIOR AND REPRODUCTION

Sheep are seasonal breeders, termed seasonally polyestrous, even though some tend to be less affected by season, and breeding activity may occur during all months of the year. Combinations of changing daylight hours and changes in temperature are involved in the seasonal pattern. Likewise, lactation contributes to seasonality, since lactating ewes are not likely to exhibit estrus (come into heat) without hormone therapy. Usually the time of lactation coincides with the time of seasonal anestrus, at least in part.

Estrus

The phenomenon of ewes being in heat, the period during which ewes are receptive to the male, is called estrus or the estrous period. It is first experienced in ewes at ages ranging from 6 to 10 months, or in many slow developing ewes at 18 to 20 months of age. The time of first estrus accompanied by ovulation is known as *puberty,* but this does not denote complete sexual maturity. The estrous period lasts for an average of 29 hours in ewes, with much individual variation. It is logical that the number of ova shed has an influence on length of the period. Ewes with multiple ovulations are probably in heat longer than those with only one ovum shed unless the multiple ovulations occur simultaneously, which is unlikely.

The *estrous cycle* indicates the length of time between estrous periods and averages nearly 17 days in ewes. Unless fertilization occurs, ewes will continue the estrous cycle with recurring periods throughout the breeding season. The cycle is regulated by a balance of hormones. Gonadotropic hormones from the anterior pituitary, which are protein in nature, interact with steroid hormones produced by the ovaries and prostaglandins from the uterus.

Progesterone is produced by the corpus luteum and is the dominant hormone in controlling the estrous cycle. During seasonal anestrus and during most of the time between estrous periods, the progesterone level in the blood is relatively high and levels of all other hormones are relatively low. Just prior to the start of a new period, prostaglandins cause regression of the corpus luteum, with a subsequent reduction in the level of progesterone. Then levels of luteinizing hormone (LH), follicle-stimulating hormone (FSH), and prolactin produced by the anterior pituitary all increase. This increase is followed by a surge of estrogen produced by the growing follicle. Surges of LH and FSH occur early in estrus and last for 8 to 10 hours. Prolactin surges and remains high during estrus. Ovulation occurs late in the estrous period, and it is after ovulation that the corpus luteum develops at the site of the ruptured follicle. This happens rapidly, aided by both prolactin and LH, and in four or five days the new corpus luteum is producing progesterone. If fertilization occurs, the corpus luteum is maintained and the level of progesterone in the blood remains high throughout pregnancy and continues as the dominant hormone. If fertilization does not occur, a new cycle begins, with the progesterone level remaining high only until the corpus luteum regresses. It is common for ewes to have silent estrus (without ovulation) at the beginning and end of the season.

Ewes do not show visible signs of estrus if a ram is not present. When a ram is present, the ewe in heat will exhibit mating behavior by rubbing against the ram, circling him and sniffing his genitalia, and shaking her tail. A sterilized ram is needed to detect ewes in heat when hand mating or artificial insemination is being used.

Fertilization

The union of ova and sperm occurs in the oviduct, in the upper part near the ovary. Ova descend toward the uterus and sperm ascend from the vagina through the uterus and into the oviducts. When conditions are right, the union occurs. The life of an unfertilized ovum is short (5 to 8 hours), and sperm cells can live up to 24 hours in the female tract. Thus correct time of mating, as well as a favorable environment, is essential to assure success of fertilization.

Artificial Insemination

Fertilization can be achieved by artificial insemination, but the use of AI in sheep has not been widespread in commercial flocks in the United States. According to Inskeep, Stevens, and Peters,[10] several problems need to be solved before AI becomes practical for use in sheep. Two of the main problems are the lack of satisfactory methods for frozen storage of ram semen and the lower fertility when AI is used at controlled estrus.

Ova Transfer

Modern technology has been developed that allows for altering the reproductive process in ewes. The use of superovulation and transfer has been accomplished, and this provides the potential for more efficient utilization of superior germ plasm.[1] Not only is this important in the introduction of new genetic material, but it can also be helpful in more rapid improvement by selection.

Gestation

Gestation, or the period of pregnancy, begins at the time of fertilization and lasts for an average of about 147 days in ewes, with considerable variation in gestation length. During the early part of gestation, the embryo is free floating and some migration occurs between the right and left horns of the uterus.[2] For the first 10 days the embryo undergoes a period of cleavage during which cell division occurs without growth. Following this, differentiation occurs, and germ layers, extraembryonic membranes, and body organs begin to develop. Size of the embryo increases

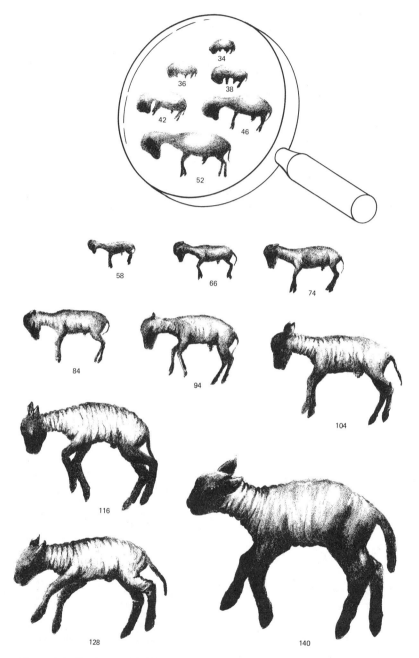

Figure 6-7 Fetal growth in sheep. The specimens at the top (from 34 through 52 days) were photographed at a different magnification that were the others. Age in days is indicated by the number below the specimen. (Adapted with permission from Winters[17])

Factors Influencing Productivity

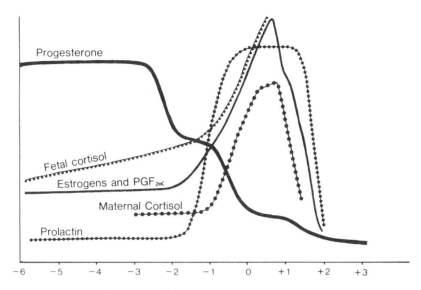

Figure 6-8 Relative changes in hormone concentration near the time of parturition in the ewe. (Reproduced with permission from Bearden and Fuquay[1])

rapidly during this state. By this time, placental attachment to the uterus has been completed and the embryo can obtain nourishment and dispose of wastes through the maternal blood supply. When differentiation has been completed, terminology changes and the embryo is now a fetus. From this stage until parturition the fetus grows and develops into a complete individual. Relative growth and development of the fetus can be observed in Figure 6-7.

A proper hormone balance is essential during gestation, with progesterone being the dominant hormone. During gestation, progesterone is produced by both the corpus luteum and the placenta surrounding the fetus. Near the end of gestation the hormone balance changes markedly in preparation for parturition. Changes in hormone levels during the preparturition period are shown in Figure 6-8. Growth of the mammary glands occurs during late gestation, sometimes beginning as much as three to four weeks ahead of parturition. As parturition approaches, the udder becomes large and distended as it fills with milk.

FACTORS INFLUENCING PRODUCTIVITY

Season, Light, and Temperature

The normal season for breeding sheep in the northern hemisphere is during the fall and extending into early winter, often stated as August through January, with the peak of the season during October and November (see Figure 6-9 and Table 6-3). Sexual activity is primarily controlled by the light–dark ratio. The incidence of es-

TABLE 6-3 PERCENTAGE OF EWES SHOWING AT LEAST ONE ESTRUS IN A FOUR-WEEK INTERVAL

Breed Group	Mean for Breed Group	Month of Breeding[a]											
		Jan	Feb	Mar	Apr	May	Jun	Jul	Aug	Sep	Oct	Nov	Dec
Wisconsin Hampshire	53	76	72	72	36	28	4	0	0	72	96	96	80
Texas Rambouillet	52	69	54	69	54	42	8	0	31	65	81	85	69
Suffolk	52	71	75	71	13	33	4	0	16	75	88	96	83
Polled Dorset	52	70	59	78	26	30	4	0	11	74	93	93	81
Beltsville Hampshire	48	64	64	60	28	8	0	4	20	68	84	96	76
Columbia	45	57	39	61	35	22	4	0	30	56	78	83	78
Montana Rambouillet	45	77	50	42	38	23	4	0	12	50	81	81	77
Targhee	43	40	52	52	28	16	0	0	4	80	88	92	68
Mean for month		66	58	63	32	25	4	0	16	68	86	90	76

Reproduced with permission from Lax and others.[12]

[a] The two-week periods beginning August 22 and December 27 were omitted and the remaining 24 two-week periods were combined to make four-week intervals that corresponded roughly to the 12 calendar months.

Factors Influencing Productivity

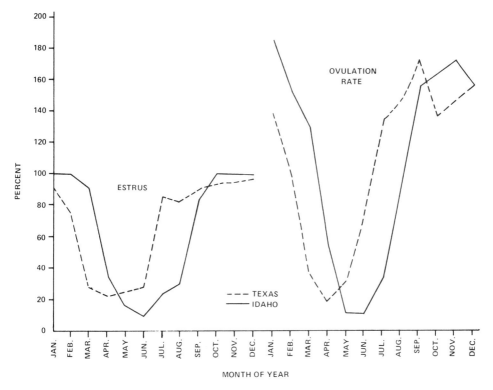

Figure 6-9 Estrus and ovulation rate by month at different locations. (Reproduced with permission from Hulet and others[9])

trus increases as days become shorter in the fall. Both fertility and ovulation rate reach their peak when daylight ranges from 10 to 12 hours, which occurs during October and November. In most areas, temperature is also getting lower at this time of year. Even though fertility and ovulation rate may peak in these two months, the optimum breeding time may be later in the year in warmer climates due to greater probability of embryo survival when temperatures are lower, as suggested in Table 6-4.

High temperatures are detrimental to reproduction, reducing both fertility and embryo survival. Extreme low temperatures during breeding, particularly when accompanied by blizzard conditions, have a detrimental effect on all body functions, including fertility. The most serious aspect of such conditions is that feed gets covered and physical discomfort prevents sheep from getting out and grazing, resulting in temporary nutritional deficiencies. These conditions are fairly common in the western range area when sheep are bred during December and January.

Rams too are affected by extremes in temperature. Although rams are capable of the reproductive process all year, both sperm concentration and spermatogenic activity in the testes are at their highest in the fall and gradually decrease to their

TABLE 6-4 OVULATION RATE AND LAMBING PERFORMANCE OF RAMBOUILLET EWES BRED DURING FOUR DIFFERENT SEASONS IN TEXAS

	Breeding Intervals			
Item	Mar 21–May 2	Jun 21–Aug 2	Sep 21–Nov 2	Dec 21–Jan 1
Ovulation rate (%)	106	141	175	152
Lamb production (%)	84	97	127	135[a]

Reproduced with permission from *Sheepman's Production Handbook*.[14]

[a]Higher lamb production in December believed due to cooler temperatures (better embryo survival).

lowest levels in summer.[4,6] Also, semen quality (viability and motility) and the amount of fructose in seminal plasma follow the same seasonal pattern. These changes sometimes produce a temporary sterility during late summer months after prolonged periods of exposure to high temperature. Fortunately, rams return to normal with no apparent harmful effects on their breeding ability.

Age of Ewe

The age of the ewe influences both lamb and wool production. In Figure 6-10 and Table 6-5, lamb production of a purebred flock comprised of sheep of four breeds is plotted by age. It is typical for production to increase with increasing age up to three years, to level off for two or three years, and gradually to decrease with further increases in age. Under range conditions, the curve of production is similar, but covers a shorter span of years.

Differences in wool production as influenced by age are shown in Figure 6-11, plotted by age at first lambing. There are expected differences, because wool

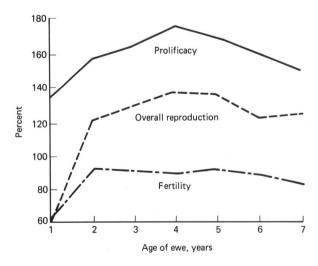

Figure 6-10 Fertility and reproductive rate by age of ewe. (University of Wyoming photo)

Factors Influencing Productivity

TABLE 6-5 INFLUENCE OF AGE OF DAM ON LAMBING PERCENT

Age	Number of Observations	Lambs Born per Ewe Lambing	SE
2	545	1.32	0.04
3	528	1.44	0.04
4	432	1.50	0.04
5	332	1.53	0.04
6	241	1.54	0.04
7	122	1.57	0.05
8	27	1.49	0.10
9	7	1.34	0.17

Reproduced with permission from Vakil, Botkin, and Roehrkasse.[16]

production competes with growth during the first year and later with lamb production and lactation. For ewes bred to lamb first at two years of age, the second year is relatively free of competition, and it is logical for the two-year-old ewes to shear the heaviest fleeces. If total production, including both lamb and wool, is plotted by age of ewe, the curve is nearly parallel to that for overall reproduction shown in Figure 6-10.

Age of Ram

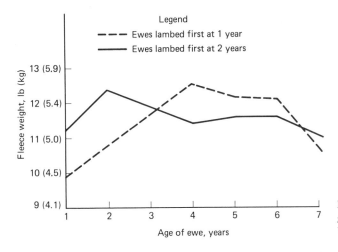

Figure 6-11 Influence of age of ewe on grease fleece weight. (University of Wyoming photo)

The age of the ram also influences its producing ability, with a reduction with advancing age, although there are not many reports of such. In Table 6-6 there is an indication that rams become less productive with advancing age. Observation of ram flocks lead to the conclusion that older rams do not maintain as much body condition, and testicle or epididymis problems are more prevalent if affected rams have not been removed. As long as rams are sound, they are capable of breeding up to six or more years of age.

TABLE 6-6 INFLUENCE OF AGE OF SIRE ON LAMBING PERCENT

Age	Number of Observations	Lambs Born per Ewe Lambing	SE
1	953	1.52	0.04
2	752	1.45	0.04
3	374	1.44	0.05
4	103	1.49	0.06
5	53	1.41	0.07

Reproduced with permission from Vakil, Botkin, and Roehrkasse.[16]

Ram lambs usually reach puberty at four to seven months of age when raised under good conditions and are capable of being used for breeding while still lambs. They are not fully mature yet and their sexual development and body growth will continue if they are not subject to too much stress as lambs.

Age of Ewes at Puberty

As previously indicated, the time of first estrus accompanied by ovulation is known as puberty in ewes. It is influenced by both heredity and environment. Anything that keeps ewe lambs from growing or developing at or near their potential, such as poor health or limited nutrition, will delay puberty. The age at which ewes reach puberty has a marked influence on lifetime production, in that ewes bred to lamb first as yearlings have a greater lifetime production than do ewes bred to produce their first lamb at two years of age. Whether bred as lambs or not, ewe lambs that exhibit estrus their first year are more productive than ewes that do not reach puberty as lambs.[8]

Many factors influence the ability to reach puberty early in life. Early-born lambs are much more likely to reach puberty as lambs, and well-fed, well-grown

TABLE 6-7 INFLUENCE OF WEIGHT AT BREEDING TIME ON OVULATION RATE AND LAMBING RATE[a]

	Year			
	1974		1975	
Weight Range (lbs)	n	Ovulation Rate[b]	n	Lambing Rate
100–129	48	1.08	36	1.03
130–139	29	1.34	29	1.07
140–149	39	1.36	29	1.10
150–159	26	1.42	19	1.19
160–169	18	1.44	13	1.23
170 and above	18	1.56	16	1.44

[a] Experimental flock, University of Wyoming.
[b] Ovulation rate determined by laparotomy.

Factors Influencing Productivity

TABLE 6-8 INFLUENCE OF WEIGHT PRIOR TO LAMBING TIME ON SUBSEQUENT FLEECE WEIGHTS

Weight Range (lbs)	n	Grease fleece weight (lbs)
100–129	44	10.4
130–139	40	10.6
140–149	38	11.3
150–159	31	11.6
160–169	19	12.8
170 and above	27	12.4

lambs have an advantage over lambs on limited nutrition. Breed has a significant effect on early puberty also.[3] Under range conditions in which lambs are born in May or June and weaned in the fall when weighing about 70 lb (34 kg), it is unusual for ewe lambs to reach puberty during their first year.

Influence of Size

Big sheep tend to produce more than do small sheep.[7,13] Figure 6-12 and Tables 6-7 and 6-8 show examples of comparisons within groups similar in breed or breed cross. Even though these data clearly support the original statement, there is much controversy over the matter of size. One notable exception is the prolificacy of Finn sheep, which are small in size. Probably the most controversial aspect of size in sheep and its influence on production relates to efficiency. Big sheep eat more and

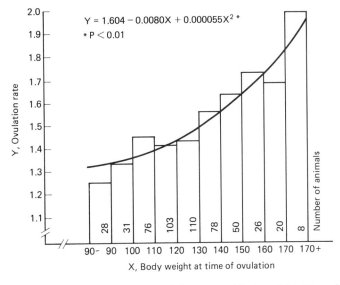

Figure 6-12 Relationship between ovulation rate and body weight. (Reproduced with permission from Gallagher and Shelton[7])

it is obvious that their maintenance requirements are relatively high. Thus, if efficiency is expressed as production per unit of body weight, medium-sized sheep are likely to be more efficient than are those at either extreme. If efficiency is based on production per unit of cost, big sheep have the advantage because most costs are on a *per head* basis. Even feed costs are the same for all sheep when grazing on public lands under lease agreements, a management system common in the western range area. When all costs are on a per head basis, this results in production and efficiency being synonymous. The best producers are the most efficient. This is one of the reasons that many sheep producers favor big sheep.

Measures of size are often difficult, since weight is commonly used to measure size and weight is variable during different times of the year or during different stages of production. Also, weight is often confounded with condition. Furthermore, desired market weight is still in the range of 105 to 110 lb (50 kg), and extremes in size are not necessary in order to produce desirable lambs at this weight.

It is likely that the controversy over size will continue and not be resolved completely. The logical solution, it would seem, is a direct approach, that of identifying and maintaining the best producers, regardless of their size.

Breed Influence

The breeds of sheep have been described in the previous chapter, but not compared for all productive traits. Breed is one of the greatest sources of variation. For any particular trait, there usually are several breeds that are similar. For example, Merinos, Rambouillets, and Dorsets are the breeds in which ewes are least seasonal in their breeding habits. This makes sheep of these breeds and their crosses most adaptable to out-of-season breeding and to accelerated management programs. Rambouillets, Merinos, and crosses with either of these in their ancestry are noted for flocking instinct and for hardiness (the ability to withstand adverse climatic conditions). They are less affected by extreme high or low temperatures and by arid conditions than are sheep of the meat breeds or long-wooled breeds. They are active grazers and are capable of traveling great distances for feed and water. They are adaptable to a wide variety of conditions and these are the ones that adapt to western range conditions. Sheep of these breeds are good wool producers, both in quantity and quality, their characteristic of fleece density in particular being important in affording insulation or protection from adverse weather. Sheep of what are sometimes termed meat breeds (Suffolk, Hampshire, Shropshire, Oxford, Dorset, Southdown, Cheviot) and Finns are relatively poor wool producers, while sheep of the long-wooled or coarse-wooled breeds produce heavy fleeces that are low in quality from the standpoint of clothing manufacture.

Reproductive rate differs among breeds, with Finn sheep in a class by themselves compared to other breeds in the United States. Not only are they more prolific, but also sexual maturity is earlier, and Finn rams are usually considered the most aggressive breeders. The other breeds are difficult to compare with regard to lambing rate due to differences in location and conditions in which they are found.

Factors Influencing Productivity

Relatively high lambing rates, 150% and above, have been obtained by breeders of several different breeds.

Growth and carcass desirability are related traits, and Suffolks are the outstanding sheep in this respect, with Hampshires a close second. Sheep of the fine-wooled breeds, coarse-wooled breeds, fine wool × coarse wool crosses and smaller sheep are not usually as desirable in carcass meatiness or cutability. This advantage becomes even more distinct when lambs are carried to heavy weights.[11]

In spite of the variations among breeds, there is much overlap, due to individual differences within every breed. No breed is outstanding in all desirable traits, and thus crossbreeding is common as a means of combining desirable traits from two or more breeds.

Birth Type

Birth type is a hereditary trait that influences productivity, as twins are more productive than are singles, even among breeds not noted for high reproductive rate. This seems to be as evident in rams as in ewes (Tables 6-9 and 6-10). Further studies with triplets at the University of Wyoming have indicated that triplet rams are superior to both twins and singles with regard to lambing percent of ewes to which they

TABLE 6-9 INFLUENCE OF TYPE OF BIRTH ON LAMBING PERCENT

Birth Type	Number of Observations	Lambs Born per Ewe Lambing	SE
Ewe			
Single	921	1.38	0.04
Twin	1275	1.52	0.04
Unknown	39	1.48	0.11
Ram			
Single	630	1.39	0.04
Twin	717	1.51	0.04
Unknown	887	1.48	0.04

Reproduced with permission from Vakil, Botkin, and Roehrkasse.[16]

TABLE 6-10 INFLUENCE OF MATING SYSTEM ON LAMBING PERCENT

Mating System	Number of Observations	Lambs Born per Ewe Lambing	SE
Single ewe, single ram	296	1.39	0.06
Single ewe, twin ram	242	1.47	0.06
Single ewe, unknown ram	382	1.40	0.06
Twin ewe, single ram	335	1.42	0.05
Twin ewe, twin ram	440	1.54	0.06
Twin ewe, unknown ram	500	1.57	0.04

Reproduced with permission from Vakil, Botkin and Roehrkasse.[16]

were mated. Testes size of multiple-birth rams has been reported to be larger than that of single rams.[15] Triplet ewes were not quite as prolific as were twin ewes out of triplet ewes and sired by triplet rams, but this difference was attributed to differences in mature size.

These data tend to refute the often quoted statement that "the number of lambs produced by a ewe is determined by the ewe." Even though the potential is entirely a function of the ewe and maternal environment is provided by the ewe, the ram or rams do influence lambing percent of ewes.

Health, Soundness, and Freedom from Defects

Ewes and rams that are healthy and free of parasites or disease or defects are more productive and are easier to manage than are those with health problems. Teeth defects, feet and leg defects, udder defects in ewes, testicle defects in rams, and wool blindness are commonly occurring problems in commercial sheep operations. Any of these can limit the usefulness of sheep either by directly reducing production or by limiting their grazing ability. Both selection and good management are involved in keeping these problems at a minimum.

REFERENCES CITED

1. Bearden, H., Joe and John Fuquay. 1980. *Applied Animal Reproduction.* Reston Publishing Co., Reston, Va.
2. Casida, L. E., C. O. Woody, and A. L. Pope. 1966. Inequality of function of right and left ovaries and uterine horns of the ewe. *J. Animal Sci.* 25:1169.
3. Dickerson, Gordon, and Danny Laster. 1975. Breed, heterosis and environmental influences on growth and puberty in ewe lambs. *J. Animal Sci.* 41:1.
4. Dufour, J. J., M. H. Fahmy, and F. Minvielle. 1984. Seasonal changes in breeding activity, testicular size, testosterone concentration and seminal characteristics with long or short breeding season. *J. Animal Sci.* 58:416.
5. Ensminger, M. E. 1970. *Sheep and Wool Science.* Interstate Printers and Publishers, Danville, Ill.
6. Frandson, R. D. 1981. *Anatomy and Physiology of Farm Animals.* Lea and Febiger Publishing Co., Philadephia.
7. Gallagher, J. R., and Maurice Shelton. 1972. Relationship of body weight to ovulation rate in mature fine-wool ewes. *Texas Agr. Exp. Sta. Progress Report 3017.*
8. Hulet, C. V., E. L. Wiggins, and S. K. Ercanbrack. 1969. Estrus in range ewe lambs and its relationship to lifetime reproductive performance. *J. Animal Sci.* 28:246.
9. Hulet, C. V. and others. 1974. Effects of origin and environment on reproductive phenomena in the ewe. *J. Animal Sci.* 38:1210.
10. Inskeep, E. K., J. T. Stevens, and J. B. Peters. 1974. Artificial insemination in sheep. *West Virginia Agr. Exp. Sta. Bulletin 629.*

11. Jacobs, J. A., and others. 1981. Increased efficiency in marketing lamb and mutton. *Idaho Agr. Exp. Sta. Research Bulletin 125.*
12. Lax, J., and others. 1979. Length of breeding season for eight breed groups of sheep in Wisconsin. *J. Animal Sci.* 49:939.
13. Nichols, C. W., and J. V. Whiteman. 1966. Productivity of farm flock ewes in relation to body size. *J. Animal Sci.* 25:360.
14. Sheep Industry Development Program. 1975. *Sheepman's Production Handbook.* Denver, Colo.
15. Snowder, G. D., M. Shelton, and D. Spiller. 1981. Factors influencing testes size of yearling Rambouillet rams. *Texas Agr. Exp. Sta. CPR 3898.*
16. Vakil, D. V., M. P. Botkin, and G. P. Roehrkasse. 1968. Influence of heredity and environmental factors on twinning in sheep. *J. Heredity* 59:256.
17. Winters, Lawrence M. 1939. *Animal Breeding,* 3rd. ed. John Wiley & Sons, New York.

7 Sheep Breeding and Applied Genetics

Breeding consists of both reproduction and control of inheritance. Reproductive aspects have been discussed in Chapter 6. Not only is it essential for producers to understand the reproductive process, but they also need to know the mechanism of inheritance and influences of heredity and environment on sheep production. The control of when sheep are bred and deciding which ewes are mated to which rams are genetic engineering, although this is only a minor part of what is included in the modern concept of the term.

MECHANISM OF INHERITANCE

Transmission of characteristics from parent to offspring begins at the same time as the reproductive process begins. Ordinary body cells, many millions in number, have 27 pairs of chromosomes (54 is the chromosome number for sheep). These chromosomes each have many genes, which are the units of inheritance, and hereditary traits are determined by these genes. Germ cells (ova and sperm), in contrast to ordinary body cells, during a process of reduction division contain only 27 chromosomes or one of each pair (Figure 7-1). The half that any germ cell contains is at random. It is a sample half and this is responsible for individual variation. When fertilization occurs, the chromosomes from the ovum and from the sperm unite to form new pairs, and as the new individual develops, its body cells contain 27 of the newly formed pairs. Thus the ova and sperm are responsible for transmitting traits from parent to offspring.

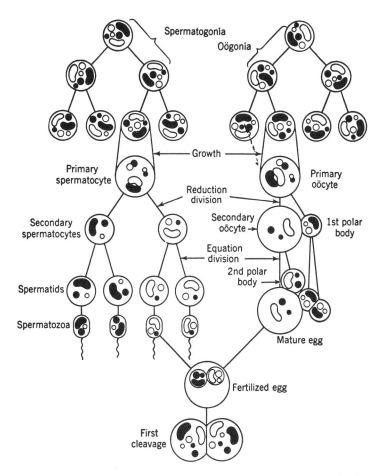

Figure 7-1 Diagrammatic representation of cell divisions during formation in animals. Bodies within the cells represent chromosomes, with those of the same sizes and shapes being of the same pair (homologous). Chromosomes are shown in black and white to represent paternal and maternal origin, respectively. Only a portion of the chromosome combinations possible in the gametes are shown. (After Shull.) (Reproduced with permission from Warwick and Legates [29])

Since there are many millions of possible combinations of genetic material, it is easy to see why nearly all individuals are different. Multiple births normally result from fertilization of separate ova, so there is no reason for twins to be any more alike in genetic makeup than two lambs from the same mating born in different years. Maternal environment may cause a pair of twins to be more alike at birth or up to weaning, but could just as well have the opposite effect because of competition between the twins for limited space or limited milk supply.

TYPES OF GENE ACTION AND TYPES OF TRAITS

The simplest form of inheritance is that in which a trait is influenced by one or only a few pairs of genes. These traits are almost entirely influenced by heredity, with only slight modification due to environment. They are known as *qualitative traits* and exhibit discrete variation, such as either the presence or absence of horns or black versus white. Several lethal genes are in this category. Apparently they are inherited as simple recessives that are expressed as presence or absence of the defect.

Color Inheritance in Sheep

Color, as in the case of black sheep occurring in a population of white sheep, can be explained by one pair of recessive genes. However, color is much more complicated than that and other color variations are common. Sometimes the black sheep are solid color, but some have white faces or white spotting on their faces, probably as a result of a different pair of genes. It is fairly common to see a black spot on the ear, eye, legs, or even on the body of otherwise white sheep, and these variations seem to be inherited. The other common variation in color is when white sheep have black faces and black legs, as seen particularly in Suffolks and Hampshires. This color pattern is due to a different pair or pairs of genes than those which cause black sheep, and the black extremities are dominant. Even though all the details of inheritance of color are not clear, the variations seen are due to differences in genetic makeup. (Figures 7-2, 7-3, and 7-4).

The only color differences that seem to be environmental in nature are the gradual browning and graying of black sheep as they age. It is common for Suffolk or Hampshire lambs to be born with black on all or part of the body and then turn white gradually so that at 3 to 4 months of age they have the same color as mature sheep (Figure 7-5). In a few cases, black sheep occur in the black-faced breeds (Figure 7-6). For a more complete discussion of color inheritance, see Rae.[15]

Figure 7-2 Spotted lamb. (University of Wyoming photo, Courtesy of Warren Livestock Company, Cheyenne, Wyoming)

Types of Gene Action and Types of Traits

Figure 7-3 Black lamb with white face. (University of Wyoming photo, Courtesy of Warren Livestock Company, Cheyenne, Wyoming)

Figure 7-4 Lamb with black leg. (University of Wyoming photo, Courtesy of Warren Livestock Company, Cheyenne, Wyoming)

Figure 7-5 Suffolk ewe with newborn lamb. (University of Wyoming photo)

Figure 7-6 Suffolk lamb that did not turn white. (University of Wyoming photo)

Inheritance of Horns

There is considerable variation in the size, shape, and rate of development in horns on sheep (Figure 7-7), but the characteristic is highly heritable. The polled condition is dominant to horns, but the trait is sex influenced in Merinos or Rambouillets or their crosses, being expressed in rams only.[15] In Dorsets, horns are common in both ewes and rams. Even though polled is dominant to horns, there is a tendency toward scurs or short horns in polled rams. Also, it is common for ewes of horned strains to have horn knobs and sometimes scurs growing out from these (Figure 7-8). Thus there likely are several pairs of modifying genes that can change the conditions of polled or horned.

Castration completely stops horn growth in Rambouillet or Merino lambs but may only reduce horn growth in Dorsets. In the early development of polled strains of Rambouillets, there was enough incidence of chryptorchidism to suspect the linkage of two pairs of genes, but in recent years this has become less of a problem. A more complete discussion of the inheritance of horns can be found in an article by Rae.[15]

Quantitative Inheritance

Traits that are influenced by many genes in which variation is continuous are called quantitative inheritance traits. These are the common measures of production, such as body weights and measurements and fleece weights. An example of the continuous variation is the array of different weights at weaning time in a group of lambs. The most difficult trait to classify is that of multiple birth, which is influenced by many genes but still expressed as a discrete variable (twins versus singles). It is easier to think of it as a continuous variable when comparing different groups or flocks within the sheep population. It is common to express lambing rate as 1.25, 1.40, or 1.52, and thus almost any degree of variation is possible.

Additive gene action is that in which two or more genes have equal effects on

Types of Gene Action and Types of Traits   113

Figure 7-7 Comparison of Rambouillet rams of similar age in horn shape. (University of Wyoming photo)

Figure 7-8 Scurs on Rambouillet ewe in foreground. (University of Wyoming photo)

a trait, and one gene does not influence the expression of another. Nonadditive gene effects are just the opposite, in that one gene influences the expression of another. One gene may mask the expression of another either partially or completely. If the gene that masks another is at the same locus on the chromosome or of the same gene pair, this is called *dominance;* if it is at a different locus, it is *epistasis.* In either case, the recessive gene is present but its influence is not noticeable. If two recessives are in combination as a pair, their influence is obvious. Thus, in cases of a pair of recessive genes resulting in an undesirable trait or defect, both parents have contributed equally.

TRAITS INFLUENCING PRODUCTIVITY AND THEIR INHERITANCE

Many different traits influence either the quantity or quality of production by sheep. These are divided into reproductive, growth, fleece, and carcass traits and defects.

Reproductive Rate

Several terms are used in describing reproductive rate, as follows:

- Ovulation rate. The number of ova shed per ovulation. This represents potential.
- Fertility. Indicates that a ewe shows estrus and ovulates or that a ram produces viable sperm.
- Prolificacy. The number of lambs born per ewe lambing.
- Survival rate. Percent of lambs born that are alive at weaning time.
- Overall reproduction. The number of lambs raised to weaning per ewe exposed for breeding. This combines fertility, prolificacy, and survival.
- Lambing percent. A loosely used term; it may mean lambs born or lambs raised per 100 ewes bred or per 100 ewes lambing. Or it may mean number of lambs per 100 ewes at any time count is made. It should be synonymous with overall production.

All these traits related to reproductive rate are relatively low in heritability (Table 7-1). Thus, environmental factors have a greater influence on reproductive rate than does heredity. However realistic "lambing percent," the lambs raised per ewe exposed for breeding, is the most important trait influencing either gross sales or net income to a sheep operation. It makes up a greater portion of total value of production than all other traits combined. For example, a ewe that raises a pair of twin lambs weighing 65 lb each (130 lb or 59 kg of lamb at weaning time) is a better producer than a ewe producing a single lamb weighing 80 lb (36.3 kg) (see Figures 7-9 and 7-10) regardless of growth rate of the lambs or differences in wool produc-

TABLE 7-1 HERITABILITY ESTIMATES FOR TRAITS OF SHEEP

Trait	Estimated Range of Heritability	
Reproductive traits		
Fertility	Low	10–20
Prolificacy	Low	10–20
Overall reproduction	Low	10–20
Semen concentration and motility	Low	10–20
Scrotal circumference	Low	10–20
Growth traits		
Birth weight	Low to medium	10–30
Weaning weight	Medium	20–30
Yearling weight	Medium	30–40
Mature weight	High	40–50
Rate of gain, postweaning	Medium	30–40
Feed efficiency, postweaning	Medium	20–40
Fleece traits		
Grease fleece weight	Medium	30–40
Clean fleece weight	Medium	30–40
Yield of clean wool	Medium	30–40
Staple length	High	40–50
Fiber diameter	High	30–50
Crimp	Low to medium	20–30
Luster	Low to medium	20–30
Carcass traits		
Carcass weight	High	40–50
Fat measures	High	40–50
Loin eye area	Medium to high	30–50
Percent lean	Medium	25–40
Tenderness	Low to medium	10–25
Color of lean	Low to medium	10–25
Marbling		

Adapted with permission from Warwick and Legates.[29]

Figure 7-9 Twin Rambouillet lambs weighing 69 and 70 lb at 110 days (weaning). (University of Wyoming photo)

Figure 7-10 Twin Hampshire lambs weighing 91 and 94 lb at 110 days (weaning). (University of Wyoming photo)

tion. This is an objective trait; that is, twins are readily identifiable in shed lambing operations. But in many range lambing operations it is not possible to distinguish twins from singles. Reproductive rate as such is not measurable in rams but is influence by fertility, semen quality, and libido or sexual drive. There is not as much information as to the heritability or mode of inheritance of these traits in males as for reproductive rate of females. What studies have been made indicate that heritability is relatively low (Table 7-1).

Growth

Several measurements denote growth in some manner: birth weight, weaning weight, rate of gain from birth to weaning, postweaning rate of gain, yearling weight, mature weight, body length or height, efficiency of gain, and age at puberty. These different measures of growth have higher heritabilities than do the reproductive traits (Table 7-1), but they have a lesser influence on productivity. It is not easy to evaluate these growth traits in terms of economic value to the producer. A fast-growing lamb is not necessarily worth more at market weight than a lamb growing at a slower rate, but he reaches market weight sooner. There is a positive correlation between growth rate and feed efficiency. Likewise, feed efficiency decreases as age and size increase. Thus, even though it is difficult to put a monetary value on average daily gain, there is no doubt that growth rate is valuable to producers.

Fleece Traits

Fleece traits rank third in importance following reproductive rate and growth, although all sheep producers do not agree. Range producers are likely to consider fleece traits of greater importance than growth. This is due to the importance of wool as an insulator essential for adaptation to undesirable climatic conditions, as well as the relative proportion of income from wool. Regardless of viewpoint, wool is a valuable product, although tremendously variable, and several measurements are used to indicate wool value. Fleece weight is the most indicative of value, but staple length, fiber diameter, and clean wool yield are major variables influencing value (see Chapters 18 and 20). Other fleece traits that influence value to a lesser extent are crimp, luster, uniformity of grade, strength, and freedom from contamination by dark fibers or hair or kemp. Most fleece traits are relatively high in heritability, as seen in Table 7-1.

Carcass Traits

Many factors are involved in carcass merit, which is hard to define and difficult to measure. Several of the traits that contribute to carcass merit are measurable and are used in attempts to measure differences in carcass value. The traits most directly related to carcass value are measures of fat and lean: fat depth, loin eye area, percent fat, and percent lean. Percent cutability and dressing percent are both influenced by fat and lean and are used to evaluate carcasses. Measures of fat or lean are highly heritable (Table 7-1), whereas quality traits such as tenderness, juiciness, color of lean, and flavor have lower heritability values. Since there are so many aspects of carcass merit and most are not easy to measure on live lambs, these are often given little attention in selection and breeding programs.

Defects

As expected, there are more defects in sheep than there are in other species, since some of them are fleece defects. Defects that are problems in sheep production are discussed next, with brief comments as to their mode of inheritance.[18] Deformities are not always lethal, as some develop later either as an inherited weakness and deformity resulting from stress of rapid growth (Figure 7-11) or resulting from min-

Figure 7-11 Ram lamb with deformed front legs. (University of Wyoming photo)

Figure 7-12 Ram with bent leg, a deformity acquired due to mineral imbalance. (University of Wyoming photo)

eral imbalance (Figure 7-12). Selection against defects tend to reduce selection pressure on other traits.[1]

Lethals. Several deformities in this category have been observed: cleft palate, dwarfism, hairlessness, muscle contracture, paralysis, and various degrees of amputated legs. These conditions are not widespread and are inherited as simple recessives. Thus removal of offending sires and dams can keep these at a low incidence.

Jaw defects. Failure of the incisor teeth in the lower jaw to meet the pad in the upper jaw properly is a problem occurring in most breeds, with incidence of occurrence reported as high as 2% of a flock. In some, the lower jaw is long with teeth extending beyond the pad, and in others the lower jaw is short and the teeth do not extend out to the end of the pad (parrot mouth), as shown in Figure 7-13. Care should be taken to cull lambs with either problem.

Wool blindness. Open-faced sheep are better producers than are wool-blind sheep, even if the face wool is sheared off in between regular shearings. The trait is highly heritable, but in spite of continuous selection pressure against it, there still remains a problem in some breeds and crosses (see Figure 7-14).

Figure 7-13 Lamb with several problems: parrot mouth, sore mouth, and hairiness. (University of Wyoming photo)

1. Open faced, clear channel below eyes and no impairment of vision.

2. Some wool on face below eyes, but not enough to cause blindness.

3. Wool covers most of face, and even though eyes seem clear, wool blindness is likely.

4. Wool blind, wool on face nearly covers eyes, and extends down over face to nose.

Figure 7-14 Suggested guide for scoring face covering. (University of Wyoming photo)

1. Smooth-bodied; neck, brisket, thigh all free of folds.

2. Folds under throat, on neck, slight amount of folds on thigh, shoulder and side smooth.

3. Heavy folds both front and rear, fairly smooth along side.

4. Heavy folds front and rear, with folds along side also. Nearly the entire body covered with folds.

Figure 7-15 Suggested guide for scoring skin folds. (University of Wyoming photo)

Skin folds or wrinkles. This is a problem in Rambouillet, Merino, and western white-faced sheep. Although not highly related to production, it creates problems at shearing. Heritability of skin folds is fairly high (Table 7-1). Variations are shown in Figure 7-15.

Chryptorchidism. This is the retention of one or both testicles in the abdominal cavity. It is a problem for purebred breeders that raise rams, but not a serious problem in commercial flocks. It is inherited as a simple recessive and has been found associated with the polled characteristic in Rambouillets and Merinos.

Entropion. This is turned in eyelids; either the upper lid or lower lid can be involved (Figure 7-16). Indications are that it is a heritable trait, but the mode of inheritance is not understood. The problem can be corrected by clipping out a small piece of skin above the lid or below the lower lid that is involved, by using surgical clips, or by sewing the lid back. If surgical clips are used, they should be removed in a few days. Lambs treated for the problem should be marked for future culling.

Hernia. Both umbilical and scrotal hernias are seen in sheep, but not at a high incidence. Although not inherited as such, it is likely that a weakness or susceptibility is inherited, but its mode of inheritance is not known.

Rectal prolapse. This is most often seen in black-faced sheep or among young lambs of several breeds that are highly fitted or on a heavy feeding program. The susceptibility seems to be inherited, although the method of inheritance is not known. It is unsightly and difficult to correct. Recurrences of prolapse following surgical correction are common. Most affected animals are slaughtered as soon as possible.

Horns or scurs. Although horns or scurs are commonly seen and are acceptable in some breeds, either can be considered a defect in other breeds. Production is not necessarily influenced, but selection pressure is reduced when horns or scurs are used for disqualification. Inheritance is explained in an earlier part of this chapter.

Color in fleeces. Colored fleeces or white fleeces contaminated with brown or black fibers are less valuable to large textile manufacturers than are pure white fleeces. In some case, people involved in home spinning or home weaving prefer dark or multicolored fleeces. Inheritance of color is explained in an earlier part of this chapter.

Figure 7-16 Yearling ram with inverted eyelid (entropion). (University of Wyoming photo)

Figure 7-17 Hairy lamb at three weeks of age. (University of Wyoming photo)

Hairiness. The appearance of hairiness is common in sheep of several breeds. Many lambs show an extremely hairy appearance at birth, but most of the hair is gone by the time of weaning. In others the hairy appearance persists. Also, it is common to see the appearance of hairiness on the thigh (britchiness). This is a contaminant and lowers the quality of fleeces. Method of inheritance is not clear, but obviously this is a heritable defect (Figure 7-17).

Belly wool. The wool that normally grows on a sheep's belly is shorter, is less dense, and has a different appearance in color and crimp than wool on the shoulder, side, and thigh. It is when this type of wool extends too high up the side or sometimes over the entire body that it is a defect. It is a problem to ram producers of several breeds, but only as it has a detrimental influence on fleece weight is it a problem to commercial producers. From limited information, it seems to be about medium in heritability.

Most defects, whether lethal, semilethal, or nonlethal, can be kept at a low level by selection or culling of breeding stock. Many are discriminated against in show ring judging and in visual inspection at sales. The most persistent defects and those that create the most difficulty for range sheep producers are wool blindness and body folds or wrinkles. Performance tests for white-faced sheep, Rambouillets in particular, emphasize these in their index in an attempt to reduce the problems. Scoring charts to aid in evaluating these problems are shown in Figures 7-14 and 7-15.

CORRELATIONS AMONG TRAITS

In many cases, traits are correlated to each other, and several are negatively correlated. Positively correlated traits usually have only moderate correlation coefficients, not high enough that one trait can be accurately estimated by measuring another. Examples of positive correlations are ovulation rate and prolificacy, birth weight and weaning weight, rate of gain and feed efficiency, grease fleece weight and clean fleece weight. Whether high, medium, or low, positive correlations do not create any problems in selection.

Negative correlations do cause extra problems. Since there is competition for available nutrients, the correlations are negative between milk, meat, and wool production. Thus there is a negative correlation between reproductive rate and early growth. Twins grow slower than do singles. Wool production is negatively correlated to reproductive rate in that ewes raising twins shear lighter fleeces than do ewes raising singles or dry ewes. Wool production and growth rate tend to be negatively correlated, as are wool growth and carcass meatiness. Correlations among carcass traits are variable in direction because of differences in methods of slaughter. For instance, if lambs are all slaughtered at a constant time but variable in age and weight, the correlation between fat weight and lean weight is likely to be positive, whereas for lambs slaughtered at a constant weight, fat and lean are negatively correlated.

Fortunately, all the correlations are apparently environmental, and there is no evidence that negative correlations are genetic.

SELECTION

The fact that sheep are dual purpose is often cited as an advantage over other species of livestock. This becomes a distinct disadvantage from the standpoint of improving production by selection. The process of improvement by selection is more difficult in sheep than in other species for the following reasons:

1. There are more traits influencing productivity and traits are not all measurable at the same time.
2. There are more defects that reduce production or reduce selection pressure on other traits.
3. Several of the important traits are negatively correlated.
4. Common methods of management in large bands do not allow individual identification.

Animal breeders do not all agree with the foregoing statements. Some cite realistic problems in other species, such as the inability to measure milk production in dairy bulls or for poultry producers to measure egg production in males. These differences in viewpoint merely help to emphasize the point that improving production by selection is not easy.

Traditional methods of selecting sheep are still in use, as many ewe lambs kept for replacements are selected by visual appraisal. This applies to selection of rams for use in commercial production also. It is suspected that such selection has been detrimental to productivity since single lambs tend to be chosen ahead of twin lambs. Even though some improvement can be realized in weaning weight, confor-

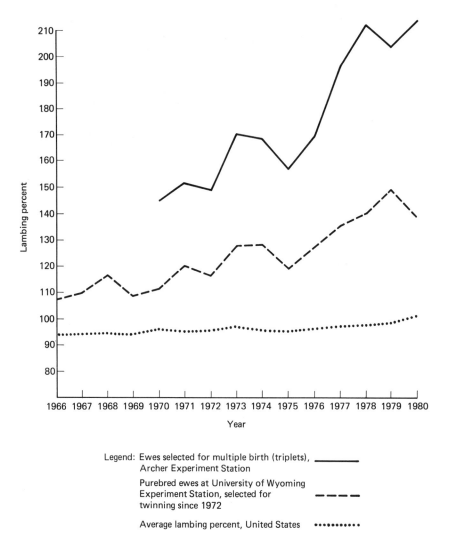

Figure 7-18 Lambing percent by year for two groups of ewes selected for multiple birth compared to the U.S. average.

mation, or wool production by visual selection, overall improvement in productivity is likely to be slow if any occurs at all.

Most authorities on animal breeding indicate that selection based on an index of overall merit is more efficient than selecting for one trait at a time (tandem selection) or selecting for certain levels of two or more traits simultaneously (independent culling levels). The index tends to adjust for individuals that are extremes, either far below or far above average in any one trait. This is easy to visualize when selecting for only a few traits measured at the same time. However, it is nearly impossible to

Selection   **125**

imagine an index of overall merit for sheep that is practical for selection at an early age.

Dwelling on the previously mentioned difficulties in sheep selection is not a valid reason for a "do-nothing" attitude. There are ways to make selection effective. Many examples can be cited to show success in improvement of traits, two of which are illustrated in Figures 7–18 and 7–19. These illustrations demonstrate that improvement is possible through concerted efforts.

Other examples of changes in production levels, not necessarily all by selection, are shown in Figures 7–20 and 7–21. These show a large improvement in two traits over a period of years, but this is not all a result of selection pressure because improvements were made in environmental conditions.[19]

Breeders tend to be more selective as to which rams are taken to central tests than in previous years. Figure 7–22 charts the average lambing percent and grease fleece weight in the United States by year. Comparing the previous figures with Figure 7–22 does lend encouragement by showing that improvement can be made; but the comparison also points out a need to change present methods of selection for the industry as a whole.

Rate of Change from Selection

Expected change per generation in any trait is estimated by the formula *selection differential × heritability*. Selection differential is merely the difference between the average of those chosen as replacements and the average of the flock from which

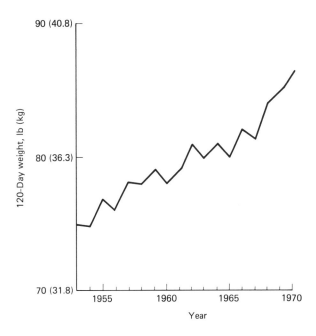

Figure 7–19 Changes in average weaning weight in a purebred flock in which selection was based on weaning performance.

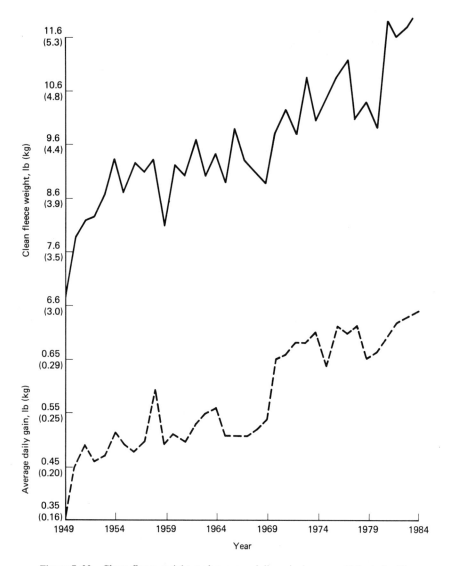

Figure 7-20 Clean fleece weight and average daily gain by year. (Adapted with permission from Texas Agr. Exp. Sta. Research Center Technical Report[23])

they came. Heritability is that portion of variation of a trait that is hereditary (estimates of heritability for various traits can be found in Table 7-1). Rate of progress per year is calculated by dividing by generation length. In Table 7-2 the average generation length is shown for both rams and ewes.[29]

Generation length can be changed, and it is not necessarily the same as shown in Table 7-2 for any particular flock; but the degree to which it can be controlled is limited. Heritability is already predetermined and cannot be controlled. Thus the

Selection   **127**

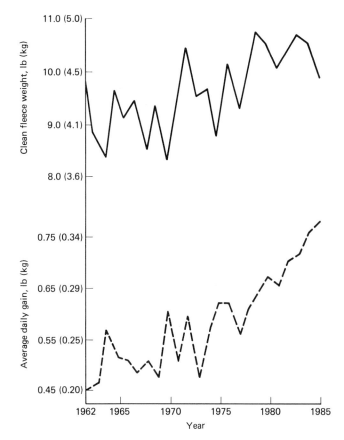

Figure 7-21 Clean fleece weight and average daily gain by year. Wyoming ram performance test.

key to success in making changes by selection is the size of the selection differential. The selection differential should be much greater for males than for females because a smaller proportion of them need to be kept for replacements. Thus a few outstanding rams for a particular trait can be kept. The reproductive rate of a flock also influences the proportion needed for replacements to maintain a constant flock number both by sex and by lambing percent.[29] Proportions needed for replacements to maintain constant numbers are shown in Table 7-3.

TABLE 7-2 GENERATION INTERVAL FOR SHEEP

Sex and Selection Method	Average Generation Length (years)
Ewes, mass selection	4.0
Rams, mass selection	2.0
Rams, with progeny testing	4.0

Adapted with permission from Warwick and Legates.[29]

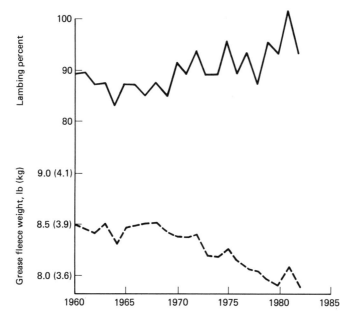

Figure 7-22 Lambing percent and grease fleece weight by year in the United States.[25]

The number of traits that are considered in making selections is likely to be the most important factor of all in determining selection effectiveness; this is simply because of a reduced selection differential for any trait when many are included. All the traits previously mentioned do have importance to the sheep industry. Since it is not impossible to find a practical index using all traits, some compromise in selection seems the logical approach. First, the most important traits should receive greatest emphasis and the greatest portion of selection pressure should be applied to rams. It has already been noted that reproductive rate is the most important trait

TABLE 7-3 PERCENT REQUIRED AS REPLACEMENTS TO MAINTAIN POPULATION NUMBERS

Percent Lamb Crop	Percent Kept for Replacement[a]	
	Ewes	Rams
75	66.7	5.3
100	50.0	4.0
125	40.0	3.2
150	33.3	2.7
175	28.6	2.3
200	25.0	2.0

[a]Replacements kept at the rate of 2 males and 25 females per 100 ewes in the flock.

and that, as reproductive rate improves, selection pressure can be increased for other traits.

Selection for Multiple Birth

Due to the nature of variability in lambing rate, it is easy to attain a relatively large selection differential. Even in a flock made up of single ewes producing all single lambs, a large selection differential can be achieved by purchasing and using twin rams. The average selection differential from this approach is 0.5 (1 for males, 0 for females). This is exactly the same as what could be achieved in a flock producing a 150% lamb crop in which only twins were kept for replacements. Using the selection differential of 0.5 and multiplying it by a heritability of 15% (or 0.15) results in expected progress per generation of 0.075, or a 7.5% increase in lambing rate. When considering the present reproductive rate in the United States, this represents quite a significant rate of improvement and is worth striving for. Use of triplets or quadruplets can speed up the rate of improvement, but this is dependent on availability. A group of ewes with triplets is shown in Figure 7–23.

Just why twin rams have an advantage over single rams in lambing rate of ewes to which they are mated [9,26] is not completely understood. A hint is found in different testes size of twins and singles.[21] Speculation and observation of the breeding behavior of Finn rams lead to the belief that at least part of the reason for a higher lambing rate is because Finn rams breed the ewes more times than do rams of other breeds. Finn rams are known to be sexually aggressive and perhaps this same trend is found in twins or triplets of the other breeds. Just because a ram is a twin, though, does not assure that he is a good breeder. There is great variation among twins in all traits, and they are just as susceptible as are single rams to reproduction problems.

Figure 7–23 Group of 15 ewes, all raising triplets. (University of Wyoming photo)

Once this approach is adopted and expanded it can be visualized that birth type can be used as an independent culling level. For example, only twin lambs can be kept for replacements, and then a selection index can be made up of other traits and performance or progeny testing used for twins only. Eventually, all single lambs could be marketed for slaughter, and then the problem of selection would become easier because of one less trait to consider. This is truly an optimistic outlook as to potential, and it cannot happen very fast or without the combined efforts of a lot of producers, but it is a realistic goal.

Selection for Growth Traits

Selection for growth traits (either growth rate or weight) is relatively easy in both rams and ewes, but care needs to be taken to avoid confounding with effects of environment. Differences in age, sex, birth type, and rearing all influence weaning weight or weight at any time early in the lambs' lives. Proper adjustments need to be made for these variables or selections made within sex or birth type or within groups similar in age. When comparing productivity by sire group, such as in progeny testing, it is particularly important to have appropriate adjustments for these nongenetic influences. These adjustments should be obtained from data under conditions similar to those where they are to be applied, because climate, feed conditions, breed, season of year, and perhaps other factors can influence the size of corrections that are needed.

Selection for Fleece Traits

This is the area of controversy and the beginning of great difficulty in selecting for improvement. Part of the controversy stems from different emphasis due to different breeds or to different areas of the country. Part of the reason, though, is in which of the fleece traits in most important. If comparisons of fleece traits are made within a breed or within a certain grade of wool, then fleece weight is the most important trait to consider (Chapter 21). But even if it is agreed on that fleece weight is the primary trait, there are still problems. Clean fleece weight is a better measure of value than is grease fleece weight, but it is much more difficult to obtain. For large flocks, individual clean fleece weight is simply not available. Furthermore, fleece weight is not directly measurable very early in a sheep's life when selection for replacements is made. Other traits, such as staple length at weaning or density, are used as an indirect measure of fleece weight. This is likely the trait, more than any other, in which the major selection emphasis is on the male and performance testing is essential.

Selection for Carcass Traits

In terms of difficulties, selection for carcass traits in sheep ranks first. The most used measures of carcass value are not available while lambs are alive. Estimates of fatness can be made visually or by use of ultrasonic devices to help in carcass evaluation of live lambs. Also, estimates of conformation can help in evaluating meatiness. However, to try to incorporate these estimates along with the numerous other traits allows a very small chance of a significant selection differential. Progeny testing is resorted to in some cases but is of necessity a slow process of improvement. Progress has been made in some breeds of sheep, particularly by reducing fatness by a combination of selection for growth and against fat.

Stratification and Raising Rams

Raising rams for one's own use can be done by any producer, with or without stratification. The market price differential for rams compared to wethers and the penalty for overweight lambs are probably the main incentives for the stratified flock approach. The purpose of the stratified program is to be able to select the best ewes, mate them to the best rams, and establish a "super herd" for production of rams that are above average. These rams are then mated to ewes in the second tier for producing replacement ewe lambs. Ewes in the third tier are then used for production of market lambs with no replacements coming from that group. This system of selective breeding requires extra care in management and record keeping in order to accomplish its purpose of improving productivity. It is often used as a combination of selection and crossbreeding program, with ewes in the lower tier being mated to rams of a different breed for production of crossbred market lambs.

Performance and Progeny Testing

There is nothing new or magical about performance testing or progeny testing of sires. Many breeders routinely measure performance traits such as reproductive rate, weaning weight, and fleece weight and use these in their selection programs. Likewise, it is common to compare progeny of different sires with regard to defects or productive traits. It is an increased emphasis on central performance test stations and the spread of these throughout the country that has made performance testing take on a new significance. Many different approaches are taken by the various stations. Some are conducted in the spring and summer and others during fall and winter. Some are breed specific and others open to rams of any breed. In some of the tests, growth rate is the primary concern, while in other tests wool production is emphasized as much as is growth. Whatever the goal as far as traits emphasized, the tests are designed to provide comparable information under similar environmental conditions. In this respect the tests can serve as a basis for sire selection both within a flock and between flocks. Central tests have been underway for many years at Texas Agricultural Experiment Station (Figure 7-24) and at the University of Wyoming (Figure 7-25).

Test stations provide a service to breeders simply by providing information that would not be available otherwise. Some breeders benefit from the use of such information, while others tend to use it more for merchandising purposes. Thus,

Figure 7-24 Part of facilities for performance testing. (University of Wyoming photo, Courtesy of Texas Agricultural Experiment Station, Sonora, Texas.)

even though there have been many cases of increased use of performance-tested rams and likely improved production as a result, much of the benefit has not extended to the sheep industry as a whole.

Performance testing on the farm or ranch is encouraged by many experiment stations, and their services are available to help producers in obtaining information. On-the-farm testing combined with testing at central stations should allow for the greatest benefit. If reproductive rate is not stressed in the index of evaluation at the station, it can be stressed in the flock. As an example, those chosen for the central test can be the heaviest twins at weaning time. At the same time, selection pressure can be applied to the ewe lambs being selected as replacements. Tests for rams of the meat breeds are relatively simple in that growth rate can be the single trait to

Figure 7-25 Rams on Wyoming performance test. (University of Wyoming photo)

measure. For rams of the breeds in which wool production is more important, tests need to emphasize both growth rate and fleece weights.

Progeny testing and subsequent use of progeny-tested rams are not as easy as performance testing. The length of the generation interval is extended, and it is time consuming and costly to obtain progeny information. Adequate test flocks or testing facilities are not usually available. Progeny testing can be used effectively in screening sires for freedom from defects and could be used for determining the effectiveness of indirect measures of traits hard to obtain, such as carcass traits.

No method of selecting replacements or of measuring productive traits can have a lasting influence in changing levels of production in the sheep industry as a whole unless a way is found to get the poor-producing sheep out of production. Measuring performance on the farm or ranch or at test stations is of little use to the industry if low-performing rams still find their way into commercial flocks. This sort of problem is hard to overcome because of the current market structure, with prices on heavy rams lower than for market-weight wethers. One goal of animal breeders and producers is to strive for methods of earlier detection of rams' producing ability so that when culled they are still relatively high in market value.

SYSTEMS OF BREEDING

Inbreeding

Inbreeding results from mating individuals more closely related to each other than the average of the population. It can vary from mild inbreeding to intense. Generally, inbreeding is not a recommended practice for commercial sheep production since it has a depressing effect on most productive traits. In past years, inbreeding was used to help stabilize the genetic makeup of populations in the development of new breeds. Likewise, in the past many experimental breeding programs included the development of inbred lines and crosses of these, patterned after the early development of hybrid corn. Most of the work with inbred lines has been discontinued due to difficulties in adequate testing facilities or due to failure of lines to withstand high levels of inbreeding. Sheep producers with small numbers of ewes are sometimes faced with the problem of increased inbreeding because of the high cost of purchasing a replacement ram for only a few ewes. Inbreeding can be useful in helping to identify carriers of undesirable recessive genes.

The depression due to inbreeding is a direct result of increased homozygosity or likeness of genes. The greatest reduction due to inbreeding is on the reproductive traits, including survival, and growth.[11,29]

The physiological basis for decreases in productivity due to increases in inbreeding is a combination of reduction at several stages: reduced ovulation rate, increased embryonic deaths, delayed puberty in both sexes, and delayed testicular development in rams resulting in inhibited or slow breeders.

Linebreeding

This is a practice involving mild inbreeding in an attempt to increase the relationship within a group to an outstanding individual or to several individuals. This is often seen in purebred flocks. In any closed flock, linebreeding is automatic, the rate of increased relationship or inbreeding being dependent mostly on the size of flock. In a flock numbering 200 or more, the rate of increase in inbreeding is minor even if the flock is closed to outside breeding unless there is intentional mating of close relatives. For flocks in which selection is based on such things as early growth and survival, the highly inbred individuals are culled, thus reducing the possibility of a rapid increase in inbreeding of the flock.

Crossbreeding

Crossing of two or more breeds has been common in the sheep business for many years. It has long been used in the development of new breeds or to combine desirable traits of earlier existing breeds, and it continues to be useful for introduction of new genetic material. When crossing breeds, the resulting crossbred offspring tend to be intermediate between the parent breeds, and normally the average of crossbred offspring does not exceed the averages of both parent breeds for any trait.

It is not unusual for a crossbred group to excel both parent groups for a particular trait in cases where both parent groups are nearly alike. For traits in which one parent breed greatly excels, though, the resulting crossbreds are below the better parent. For example, Finn crosses are not as prolific as Finns and Suffolk crosses are not as fast growing as Suffolks. Combinations of breeds that are each highly productive for different traits may result in the crossbred offspring exceeding the average of both parent breeds in overall merit. While inbreeding tends to bring out undesirable recessives, crossing tends to cover them up or mask the effect of the recessive genes.

Crossing two breeds results in heterosis as expressed in offspring. Additional maternal heterosis can be obtained by using crossbred ewes. Degrees of heterosis for various traits are shown in Table 7-4. These comments concerning heterosis or increased vigor of crossbreds should not be construed to mean that crossing of any two or more breeds results in improved production. First, genetic diversity is necessary and degree of diversity is related to degree of heterosis. Also, it must be kept in mind that the crossbred offspring are inferior to offspring of the better parent breed for most traits. Furthermore, adaptability to environment is an important consideration for sheep breeders and may further limit the usefulness of some crosses.

Choice of breeds to be used is the major consideration for a producer anticipating the use of a crossbreeding program. Quite often this decision is dictated by climatic conditions or by the type of operation. As examples, range sheep producers need sheep of some breed in which the sheep are hardy and gregarious, with the ability to forage in undesirable conditions, and producers who operate accelerated lambing programs need sheep with the ability to breed out of season.

TABLE 7-4 INDIVIDUAL AND MATERNAL HETEROSIS IN TRAITS OF SHEEP (AVERAGE OF ESTIMATES EXPRESSED AS PERCENTAGE OF PARENTAL MEAN)

Trait	Individual Heterosis (%)	Maternal Heterosis (%)
Growth		
Birth weight	+3.2	+5.1
Weaning weight	+5.0	+6.3
Preweaning growth rate	+5.3	
Postweaning growth rate	+6.6	
Yearling or adult weight	+5.2	+5.0
Reproduction		
Fertility (conception rate)	+2.6	+8.7
Prolificacy (litter size)	+2.8	+3.2
Survival (birth to weaning)	+9.8	+2.7
Aggregate performance		
Lambs born per ewe joined	+5.3	+11.5
Lambs reared per ewe joined	+15.2	+14.7
Weight of lambs reared per ewe joined	+17.8	+18.0
Carcass traits	0.0	
Fleece weight	+7.0	

Reproduced from chapter by A. L. Rae in *World Animal Science,* Vol. C-1.[16] Courtesy of Elsevier Science Publishers, Amsterdam, the Netherlands.

Reports of studies on crossbreeding of sheep are common. See references 2 through 8, 10, 13, 17, 20, 24, 27, 28, and 30).

Crossbreeding systems. A static two-breed cross has been the most commonly used system for crossing, particularly with western range sheep. This involves mating black-faced rams (Suffolk or Hampshire) to western white-faced ewes (Rambouillet, Targhee, Columbia, Corriedale, and crosses among these). This takes advantage of the wool-producing ability as well as the hardiness and adaptability of the ewes and heterosis as expressed in increased growth rate and carcass value of the crossbred lambs. Normally, the crossbred ewes are not kept as replacements. Wool production of the crossbred ewes is below that of the white-faced ewes. Also, there is a tendency for a reduction in hardiness and adaptability as wool is less dense and has less protective value on the crossbred ewes. These crossbred ewe lambs are often sold to producers in other areas or other types of operations as replacement ewes in crossbreeding programs. Selling all the crossbred lambs results in a problem of replacements. Either part of the flock needs to be mated to white-faced rams to raise replacements or replacement ewes need to be purchased. In either case, progress from selection is limited.

Systems of crossbreeding that allow for the use of crossbred ewes and that take advantage of maternal heterosis are as follows:

1. *Crisscross,* or *alternating backcross,* involves two breeds; crossbred ewes are mated back to sires of one of the original breeds and these backcross lambs mated to rams of the other breed.
2. *Static three-breed cross,* a system to utilize crossbred ewes mated to rams of a third breed. Offspring are sold and replacement crossbred ewes are purchased each year.
3. *Rotational crosses* involving three or more breeds are used to obtain maximum heterosis both in ewes and offspring. Also, selection can be included in choosing replacement ewes. These rotational systems can become complicated and require that rams of all breeds involved are available every year.

Any of these systems requires careful management and the use of superior breeding sheep to be of benefit. Crossbreeding is not a panacea that in itself solves all of a producer's problems; neither does it eliminate the need for selection if improvement in production is desired.

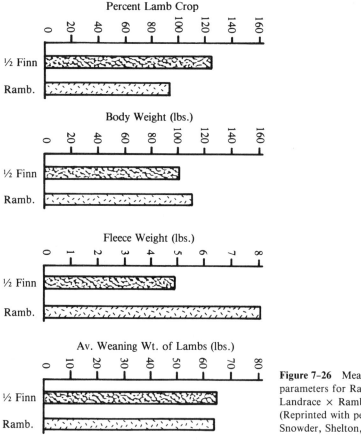

Figure 7-26 Mean lifetime production parameters for Rambouillet and Finnish Landrace × Rambouillet ewes. (Reprinted with permission from Snowder, Shelton, and Thompson[22])

Systems of Breeding

TABLE 7-5 INFLUENCE OF BREED OR BREED CROSS ON FERTILITY AND LAMBING RATE OF EWES BRED TO LAMB AS YEARLINGS

Breed	Percent Ewes Lambing	Lambs Born per Ewe Exposed (%)	Lambs Weaned per Ewe Lambing (%)
Suffolk	59	72	57
Hampshire	55	55	43
Rambouillet	31	32	80
Dorset	56	55	58
Targhee	38	44	77
Corriedale	33	36	49
Coarse Wool	80	83	73
Rambouillet × Dorset	92	99	89
Rambouillet × Targhee	51	51	42
Rambouillet × Corriedale	62	62	79
Rambouillet × Coarse Wool	87	88	70
Finn × Rambouillet	83	118	118
Finn × Dorset	74	138	149
Finn × Targhee	95	147	126
Finn × Corriedale	90	122	112
Finn × Coarse Wool	78	129	122
Finn × Fine Wool	101	172	138
Finn × Columbia	90	120	131
Finn × Navajo	102	149	124
Average of all	71	93	91

Reprinted with permission from Laster, Glimp, and Dickerson.[(12)]

During recent years, the majority of research in crossbreeding has involved the Finnish Landrace or Finn sheep as it is commonly called. Finns themselves are at a disadvantage compared to sheep of domestic breeds in size, growth rate, carcass merit, and wool production. In spite of this, Finns do have a lot to offer in reproductive rate and early puberty. They have been used with many breeds with favorable results. Examples of comparisons between Finn crosses and various breeds are seen in Tables 7-5 through 7-8 and in Figure 7-26. In development of composites made up of several breeds, or what are called "synthetics," the Finn is nearly always one of the breeds included.

Finn cross rams are also being used in commercial production but records are not always available to show the comparison between them and rams of other breeds. In some cases an attempt is made to produce a ewe flock with one-quarter Finn breeding. This is an area of much needed research, not only for extended and controlled studies in extensive sheep production, but also in evaluation of other foreign breeds and introduction of new genetic material. Import regulations have made it difficult to introduce new breeds. Knowledge of some breeds is greater than of others and the combining ability of different breeds is often not known. Perhaps development in future research can help to ease the problems.

TABLE 7-6 LEAST-SQUARES MEANS AND STANDARD ERRORS FOR NUMBER OF LAMBS WEANED (LITTER SIZE) PER 100 EWES LAMBING, 1971 TO 1973

Breed of Ewe[a]	Age of Ewe[b]						
	1 Year		2 Years		3 Years		
	N	Mean (%)	N	Mean (%)	N	Mean (%)	
Finn Sheep (F)	45	146.1 ± 9.6 (4)[c]	28	231.6 ± 12.2 (19)	17	243.9 ± 15.6 (14)	
Minnesota 100 (M)	28	62.4 ± 12.2	42	102.4 ± 9.9	46	103.3 ± 9.5 (1)	
Suffolk (S)	35	69.7 ± 10.9	58	109.8 ± 8.4 (2)	22	110.7 ± 13.7	
Targhee (T)	19	75.2 ± 14.8	18	115.3 ± 15.2 (2)	7	116.2 ± 24.3	
Std mean	82	69.1 ± 7.3	118	109.2 ± 6.7	75	110.1 ± 9.8	
F × M	62	95.2 ± 8.2 (1)	52	163.3 ± 8.9 (8)	27	176.2 ± 12.4 (1)	
F × S	55	101.3 ± 8.7	54	169.4 ± 8.8 (4)	27	182.4 ± 12.4 (3)	
F × T	72	89.8 ± 7.6 (2)	50	157.9 ± 9.1 (10)	17	170.8 ± 15.6	
F₁ mean	189	95.4 ± 4.7	156	163.6 ± 5.2	71	176.5 ± 7.8	
F₂ (F × M)	24	76.7 ± 13.1	8	101.7 ± 22.8 (2)	3	188.2 ± 37.2 (1)	
F₂ (F × S)	18	73.4 ± 15.2	7	98.4 ± 24.3	3	184.9 ± 37.2	
F₂ (F × T)	15	103.9 ± 16.6 (1)	2	128.9 ± 45.5			
F₂ mean	57	84.7 ± 8.7	17	109.6 ± 18.8	6	186.5 ± 26.3	
F × (F × M)	28	121.5 ± 12.2 (2)	16	185.9 ± 16.1 (4)	6	193.2 ± 26.3	
F × (F × S)	26	111.6 ± 12.6 (1)	13	176.1 ± 17.8 (3)	5	183.3 ± 28.8 (2)	
F × (F × T)	8	110.8 ± 22.8	2	175.2 ± 45.5 (1)			
BCF mean	62	114.6 ± 9.6	31	179.1 ± 17.2	11	188.3 ± 19.5	
M × (F × M)	16	92.4 ± 16.1	5	117.5 ± 28.8			
S × (F × S)	21	68.5 ± 14.0	8	93.7 ± 22.8			
T × (F × T)	12	86.9 ± 18.6	3	112.1 ± 37.2			
BCStd mean	49	82.6 ± 9.4	16	107.8 ± 17.4			

Reprinted with permission from Oltenacu and Boylan.[14]

[a] Crosses are designated by letters, with breed of sire listed first.
[b] Differences observed for age of ewe, class of ewe, breed of lamb (i.e., BCF versus BCStd) within F × S ewes ($P < 0.01$) and year and breed of lamb (i.e., S versus F × S) within S ewes ($p < 0.05$).
[c] Values in parentheses are the numbers of lambs raised artificially.

TABLE 7-7 MEANS (±SE) FOR COMPONENTS OF EWE PRODUCTION BY COMPOSITE 1 AND PARENTAL BREEDS, AND HETEROSIS LEVELS BY GENERATIONS OF CROSSING

				Component Traits[a]					
					Lamb Survival (%)			Weaning Weight/Lamb (kg)	
						Preweaning			
Ewe Breed[b]	Number of Sires/Ewes/ Records	Fertility (EL), %	Lambs Born (LB), No.	To 1 d (LA/B)	On Ewe (LW/A)	Total (LWN/A)		On Ewe (WW)	Total (WWN)
Finn Sheep (F)	89/ 599/1431	70.0 ± 1.2	2.55 ± 0.02	86.0 ± 0.9	57.8 ± 1.3	67.8 ± 1.3		10.3 ± 0.01	9.8 ± 0.1
Rambouillet (R)	72/ 628/1960	62.7 ± 1.0	1.58 ± 0.02	92.5 ± 0.8	76.5 ± 1.2	79.8 ± 1.1		13.1 ± 0.01	12.9 ± 0.1
Dorset (D)	66/ 599/1768	71.5 ± 1.1	1.54 ± 0.02	91.8 ± 0.8	85.1 ± 1.2	86.2 ± 1.1		12.9 ± 0.01	12.8 ± 0.1
F × R }$\overline{FX}$	10/ 79/212	75.1 ± 3.2	2.05 ± 0.05	88.6 ± 2.0	78.2 ± 3.0	83.5 ± 2.8		12.0 ± 0.02	11.7 ± 0.2
F × D	9/ 53/133	85.4 ± 4.0	2.17 ± 0.06	94.6 ± 2.4	75.7 ± 3.5	80.6 ± 3.4		11.3 ± 0.03	11.1 ± 0.3
½F¼R¼D (C1$_1$)	12/ 401/1247	82.2 ± 1.3	2.13 ± 0.02	93.4 ± 0.8	79.6 ± 1.2	83.9 ± 1.1		11.7 ± 0.01	11.5 ± 0.1
C1$_2$	27/ 355/711	80.0 ± 1.7	2.01 ± 0.03	92.5 ± 1.3	80.2 ± 1.9	83.8 ± 1.8		11.7 ± 0.02	11.6 ± 0.2
Parental means ($\overline{P}$)[c]	227/1826/5159	68.6 ± .71	2.06 ± 0.012	89.1 ± 0.53	69.3 ± 0.78	75.4 ± 0.76		11.6 ± 0.06	11.3 ± 0.06
Heterosis, %									
100($\overline{FX} - \overline{P}$)/$\overline{P}$[d]		17.1[e]	2.4	2.9	11.0[e]	8.9[f]		0.2	0.1
100(C1$_1$ − $\overline{P}$)/$\overline{P}$		19.9[e]	3.6	4.9[f]	14.9[e]	11.3[e]		0.7	1.5
100(C1$_2$ − $\overline{P}$)/$\overline{P}$		16.7[e]	−2.2	3.9[f]	15.7[e]	11.2[e]		0.9	2.6

Reprinted with permission from Fogarty, Dickerson, and Young.[(8)]

[a] Adjusted for age of ewe and for year-season. See text.
[b] C1$_1$, ewes were reciprocal crosses of F · R × F · D. C1$_2$ ewes were from inter SE matings of C1$_1$.
[c] $\overline{P} = ½F + ¼R + ¼D$.
[d] Assuming that FR = RF in ewe performance.
[e] $P < 0.01$. } after adjusting the within-breed SE shown to a sire/breed basis.
[f] $P < 0.05$.

TABLE 7-8 MEANS (±SE) FOR COMPONENTS OF EWE PRODUCTION BY COMPOSITE 2 AND PARENTAL BREEDS, AND HETEROSIS LEVELS BY GENERATIONS OF CROSSING

Ewe Breed[b]	Number of Sires/Ewes/Records	Fertility (EL), %	Lambs Born (LB), No.	Component Traits[a] To 1 d (LA/B)	Lamb Survival (%) Preweaning On Ewe (LW/A)	Total (LWN/A)	Weaning Weight/Lamb (kg) On Ewe (WW)	Total (WWN)
Finn Sheep (F)	89/599/1431	70.0 ± 1.2	2.55 ± 0.02	86.0 ± 0.9	57.8 ± 1.3	67.8 ± 1.3	10.3 ± 0.01	9.8 ± 0.1
Suffolk (S)	50/103/164	62.4 ± 3.6	1.62 ± 0.06	90.8 ± 2.2	65.5 ± 3.2	69.7 ± 3.1	15.1 ± 0.3	14.5 ± 0.3
Targhee (T)	26/219/591	51.7 ± 1.9	1.51 ± 0.04	91.1 ± 1.4	75.9 ± 2.1	79.0 ± 2.0	12.7 ± 0.2	12.6 ± 0.2
½F½S }$\overline{FX}$	11/57 /173	77.4 ± 3.5	2.21 ± 0.06	85.4 ± 2.2	69.0 ± 3.2	74.3 ± 3.1	12.5 ± 0.3	12.3 ± 0.3
½F½T	10/212/680	76.7 ± 1.8	2.06 ± 0.03	94.2 ± 1.1	77.2 ± 1.5	82.5 ± 1.5	12.4 ± 0.1	12.1 ± 0.1
½F¼S¼T ($C2_1$)	17/373/ 560	80.7 ± 2.0	2.10 ± 0.03	91.7 ± 1.2	80.5 ± 1.8	84.6 ± 1.7	12.3 ± 0.1	12.1 ± 0.1
$C2_2$	15/104/135	79.2 ± 4.0	2.04 ± 0.06	95.2 ± 2.4	76.8 ± 3.5	82.4 ± 3.3	12.0 ± 0.03	11.8 ± 0.3
Parental means ($\overline{P}$)[c]	165/921/2186	63.5 ± 1.18	2.06 ± 0.021	88.4 ± 0.79	64.3 ± 1.16	71.1 ± 1.13	12.1 ± 0.10	11.7 ± 0.1
Heterosis, %								
$100(\overline{FX} - \overline{P})/\overline{P}$[d]		21.4[e]	3.6[f]	1.5	13.7[e]	10.3[e]	3.0	4.2[f]
$100(C2_1 - \overline{P})/\overline{P}$		27.0[e]	1.9	3.6	25.2[e]	19.0[e]	1.4	3.3[f]
$100(C2_2 - \overline{P})/\overline{P}$		24.7[e]	−0.8	7.6[f]	19.5[e]	15.9[e]	−0.9	0.6

Reproduced with permission from Fogarty, Dickerson, and Young.[(8)]

[a] Adjusted for age of ewe and for year-season. See text.
[b] $C2_1$ ewes were reciprocal crosses of F · S × F · T. $C2_2$ ewes were from inter se matings of $C2_1$.
[c] $\overline{P} = \frac{1}{2}F + \frac{1}{4}S + \frac{1}{4}T$.
[d] Assuming that FS = SF in ewe performance.
[e] $P < 0.01$. } after adjusting the within-breed SE shown to a sire/breed basis.
[f] $P < 0.05$.

REFERENCES CITED

1. Botkin, M. P., and P. O. Stratton. 1967. Factors limiting selection effectiveness in small flocks of Columbias and Corriedales. *J. Animal Sci.* 26:971.
2. Bradley, B. P., and others. 1972. Two- and three-way crosses estimating combining ability of Suffolk, Targhee and Shropshire breeds of sheep. *J. Animal Sci.* 34:541.
3. Dahman, J. J., and others. 1980. Lamb production of Panama and Finn × Panama ewes. *Proc. Western Section Amer. Soc. Animal Sci.* 31:5.
4. Dickerson, G. E., and D. B. Laster. 1975. Breed, heterosis and environmental influences on growth and puberty in ewe lambs. *J. Animal Sci.* 41:1.
5. Dickerson, G. E., H. A. Glimp, and K. E. Gregory. 1975. Genetic resources for efficient meat production in sheep: Preweaning variability and growth of Finnsheep and domestic crossbred lambs. *J. Animal Sci.* 4l:43.
6. Dickerson, G. E. 1977. Crossbreeding evaluation of Finnsheep and some U.S. breeds for market lamb production. *North Central Regional Pub. No. 246.*
7. Dickerson, G. E., and others. 1985. Crossing performance of Finnsheep and domestic breeds. *SIDP Research Digest,* Vol. 2 No. 1, p. 20.
8. Fogarty, N. M., G. E. Dickerson, and L. D. Young. 1984. Lamb production and its components in pure breeds and composite lines. *J. Animal Sci.* 58:285.
9. Hodgson, C. W., T. D. Bell, and E. W. Owens. 1965. Influence of sire on multiple births in sheep. *Idaho Agr. Exp. Sta. PR 107.*
10. Hohenboken, W. D., and S. E. Clarke. 1981. Genetic, environmental and interaction effects on lamb survival, cumulative lamb production and longevity of crossbred ewes. *J. Animal Sci.* 53:966.
11. Lambertson, W. R., D. L. Thomas, and K. E. Rowe. 1982. The effects of inbreeding in a flock of Hampshire sheep. *J. Animal Sci.* 55:780.
12. Lancaster, D. B., H. A. Glimp, and H. E. Dickerson. 1972. Factors affecting reproduction in ewe lambs. *J. Animal Sci.* 35:79.
13. Magid, A. F., and others. 1981. Border Leicester and Finnsheep crosses. *J. Animal Sci.* 52:1253 and 1262.
14. Oltenacu, E. A. B., and W. J. Boylan. 1981. Productivity of purebred and crossbred Finnsheep. *J. Animal Sci.* 52:989, 998.
15. Rae, A. L. 1956. Genetics of sheep. *Advances in Genetics* 8:189.
16. Rae, A. L. 1982. Breeding. In *World Animal Science C-1.* Elsevier Publishing Co., Amsterdam, The Netherlands.
17. Rastigo, R., and others. 1982. Crossbreeding in sheep with evaluation of combining ability, heterosis and recombination effects for lamb growth. *J. Animal Sci.* 54:524.
18. Sheep Industry Development Program. 1975. *Sheepman's Production Handbook.* Denver, Colo.
19. Shelton, M. 1979. Estimation of genetic change in a performance testing program for sheep. *J. Animal Sci.* 48:26.
20. Sidwell, G. M., and L. R. Miller. 1971. Production in some pure breeds of sheep and their crosses. *J. Animal Sci.* 32:1084, 1090, 1095, 1099.

21. Snowder, G. D., M. Shelton, and D. Spiller. 1981. Factors influencing testes size of yearling Rambouillet rams. *Texas Agr. Exp. Sta. CPR3898.*
22. Snowder, Gary, Maurice Shelton, and P. Thompson. 1986. Evaluation of Finn-cross and Rambouillet ewes under Texas range conditions. *S.I.D. Research Digest,* Vol. 2, No. 1, p. 31.
23. *Texas Agr. Exp. Sta. Research Center Technical Report.* 84:1.
24. Thomas, D. L., and J. V. Whiteman. 1979. Effects of substituting Finnsheep and Dorset breeding for Rambouillet breeding. *J. Animal Sci.* 48:256, 265.
25. *USDA Agricultural Statistics.* Agricultural Marketing Service, Washington, D.C.
26. Vakil, D. V., M. P. Botkin, and G. P. Roehrkasse. 1968. Influence of hereditary and environmental factors on twinning in sheep. *J. Heredity* 59:256.
27. Veseley, J. A. 1978. Performance of progeny of Finnish Landrace and Dorset Horn rams mated to ewes of various breeds. *Canad. J. Animal Sci.* 58:399.
28. Veseley, J. A., and R. F. Peters. 1979. Lamb growth performance of certain pure breeds and their 2, 3 and 4 breed crosses. *Canad. J. Animal Sci.* 59:349.
29. Warwick, E. J., and J. E. Legates. 1979. *Breeding and Improvement of Farm Animals.* McGraw-Hill Book Co., New York.
30. Zidon, J. F., and others. 1979. Carcass evaluation of two- and three-way crosses of six breeds of sheep. *J. Animal Sci.* 49:1151.

SUGGESTED READING

Botkin, M. P., and others. Heritability of carcass traits in lambs. *J. Animal Sci.* 29:251.

Bradford, G. E., and G. M. Spurlock. 1972. Selection for meat production in sheep—results of a progeny test. *J. Animal Sci.* 34:737.

Burfening, P. J., J. L. Van Horn, and R. L. Blackwell. 1971. Genetics and phenotypic parameters including occurrence of estrus in Rambouillet ewe lambs. *J. Animal Sci.* 33:919.

Clarke, S.E., and W. D. Hohenboken. 1983. Estimation of repeatability, heritability and breed differences for lamb production. *J. Animal Sci.* 56:309.

Dickerson, G. E., and H. A. Glimp. 1975. Breed and age effects on lamb production of ewes. *J. Animal Sci.* 40:397.

Dzakuma, J. M., J. V. Whiteman, and R. W. McNew. 1982. Repeatability of lambing rate. *J. Animal Sci.* 54:540.

Dzakuma, J. M., M. K. Nielsen, and T. H. Doane. 1978. Genetics and phenotypic parameter estimates for growth and wool traits of Hampshire sheep. *J. Animal Sci.* 47:1014.

Ercanbrack. S. K., and D. A. Price. 1972. Selection for weight and rate of gain in non-inbred lambs. *J. Animal Sci.* 34:713.

Ercanbrack, S.K., and A. D. Knight. 1981. Weaning trait comparisons among inbred lines and selected non-inbred and randomly bred control groups of Rambouillet, Targhee and Columbia sheep. *J. Animal Sci.* 52:977.

Gould, M. B., and J. V. Whiteman, 1975. Relationship between preweaning growth rate of female lambs and the growth of their offspring. *J. Animal Sci.* 40:585.

Hohenboken, W., and P. E. Cochran. 1976. Heterosis for ewe lamb productivity. *J. Animal Sci.* 42:819.

Lax, J., and others. 1979. Comparison of single trait and index selection in sheep. *J. Animal Sci.* 48:776.

Notter, D. R. 1981. Repeatability of conception rate and litter size for ewes in an accelerated lambing system. *J. Animal Sci.* 30:1.

Olson, L. W., G. E. Dickerson, and H. A. Glimp. 1976. Selection criteria for intensive market lamb production. *J. Animal Sci.* 43:78, 90.

Pirchner, F. 1969. Population genetics in animal breeding. W. H. Freeman and Co., San Francisco.

Quirke, J. F., and others. 1985. Ovulation rate in sheep selected for weaning or litter size. *J. Animal Sci.* 61:1421.

Shelton, M., and J. W. Menzies. 1968. Genetic parameters of some performance characteristics of range fine-wool ewes. *J. Animal Sci.* 27:1219.

Terrill, C. E. 1958. Fifty years of progress in sheep breeding. *J. Animal Sci.* 17:944.

Thrift, F. A., J. V. Whiteman, and D. D. Kratzer. 1973. Genetic analysis of preweaning and post weaning lamb growth traits. *J. Animal Sci.* 36:640.

Turner, H. N. 1978. Selection for reproduction rate in Australian Merino sheep. *J. Agric. Sci.* 29:327.

Veseley, J. A., and others. 1970. Heritabilities and genetic correlations in growth and wool traits in sheep. *J. Animal Sci.* 30:174.

8 Sheep Nutrition

The first consideration of any sheep producer with regard to sheep feeding is to provide adequate nutrition. Nutrients are constantly required for maintenance of all essential body systems, such as skeletal, nervous, circulatory, respiratory, digestive, and musculature, regardless of age or stage of production. In addition, there are specific needs for reproduction, growth, lactation, and wool production. In sheep there is never a time during which simple maintenance requirements are sufficient to meet their needs, because wool growth is a continual process. Wool is protein and this creates competition for nutrients, particularly during growth and lactation. This was pointed out previously in Chapter 7 as being responsible for negative correlations between several productive traits (wool, growth, reproduction, and lactation).

Maintenance requirements vary with body surface and length of wool and are influenced by climatic conditions, facilities, and sources of feed and water. More energy is required by sheep grazing in semiarid pasture areas than by those in confinement with feed being carried to them. More nutrients are required for maintaining body temperature immediately following shearing than when fleeces are long enough to provide protection. This is particularly true in colder climates and is still true even when sheds or windbreaks are available to provide shelter.

NUTRIENTS

To provide adequate nutrition, familiarity with the various nutrients and their functions is essential. Nutrients fall into the following categories according to their chemical, physical, or biological nature: water, proteins, carbohydrates, fats, minerals, and vitamins.

Water

Water is necessary for all body processes. It aids in the functioning of digestive processes, absorption of nutrients into the body, and elimination of wastes. Water makes up a large part of the sheep's body tissues. An adequate and constant supply is required for all body systems to function properly and without interruptions. Water also contributes to the regulation of body temperature. Energy is used to bring the temperature of water that sheep drink up to body temperature. The amount of water used for these various functions is variable with respect to age, state of production, climatic conditions, and time of year.

Protein

Sheep require protein for all aspects of life and production: growth, reproduction, lactation, and wool growth. There is a constant need for normal replacement of body tissues.[6,7,10,16] Requirements vary with age and production. Young growing lambs require higher levels of protein in their feed than do mature sheep. Ewes in late gestation, particularly those carrying more than one lamb, and ewes during lactation require higher levels of protein than do the same ewes during the interim from weaning until the next breeding season or during the early part of gestation. Since wool is largely protein, it is obvious that protein is constantly demanded for uniform wool growth. Thus, during the stages of production involving the greatest stress, the competition for available protein becomes great. Even though ewes survive and produce when receiving less than adequate protein, their level of production is not at its true potential. (See Chapter 21 for specific needs for wool production.)

Sheep do not produce protein from other nutrients, but nonprotein nitrogen may serve as supplementary protein because such nitrogen is utilized by bacteria and protozoa in the rumen. Protein from these microorganisms is then utilized by the sheep.

Excess protein above actual needs by the sheep can be used as an energy source. This is a common situation when sheep are on high-quality forage diets, such as legume pastures or legume hays. Providing more protein than required, when it is from an expensive supplement, is a mistake in terms of economical production.

Energy

Energy, as used in describing or evaluating feeds, is actually the end product, rather than an inherent characteristic of compounds found in feeds. Energy results from utilization of the absorbed nutrients through metabolic processes such as oxidation and synthesis.[8] Sheep derive energy from a wide variety of sources, as they have the capability of utilizing feeds that differ greatly in nutrient content. Energy is often the most important ingredient in sheep feeds and can be a limiting factor due to many causes. Sheep grazing on dry range areas may be subject to energy deficiency because of the low energy content of feeds or because of limited feed intake due to unpalatability. Drought may intensify the problem, and inclement weather (such as a heavy snowstorm) may make feed less readily available. Energy shortage is sometimes accompanied by deficiencies of other nutrients such as protein or minerals. Without adequate energy, proper growth and development are impossible. Weight loss, lowered production, lowered resistance to stress, and increased mortality may result from energy deficiency.

In every phase of life and production, energy is important, but other nutrients are also essential in proper amounts for the energy from feeds to be used efficiently. Energy requirements of sheep vary greatly, depending on age, size, time of year, environmental conditions, or stage of production.[3,4,7,8,9,10,13,15,16] The energy values for various feeds can be found in Table 8-12.

Minerals

Minerals function as integral parts of bones and teeth and help give rigidity as well as strength to the skeleton. Some minerals are essential parts of organic compounds that occur in muscle, blood, and various secretions. Some are necessary for regulation of pH, the acid–base balance of body fluids. Also, some minerals are necessary for wool production (see Chapter 21).

A number of minerals have been found essential for sheep. Calcium, chlorine, sodium, phosphorus, magnesium, potassium, and sulfur are macrominerals (Table 8-3). Microminerals (Table 8-4) are iodine, iron, copper, molybdenum, cobalt, manganese, zinc, selenium, and fluorine. These are also known as trace minerals and in most sheep feeds are present in adequate amounts. Minerals of most concern to sheep producers in providing adequate nutrition are calcium, phosphorus, and salt (sodium chloride). In many cases an imbalance of important minerals, such as an oversupply of one, can become a problem just as serious as a deficiency. Also, some of the essential trace minerals (selenium or copper, for example) are toxic when present at levels that are too high. According to Pope,[12] some minerals are "oversupplemented" and some that should be added to the diet are not. His feeling is that too often minerals carry the blame for inadequacies of energy or management.

Vitamins

Most of the known vitamins have specific functions in the well-being of sheep, as in other species. Vitamin A is necessary for maintaining normal epithelial tissue. Vitamin D interacts with calcium and phosphorus in the development and maintenance of bones. Vitamin E is a dietary requirement for young nursing lambs to aid in muscle development. B complex vitamins and vitamins C and K likewise have important body functions, but they are synthesized in adequate amounts and do not need to be added to normal sheep diets. Of all the vitamins, the one of most concern in the business of sheep production is vitamin A, since grazing on dry feeds or in semiarid range areas, which lack vitamin A, is so common.

REQUIREMENTS

Nutritive requirements vary with age, development, and level of production, not only in total intake but also in the proportion of the various nutrients. Body weight is often used as a basis for determining requirements, but caution should be taken to allow for normal weight changes of the same ewe. Expected weight change as influenced by the state and level of production is depicted in Figure 8-1. This is an average of what happens to ewes weighing about 140 lb (63.5 kg). Some gain more during gestation or lose more during parturition or lactation, and some may not quite return to their original weight even after the flushing period. Care should also be used when weight is used as a basis for requirements to avoid confusing extra condition with weight. A small-framed ewe that is fat and weighs 140 lb is entirely different from a large-framed ewe in normal condition that weighs the same, and for fat ewes, requirements should be lower than what is indicated for ewes of their weight.

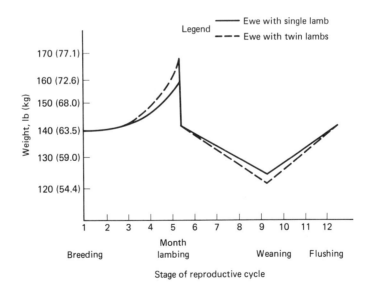

Figure 8-1 Weight changes of a ewe by stage and level of production, starting from breeding in once a year lambing program. (Adapted with permission from *Sheepman's Production Handbook*[15])

Requirements for the major nutrients by age, weight, sex, and stage of production are shown in Tables 8-1 and 8-2 as suggested by the National Research Council Subcommittee on Sheep Nutrition and published by the National Academy of Sciences.[8] Table 8-1 gives requirements per sheep per day, while Table 8-2 gives nutrient concentration in diets for sheep, expressed on a 100% dry matter basis.

The requirements listed in Tables 8-1 and 8-2 are based on available research data. Energy requirements in the diet are expressed as metabolizable energy (ME) in megacalories, and digestible energy (DE) in megacalories or as total digestible nutrients (TDN) in kilograms and pounds. Good-quality forages provide about 2 megacalories of ME per kilogram, and as concentrates are mixed with forages, the energy content of the diet increases. For these tables, it is assumed that ewes on a maintenance diet or ewes during the first 15 weeks of gestation are fed all-forage diets. Protein is listed as crude protein. The only minerals included in these tables are calcium and phosphorus, since these are critical in balancing the diets for sheep.

Mineral requirements in more detail are given in Tables 8-3 and 8-4 separated into macrominerals or microminerals.

As indicated by the Subcommittee on Sheep Nutrition, the requirements in Tables 8-1 and 8-2 are meant to be useful guides for providing adequate nutrition. Not only are sheep variable in size, shape, and condition, but there are also differences in quality of any type of feed. These recommendations apply to the average sheep receiving a feed or feeds of average quality. This is not intended to take the place of observation and sound judgment when determining needs of sheep.

Feed requirements, as listed in Tables 8-1 and 8-2, are aimed at providing the necessary nutrients for optimum production and for preventing any sign of deficiencies. A margin of safety is allowed for average or below average producers or for those with below average requirements.

Table 8-5 shows comparative requirements of net energy for ewes carrying different numbers of fetuses at various stages of gestation.

TABLE 8-1 DAILY NUTRIENT REQUIREMENTS OF SHEEP

Body Weight		Weight Change/Day		Dry Matter per Animal[a]			Energy[b]				Nutrients per Animal						
							TDN		DE	ME	Crude Protein		Ca	P	Vitamin A Activity	Vitamin E Activity	
(kg)	(lb)	(g)	(lb)	(kg)	(lb)	(% body weight)	(kg)	(lb)	(Mcal)	(Mcal)	(g)	(lb)	(g)	(g)	(IU)	(IU)	
Ewes[c]																	
Maintenance																	
50	110	10	0.02	1.0	2.2	2.0	0.55	1.2	2.4	2.0	95	0.21	2.0	1.8	2350	15	
60	132	10	0.02	1.1	2.4	1.8	0.61	1.3	2.7	2.2	104	0.23	2.3	2.1	2820	16	
70	154	10	0.02	1.2	2.6	1.7	0.66	1.5	2.9	2.4	113	0.25	2.5	2.4	3290	18	
80	176	10	0.02	1.3	2.9	1.6	0.72	1.6	3.2	2.6	122	0.27	2.7	2.8	3760	20	
90	198	10	0.02	1.4	3.1	1.5	0.78	1.7	3.4	2.8	131	0.29	2.9	3.1	4230	21	
Flushing: 2 weeks prebreeding and first 3 weeks of breeding																	
50	110	100	0.22	1.6	3.5	3.2	0.94	2.1	4.1	3.4	150	0.33	5.3	2.6	2350	24	
60	132	100	0.22	1.7	3.7	2.8	1.00	2.2	4.4	3.6	157	0.34	5.5	2.9	2820	26	
70	154	100	0.22	1.8	4.0	2.6	1.06	2.3	4.7	3.8	164	0.36	5.7	3.2	3290	27	
80	176	100	0.22	1.9	4.2	2.4	1.12	2.5	4.9	4.0	171	0.38	5.9	3.6	3760	28	
90	198	100	0.22	2.0	4.4	2.2	1.18	2.6	5.1	4.2	177	0.39	6.1	3.9	4230	30	
Nonlactating: First 15 weeks gestation																	
50	110	30	0.07	1.2	2.6	2.4	0.67	1.5	3.0	2.4	112	0.25	2.9	2.1	2350	18	
60	132	30	0.07	1.3	2.9	2.2	0.72	1.6	3.2	2.6	121	0.27	3.2	2.5	2820	20	
70	154	30	0.07	1.4	3.1	2.0	0.77	1.7	3.4	2.8	130	0.29	3.5	2.9	3290	21	
80	176	30	0.07	1.5	3.3	1.9	0.82	1.8	3.6	3.0	139	0.31	3.8	3.3	3760	22	
90	198	30	0.07	1.6	3.5	1.8	0.87	1.9	3.8	3.2	148	0.33	4.1	3.6	4230	24	
Last 4 weeks gestation (130%–150% lambing rate expected) or last 4–6 weeks lactation suckling singles[d]																	
50	110	180 (45)	0.40 (0.10)	1.6	3.5	3.2	0.94	2.1	4.1	3.4	175	0.38	5.9	4.8	4250	24	
60	132	180 (45)	0.40 (0.10)	1.7	3.7	2.8	1.00	2.2	4.4	3.6	184	0.40	6.0	5.2	5100	26	
70	154	180 (45)	0.40 (0.10)	1.8	4.0	2.6	1.06	2.3	4.7	3.8	193	0.42	6.2	5.6	5950	27	
80	176	180 (45)	0.40 (0.10)	1.9	4.2	2.4	1.12	2.4	4.9	4.0	202	0.44	6.3	6.1	6800	28	
90	198	180 (45)	0.40 (0.10)	2.0	4.4	2.2	1.18	2.5	5.1	4.2	212	0.47	6.4	6.5	7650	30	

(continued)

TABLE 8-1 (continued)

Body Weight		Weight Change/Day		Dry Matter per Animal[a]			Nutrients per Animal				Crude Protein		Ca	P	Vitamin A Activity	Vitamin E Activity
							Energy[b]									
						(% body	TDN		DE	ME						
(kg)	(lb)	(g)	(lb)	(kg)	(lb)	weight)	(kg)	(lb)	(Mcal)	(Mcal)	(g)	(lb)	(g)	(g)	(IU)	(IU)

Last 4 weeks gestation (180%–225% lambing rate expected)

50	110	225	0.50	1.7	3.7	3.4	1.10	2.4	4.8	4.0	196	0.43	6.2	3.4	4250	26
60	132	225	0.50	1.8	4.0	3.0	1.17	2.6	5.1	4.2	205	0.45	6.9	4.0	5100	27
70	154	225	0.50	1.9	4.2	2.7	1.24	2.8	5.4	4.4	214	0.47	7.6	4.5	5950	28
80	176	225	0.50	2.0	4.4	2.5	1.30	2.9	5.7	4.7	223	0.49	8.3	5.1	6800	30
90	198	225	0.50	2.1	4.6	2.3	1.37	3.0	6.0	5.0	232	0.51	8.9	5.7	7650	32

First 6–8 weeks lactation suckling singles or last 4–6 weeks lactation suckling twins[d]

50	110	−25 (90)	−0.06 (0.20)	2.1	4.6	4.2	1.36	3.0	6.0	4.9	304	0.67	8.9	6.1	4250	32
60	132	−25 (90)	−0.06 (0.20)	2.3	5.1	3.8	1.50	3.3	6.6	5.4	319	0.70	9.1	6.6	5100	34
70	154	−25 (90)	−0.06 (0.20)	2.5	5.5	3.6	1.63	3.6	7.2	5.9	334	0.73	9.3	7.0	5950	38
80	176	−25 (90)	−0.06 (0.20)	2.6	5.7	3.2	1.69	3.7	7.4	6.1	344	0.76	9.5	7.4	6800	39
90	198	−25 (90)	−0.06 (0.20)	2.7	5.9	3.0	1.75	3.8	7.6	6.3	353	0.78	9.6	7.8	7650	40

First 6–8 weeks lactation suckling twins

50	110	−60	−0.13	2.4	5.3	4.8	1.56	3.4	6.9	5.6	389	0.86	10.5	7.3	5000	36
60	132	−60	−0.13	2.6	5.7	4.3	1.69	3.7	7.4	6.1	405	0.89	10.7	7.7	6000	39
70	154	−60	−0.13	2.8	6.2	4.0	1.82	4.0	8.0	6.6	420	0.92	11.0	8.1	7000	42
80	176	−60	−0.13	3.0	6.6	3.8	1.95	4.3	8.6	7.0	435	0.96	11.2	8.6	8000	45
90	198	−60	−0.13	3.2	7.0	3.6	2.08	4.6	9.2	7.5	450	0.99	11.4	9.0	9000	48

Ewe lambs

Nonlactating: First 15 weeks gestation

40	88	160	0.35	1.4	3.1	3.5	0.83	1.8	3.6	3.0	156	0.34	5.5	3.0	1880	21
50	110	135	0.30	1.5	3.3	3.0	0.88	1.9	3.9	3.2	159	0.35	5.2	3.1	2350	22
60	132	135	0.30	1.6	3.5	2.7	0.94	2.0	4.1	3.4	161	0.35	5.5	3.4	2820	24
70	154	125	0.28	1.7	3.7	2.4	1.00	2.2	4.4	3.6	164	0.36	5.5	3.7	3290	26

Last 4 weeks gestation (100%–120% lambing rate expected)

40	88	180	0.40	1.5	3.3	3.8	0.94	2.1	4.1	3.4	187	0.41	6.4	3.1	3400	22
50	110	160	0.35	1.6	3.5	3.2	1.00	2.2	4.4	3.6	189	0.42	6.3	3.4	4250	24
60	132	160	0.35	1.7	3.7	2.8	1.07	2.4	4.7	3.9	192	0.42	6.6	3.8	5100	26
70	154	150	0.33	1.8	4.0	2.6	1.14	2.5	5.0	4.1	194	0.43	6.8	4.2	5950	27

Last 4 weeks gestation (130%–175% lambing rate expected)														
40	88	225	0.50	3.8	1.5	3.3	0.99	2.2	4.4	3.6	202	0.44	7.4	3.5	3400	22
50	110	225	0.50	3.2	1.6	3.5	1.06	2.3	4.7	3.8	204	0.45	7.8	3.9	4250	24
60	132	225	0.50	2.8	1.7	3.7	1.12	2.5	4.9	4.0	207	0.46	8.1	4.3	5100	26
70	154	215	0.47	2.6	1.8	4.0	1.14	2.5	5.0	4.1	210	0.46	8.2	4.7	5950	27
First 6–8 weeks lactation suckling singles (wean by 8 weeks)														
40	88	−50	−0.11	4.2	1.7	3.7	1.12	2.5	4.9	4.0	257	0.56	6.0	4.3	3400	26
50	110	−50	−0.11	4.2	2.1	4.6	1.39	3.1	6.1	5.0	282	0.62	6.5	4.7	4250	32
60	132	−50	−0.11	3.8	2.3	5.1	1.52	3.4	6.7	5.5	295	0.65	6.8	5.1	5100	34
70	154	−50	−0.11	3.6	2.5	5.5	1.65	3.6	7.3	6.0	301	0.68	7.1	5.6	5450	38
First 6–8 weeks lactation suckling twins (wean by 8 weeks)														
40	88	−100	−0.22	5.2	2.1	4.6	1.45	3.2	6.4	5.2	306	0.67	8.4	5.6	4000	32
50	110	−100	−0.22	4.6	2.3	5.1	1.59	3.5	7.0	5.7	321	0.71	8.7	6.0	5000	34
60	132	−100	−0.22	4.2	2.5	5.5	1.72	3.8	7.6	6.2	336	0.74	9.0	6.4	6000	38
70	154	−100	−0.22	3.9	2.7	6.0	1.85	4.1	8.1	6.6	351	0.77	9.3	6.9	7000	40
Replacement ewe lambs[e]														
30	66	227	0.50	4.0	1.2	2.6	0.78	1.7	3.4	2.8	185	0.41	6.4	2.6	1410	18
40	88	182	0.40	3.5	1.4	3.1	0.91	2.0	4.0	3.3	176	0.39	5.9	2.6	1880	21
50	110	120	0.26	3.0	1.5	3.3	0.88	1.9	3.9	3.2	136	0.30	4.8	2.4	2350	22
60	132	100	0.22	2.5	1.5	3.3	0.88	1.9	3.9	3.2	134	0.30	4.5	2.5	2820	22
70	154	100	0.22	2.1	1.5	3.3	0.88	1.9	3.9	3.2	132	0.29	4.6	2.8	3290	22
Replacement ram lambs[e]														
40	88	330	0.73	4.5	1.8	4.0	1.1	2.5	5.0	4.1	243	0.54	7.8	3.7	1880	24
60	132	320	0.70	4.0	2.4	5.3	1.5	3.4	6.7	5.5	263	0.58	8.4	4.2	2820	26
80	176	290	0.64	3.5	2.8	6.2	1.8	3.9	7.8	6.4	268	0.59	8.5	4.6	3760	28
100	220	250	0.55	3.0	3.0	6.6	1.9	4.2	8.4	6.9	264	0.58	8.2	4.8	4700	30
Lambs finishing; 4 to 7 months old[f]														
30	66	295	0.65	4.3	1.3	2.9	0.94	2.1	4.1	3.4	191	0.42	6.6	3.2	1410	20
40	88	275	0.60	4.0	1.6	3.5	1.22	2.7	5.4	4.4	185	0.41	6.6	3.3	1880	24
50	110	205	0.45	3.2	1.6	3.5	1.23	2.7	5.4	4.4	160	0.35	5.6	3.0	2350	24
Early weaned lambs: moderate growth potential[f]														
10	22	200	0.44	5.0	0.5	1.1	0.40	0.9	1.8	1.4	127	0.38	4.0	1.9	470	10
20	44	250	0.55	5.0	1.0	2.2	0.80	1.8	3.5	2.9	167	0.37	5.4	2.5	940	20
30	66	300	0.66	4.3	1.3	2.9	1.00	2.2	4.4	3.6	191	0.42	6.7	3.2	1410	20
40	88	345	0.76	3.8	1.5	3.3	1.16	2.6	5.1	4.2	202	0.44	7.7	3.9	1880	22
50	110	300	0.66	3.0	1.5	3.3	1.16	2.6	5.1	4.2	181	0.40	7.0	3.8	2350	22

TABLE 8-1 (continued)

Body Weight		Weight Change/Day		Dry Matter per Animal[a]			Energy[b]				Nutrients per Animal					
							TDN		DE	ME	Crude Protein		Ca	P	Vitamin A Activity	Vitamin E Activity
(kg)	(lb)	(g)	(lb)	(kg)	(lb)	(% body weight)	(kg)	(lb)	(Mcal)	(Mcal)	(g)	(lb)	(g)	(g)	(IU)	(IU)
Early weaned lambs: rapid growth potential[f]																
10	22	250	0.55	0.6	1.3	6.0	0.48	1.1	2.1	1.7	157	0.35	4.9	2.2	470	12
20	44	300	0.66	1.2	2.6	6.0	0.92	2.0	4.0	3.3	205	0.45	6.5	2.9	940	24
30	66	325	0.72	1.4	3.1	4.7	1.10	2.4	4.8	4.0	216	0.48	7.2	3.4	1410	21
40	88	400	0.88	1.5	3.3	3.8	1.14	2.5	5.0	4.1	234	0.51	8.6	4.3	1880	22
50	110	425	0.94	1.7	3.7	3.4	1.29	2.8	5.7	4.7	240	0.53	9.4	4.8	2350	25
60	132	350	0.77	1.7	3.7	2.8	1.29	2.8	5.7	4.7	240	0.53	8.2	4.5	2820	25

Reprinted with the permission of the National Academy Press, Washington, D.C.[(8)]

[a] To convert dry matter to an as-fed basis, divide dry matter values by the percentage of dry matter in the particular feed.

[b] One kilogram TDN (total digestible nutrients) = 4.4 Mcal DE (digestible energy); ME (metabolizable energy) = 82% of DE. Because of rounding errors, values in Tables 8-1 and 8-2 may differ.

[c] Values are applicable for ewes in moderate condition. Fat ewes should be fed according to the next lower weight category and thin ewes at the next higher weight category. Once desired or moderate weight condition is attained, use that weight category through all production stages.

[d] Values in parentheses are for ewes suckling lambs the last 4 to 6 weeks of lactation.

[e] Lambs intended for breeding; thus, maximum weight gains and finish are of secondary importance.

[f] Maximum weight gains expected.

TABLE 8-2 NUTRIENT CONCENTRATION IN DIETS FOR SHEEP (EXPRESSED ON 100% DRY MATTER BASIS)[a]

Body Weight		Weight Change/Day		Energy[b]			Example Diet Proportions		Crude Protein (%)	Calcium (%)	Phosphorus (%)	Vitamin A Activity (IU/kg)	Vitamin E Activity (IU/kg)
(kg)	(lb)	(g)	(lb)	TDN[c] (%)	DE (Mcal/kg)	ME (Mcal/kg)	Concentrate %	Forage %					

Ewes[d]
Maintenance

| 70 | 154 | 10 | 0.02 | 55 | 2.4 | 2.0 | 0 | 100 | 9.4 | 0.20 | 0.20 | 2742 | 15 |

Flushing: 2 weeks prebreeding and first 3 weeks of breeding

| 70 | 154 | 100 | 0.22 | 59 | 2.6 | 2.1 | 15 | 85 | 9.1 | 0.32 | 0.18 | 1828 | 15 |

Nonlactating: first 15 weeks gestation

| 70 | 154 | 30 | 0.07 | 55 | 2.4 | 2.0 | 0 | 100 | 9.3 | 0.25 | 0.20 | 2350 | 15 |

Last 4 weeks gestation (130%–150% lambing rate expected) or last 4–6 weeks lactation suckling singles[e]

| 70 | 154 | 180 (0.45) | 0.40 (0.10) | 59 | 2.6 | 2.1 | 15 | 85 | 10.7 | 0.35 | 0.23 | 3306 | 15 |

Last 4 weeks gestation (180%–225% lambing rate expected)

| 70 | 154 | 225 | 0.50 | 65 | 2.9 | 2.3 | 35 | 65 | 11.3 | 0.40 | 0.24 | 3132 | 15 |

First 6–8 weeks lactation suckling singles or last 4–6 weeks lactation suckling twins[e]

| 70 | 154 | −25(90) | −0.06 (0.20) | 65 | 2.9 | 2.4 | 35 | 65 | 13.4 | 0.32 | 0.26 | 2380 | 15 |

First 6–8 weeks lactation suckling twins

| 70 | 154 | −60 | −0.13 | 65 | 2.9 | 2.4 | 35 | 65 | 15.0 | 0.39 | 0.29 | 2500 | 15 |

Ewe lambs:
Nonlactating: first 15 weeks gestation

| 55 | 121 | 135 | 0.30 | 59 | 2.6 | 2.1 | 15 | 85 | 10.6 | 0.35 | 0.22 | 1668 | 15 |

Last 4 weeks gestation (100%–120% lambing rate expected)

| 55 | 121 | 160 | 0.35 | 63 | 2.8 | 2.3 | 30 | 70 | 11.8 | 0.39 | 0.22 | 2833 | 15 |

Last 4 weeks gestation (130%–175% lambing rate expected)

| 55 | 121 | 225 | 0.50 | 66 | 2.9 | 2.4 | 40 | 60 | 12.8 | 0.48 | 0.25 | 2833 | 15 |

First 6–8 weeks lactation suckling singles (wean by 8 weeks)

| 55 | 121 | −50 | 0.22 | 66 | 2.9 | 2.4 | 40 | 60 | 13.1 | 0.30 | 0.22 | 2125 | 15 |

(continued)

TABLE 8-2 (continued)

Body Weight		Weight Change/Day		Energy[b]			Example Diet Proportions		Crude Protein (%)	Calcium (%)	Phosphorus (%)	Vitamin A Activity (IU/kg)	Vitamin E Activity (IU/kg)
(kg)	(lb)	(g)	(lb)	TDN[c] (%)	DE (Mcal/kg)	ME (Mcal/kg)	Concentrate %	Forage %					
First 6–8 weeks lactation suckling twins (wean by 8 weeks)														
55	121	−100	−0.22	69	3.0	2.5	50	50	13.7	0.37	0.26	2292	15
Replacement ewe lambs[f]														
30	66	227	0.50	65	2.9	2.4	35	65	12.8	0.53	0.22	1175	15
40	85	182	0.40	65	2.9	2.4	35	65	10.2	0.42	0.18	1343	15
50–70	110–154	115	0.25	59	2.6	2.1	15	85	9.1	0.31	0.17	1567	15
Replacement ram lambs[f]														
40	88	330	0.73	63	2.8	2.3	30	70	13.5	0.43	0.21	1175	15
60	132	320	0.70	63	2.8	2.3	30	70	11.0	0.35	0.18	1659	15
80–100	176–220	270	0.60	63	2.8	2.3	30	70	9.6	0.30	0.16	1979	15
Lambs finishing: 4 to 7 months old[g]														
30	66	295	0.65	72	3.2	2.5	60	40	14.7	0.51	0.24	1085	15
40	88	275	0.60	76	3.3	2.7	75	25	11.6	0.42	0.21	1175	15
50	110	205	0.45	77	3.4	2.8	80	20	10.0	0.35	0.19	1469	15
Early weaned lambs: moderate and rapid growth potential[g]														
10	22	250	0.55	80	3.5	2.9	90	10	26.2	0.82	0.38	940	20
20	44	300	0.66	78	3.4	2.8	85	15	16.9	0.54	0.24	940	20
30	66	325	0.72	78	3.3	2.7	85	15	15.1	0.51	0.24	1085	15
40–60	88–132	400	0.88	78	3.3	2.7	85	15	14.5	0.55	0.28	1253	15

Reprinted with permission of the National Academy Press, Washington, D.C.[(8)]

[a]Values are calculated from daily requirements in Table 8-1 divided by DM intake. The exception, vitamin E daily requirements/head, is calculated from vitamin E/kg diet × DM intake.

[b]One kilogram TDN = 4.4 Mcal DE digestible energy. ME (metabolizable energy) = 82% of DE. Because of rounding errors, values in Tables 8-1 and 8-2 may differ.

[c]TDN calculated on following basis: hay DM, 55% TDN and on as-fed basis 50% TDN; grain DM, 83% TDN and on as-fed basis 75% TDN.

[d]Values are for ewes in moderate condition. Fat ewes should be fed according to the next lower weight category and thin ewes at the next higher weight category. Once desired or moderate weight condition is attained, use that weight category through all production stages.

[e]Values in parentheses are for ewes suckling lambs the last 4 to 6 weeks of lactation.

[f]Lambs intended for breeding; thus maximum weight gains and finish are of secondary importance.

[g]Maximum weight gains expected.

TABLE 8-3 MACROMINERAL REQUIREMENTS OF SHEEP (PERCENTAGE OF DIET DRY MATTER)[a]

Nutrient	Requirement
Sodium	0.09–0.18
Chlorine	—
Calcium	0.20–0.82
Phosphorus	0.16–0.38
Magnesium	0.12–0.18
Potassium	0.50–0.80
Sulfur	0.14–0.26

Reprinted with permission of the National Academy Press, Washington, D.C.[8]

[a]Values are estimates based on experimental data.

TABLE 8-4 MICROMINERAL REQUIREMENTS OF SHEEP AND MAXIMUM TOLERABLE LEVELS (PPM, MG/KG OF DIET DRY MATTER)[a]

Nutrient	Requirement	Maximum Tolerable Level[b]
Iodine	0.10–0.80[c]	50
Iron	30–5	500
Copper	7–11[d]	25[e]
Molybdenum	0.5	10[e]
Cobalt	0.1–0.2	10
Manganese	20–40	1000
Zinc	20–33	750
Selenium	0.1–0.2	2
Fluorine	—	60–150

Reprinted with permission of the National Academy Press, Washington, D.C.[8]

[a]Values are estimates based on experimental data.

[b]National Research Council 1985

[c]High level for pregnancy and lactation in diets not containing goitrogens; should be increased if diets contain goitrogens.

[d]Requirement when dietary Mo concentrations are <1 mg/kg DM. See text for requirements under other circumstances.

[e]Lower levels may be toxic under some circumstances. See text.

TABLE 8-5 NEPREG (NE_v) REQUIREMENTS (KCAL/DAY) OF EWES CARRYING DIFFERENT NUMBERS OF FETUSES AT VARIOUS STAGES OF GESTATION

Number of Fetuses Being Carried	Stage of Gestation (days)[a]					
	100	%[b]	120	%[b]	140	%[b]
1	70	100	145	100	260	100
2	125	178	265	183	440	169
3	170	243	345	238	570	219

Reprinted with permission of the National Academy Press, Washington, D.C.[8]

[a]For gravid uterus (plus contents) and mammary gland development only.

[b]As a percentage of a single fetus's requirement.

BALANCING RATIONS AND FORMULATING DIETS

An understanding of how rations are balanced and diets are formulated is an essential part of the ability to provide adequate nutrition. Several methods are used for balancing rations. Examples are as follows.

Percentage Method

Reprinted with Permission: This is the method of choice to be used when feeding a total mixed ration.[14] Assume the objective is to develop a high concentrate finishing ration for old crop lambs already on feed and that the price of wheat economically justifies its use in the ration.

Ingredient	Percentage of Ration	Protein (%)	CA (%)	P (%)
Wheat, ground	40	4.44	0.036	0.120
Corn, ground, shelled	40	3.56	0.008	0.124
Alfalfa hay, ground	15	2.25	0.108	0.030
Total		10.25	0.224	0.274

Using wheat at the rate of one-half of the grain portion of the ration and adding 15% alfalfa hay for roughage would provide the above ingredients on a percentage basis. Note that only 95% of the total ration is included since as estimated 5% would be required to supplement deficient nutrients. The above ration needs additional protein. Thus, the following must be added:

	Percentage
Soybean meal, 44% CP	4.0
Salt, trace-mineralized	0.5
Ammonium chloride, feed grade	0.25
Aureomycin or Terramycin	20 g/ton

The above additions increase the protein level to 12% and meet the salt and antibiotic needs. If ammonium chloride is not available, it is recommended that 1.0% feed-grade ground limestone be added to the ration as a safety factor in protection against urinary calculi.

In summary, the percentage method can be used most effectively by determining in advance the desired concentrate–roughage ratio and the grain and roughage sources to be used. Once this has been done, the difference between the nutrients provided by these ingredients and the requirements can be easily determined.

Fixed Supplement Method

Reprinted with Permission: This method is used when lambs are allowed to consume unlimited amounts of roughage and the supplement is fed to provide the needed nutrients to balance the ration.[14] Examples of such a program would be wheat pasture grazing, grazing corn or milo stubble, or a silage or haylage feeding program. As an example, assume that lambs will consume 2 lb of dry matter daily from wheat pasture. *Feeds and Feeding* gives the following analysis for wheat pasture, with the requirements for 60-lb early weaned lambs listed below:

	Dry Matter	Crude Protein	TDN	Ca	P
Wheat pasture, % composition	20	4.8	12.7	0.09	0.08
Nutrient requirements, lb/head/day	2.7 (1.22 kg)	0.38 (0.17 kg)	1.6 (0.725 kg)	0.007 (3 g)	0.006 (2.7 g)

The lambs would therefore be eating 10 lb (4.54 kg) per head per day of wheat pasture in order to consume 2.0 lb (0.91 kg) of dry matter. The wheat pasture would

Dry Matter (lb)	Crude Protein (lb)	TDN (lb)	Ca (lb)	P (lb)
2.0 (0.91/kg)	0.48 (0.22 kg)	1.27 (0.58 kg)	0.009 (4 g)	0.008 (3.6 g)

provide the following nutrients (lb intake × % composition). As most research with supplemental feeding of lambs on wheat pasture has shown, supplemental dry matter and energy (TDN) are all that would be required. This could be provided by 0.75 lb (0.34 kg) per day of good quality hay. Salt and steamed bonemeal would be fed free choice on this feeding program.

Rations for all classes of sheep may be calculated in a similar manner. As an example, the requirements of a ewe flock averaging 160 lb (73.5 kg) in the last 6 weeks of pregnancy would be calculated as follows:

1. Find the nutrient requirements of the 160-lb (73.5-kg) ewe in the table of requirements.
2. List the proposed ration and determine the nutrient content of the ration from a feed composition table.
3. Subtract content of ration from requirements and determine the deficiencies.
4. Add the additional amount necessary to fulfill requirements.

Daily Ration

Alfalfa hay, 3 lb (1.36 kg)
Shelled corn, 0.5 lb (0.23 kg)

Requirements

Body Weight, lb (kg)	Daily Feed per Animal, lb (kg)	TDN[a] lb (kg)	Total Protein, lb (kg)	Ca, g	P, g
160 (73.5)	4.80 (2.2)	2.5 (1.13)	0.37 (0.17)	4.8	3.7
Nutrient Content of Ration					
Alfalfa hay (mid-bloom)	3.0 (1.36)	1.5 (0.68)	0.46 (0.21)	16.3	2.7
Shelled corn	0.5 (0.23)	0.4 (0.18)	0.04 (0.02)	—	0.7
Total	3.5 (1.58)	1.9 (0.86)	0.50 (0.23)	16.3	3.4
Deficiencies	1.3 (0.59)	0.6 (0.27)	—	—	0.7
Addition Required to Balance Ration					
Alfalfa hay	1.0 (0.45)	0.5 (0.23)	0.15 (0.07)	5.4	0.9
Shelled corn	0.25 (0.11)	0.2 (0.09)	0.02 (0.009)		

[a]TDN = total digestible nutrients

Pearson Square Method

This is a simple, direct method of formulating rations in which balancing the protein requirements receives major consideration.[4] An example of how it can be used is as follows: Using prairie hay and soybean meal (solvent extracted) as a supplement, a ration can be balanced for protein.

The daily requirements for a 154-lb (70-kg) ewe during gestation are as follows:

DM, lb (kg)	DE, Mcal	Crude protein, lb (kg)	Ca, g	P, g
4.0 (1.8)	4.7	0.42 (0.19)	6.3	6.1

To meet the suggested DM, it requires 4.44 lb (2 kg) of total feed, since both feeds contain 90% DM. Then, dividing the protein required, 0.42 lb (193 g) by total feed, the percent protein in the ration is 9.5%. Next a square is made with the needed protein percent in the center and with the protein percent of the two feeds at the left (Figure 8-2). Subtracting diagonally gives figures at the right (44.8 − 9.5 = 35.3 and 9.5 − 7.8 = 1.7). Adding 35.3 + 1.7 gives a total of 37.0. This means that, out of 37 total parts, 35.3 (95.4% of the total) comes from prairie hay and 1.7

Balancing Rations and Formulating Diets

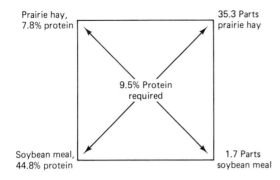

Figure 8-2 Example of Pearson square for determining portions of feeds to provide protein requirements.

parts (4.6% of the total) comes from soybean meal. Then of the total feed (4.44 lb or 2 kg), 4.19 lh (1.9 kg) should be prairie hay and 0.21 lb (0.09 kg) should be soybean meal.

The diet will provide the following:

	DE	Protein, lb (kg)	Ca, g	P, g
4.19 lg (1.9 kg) prairie hay	4.78	0.33 (0.15)	8.16	4.08
0.21 lb (0.09 kg) soybean meal	0.31	0.09 (0.04)	0.30	0.59
Total	5.09	0.42 (0.19)	8.46	4.67

Comparing the nutrients provided by these ingredients with the requirements yields the following:

	DE, Mcal	Protein, lb (kg)	Ca, g	P, g
Requirements	4.7	0.42 (0.19)	6.3	6.1
Provided by diet	5.1	0.42 (0.19)	8.5	4.7

As can be seen from this comparison, the dietary needs are met except for phosphorus. This deficiency can be corrected by feeding a high-phosphorus supplement, such as monosodium phosphate, 0.015 lb (7 g) daily either alone or in a mixture with salt. Also, phosphorus can be added in the manufacturing process to soybean meal or to another protein concentrate.

Admittedly, this is a simple example in that nearly all the nutrient requirements are met with the feeds selected, but the example is merely intended to demonstrate the use of the Pearson square in balancing the protein in the ration. It is fairly common in the sheep business for home-grown forages, when supplemented with a good protein source to provide a balanced diet. Also, the feed industry has the capability of supplying any type of supplement that may be required.

National Research Council Method

A method of balancing a ration for all ingredients is provided in reference 8 and is described, with the permission of the NRC, in the following:

By combining information in Tables 8-1 and 8-2 with data on feed composition listed in Table 8-12, rations can be formulated to provide all requirements. Rations can be balanced to provide specific amounts of nutrients per sheep daily as recommended in Table 8-1 or to provide a complete diet containing the recommended composition shown in Table 8-2, which when fed at recommended levels will provide nutrients in the amounts recommended in Table 8-1. The weight is based on normal weight for a ewe and requirements are listed for normal weight, even though there is much variation due to stage of production, as previously seen in Figure 8-1.

TABLE 8-6 RECOMMENDED NUTRIENT CONCENTRATION IN DIETS FOR 60-KG (132-LB) EWES, FIRST 6 to 8 WEEKS LACTATION SUCKLING TWINS, AND NUTRIENT CONTENT OF FEEDS, BOTH ON DM BASIS

Item	DE (Mcal/kg)	Crude Protein (%)	CA (%)	P (%)	Carotene (mg/kg)	Vitamin A (IU/kg)
Diet concentration	2.9	15.0	0.39	0.29	—	2500
Oat hay	2.38	9.2	0.26	0.24	101.0	—
Barley	3.79	13.0	0.09	0.47	—	—
Soybean meal	3.53	51.5	0.36	0.75	—	—

Sheep rations are formulated by selecting a major feed energy source, such as pasture grass or hay or silage. Next determine what nutrients the feed provides and compare these with the values in Table 8-1 or 8-2. Then determine the composition and amount of supplement that when fed with the grass or hay or silage will compensate for the nutrient shortages. In the following examples, rations are formulated for a 132-lb (60-kg) ewe suckling twin lambs during early lactation, using oat hay, barley, and soybean meal as available feeds.

Complete diet formulation. Write down the recommended nutrient composition of the diet from Table 8-2 and the nutrient content of the feeds (Table 8-6). In this example, the feeds used were analyzed by a commercial laboratory; if analyzed values are not available, use average values from Table 8-12.

Procedure

1. Compare the composition of oat hay with the ewe's requirements. The oat hay is inadequate in all nutrients except carotene (5 mg of carotene would supply the vitamin A requirement).
2. Determine the substitution value of barley for oat hay: 3.79 Mcal DE (barley) − 2.38 Mcal DE (hay) = 1.41 Mcal DE/kg. The DE deficiency in an oat hay diet is 2.90 Mcal (required) − 2.38 Mcal (hay) = 0.52 Mcal DE. Determine the percent of barley to substitute for part of the hay to provide the 0.52-Mcal DE/kg deficiency (0.52 Mcal ÷ 1.41 Mcal = 0.37). The diet at this stage becomes 37% barley and 63% oat hay.
3. Compare this combination of oat hay and barley with the dietary requirements for DM, DE, and protein (Table 8-7). The diet is now adequate in digestible energy but is 4.4% deficient in protein.
4. Substitute soybean meal for barley to provide for the protein deficiency. Determine the difference in crude protein content of the two feeds: 51.5% protein (soybean meal) − 13% (barley) = 38.5% protein. Divide the amount of protein that is deficient (4.4%) by the amount provided when soybean meal is substi-

TABLE 8-7 COMPARISON OF DIETARY REQUIREMENTS WITH AMOUNT OF NUTRIENTS PROVIDED IN HAY–BARLEY DIET

Item	DM (% of diet)	DE (meal)	Protein (%)
Dietary requirement	100.0	2.9	15.0
Oat hay	63.0	1.5[a]	5.8[a]
Barley	37.0	1.4	4.8
Total	100.0	2.9	10.6
Difference	—	—	−4.4

[a]Values obtained by multiplying energy or protein content in feeds by percent of feeds in diet.

tuted for a unit of barley: 4.4% ÷ 38.5% = 0.114, or 11.4% of the entire ration will be soybean meal. The diet then is as shown in Table 8-8 and is now adequate in both digestible energy and protein.

5. Write down the calcium and phosphorus requirement and compare these with the amounts provided by the hay–barley–soybean meal diet (Table 8-9).

The diet is adequate in phosphorus but is 0.17% deficient in calcium. Limestone is a rich (34%) and inexpensive source of calcium. Dividing the 0.17% deficiency by the 34% calcium in limestone gives 0.5 parts limestone that should be

TABLE 8-8 COMPARISON OF DIETARY REQUIREMENTS WITH ENERGY AND PROTEIN PROVIDED BY THE OAT HAY–BARLEY–SOYBEAN MEAL DIET

Item	DM (%)	DE (meal)	Protein (%)
Requirement	100.0	2.9	15.0
Oat hay	63.0	1.5	5.8
Barley	25.6	1.0	3.3
Soybean meal	11.4	0.4	5.9
Total	100.0	2.9	15.0

TABLE 8-9 COMPARISON OF DIETARY REQUIREMENTS WITH CALCIUM AND PHOSPHORUS PROVIDED BY THE OAT HAY–BARLEY–SOYBEAN MEAL DIET

Item	DM (%)	Calcium (%)	Phosphorus (%)
Requirement	100.0	0.39	0.29
Oat hay	63.0	0.16	0.15
Barley	25.6	0.02	0.12
Soybean meal	11.4	0.04	0.08
Total	100.0	0.22	0.35
Difference	—	−0.17	+0.06

added to the diet. The final diet now becomes as shown in Table 8-10 (parts, DM basis). When fed at the levels recommended in Table 8-1, it will satisfy the daily requirements for this category of sheep.

Up to this point, all calculations have been made on a DM basis; however, few feeds are 100% dry matter. To convert the portions of feed on a DM basis to an as-fed basis (Table 8-10), divide the contribution of the feed in question by the percent of dry matter it contains (oat hay: 63 parts of the diet DM ÷ 88.2% dry matter = 71.4 parts, air dry). To convert the various component parts of the diet

TABLE 8-10 COMPLETE DIET FOR 60-KG (132-LB) EWES, FIRST 6 TO 8 WEEKS LACTATION SUCKLING TWINS

Ingredient	DM (%)	Parts in Diet DM[a]	Air Dry or As-Fed Basis Parts[b]	Percent[c]
Oat hay	88.2	63.0	71.4	62.9
Barley	89.0	25.6	28.8	25.4
Soybean meal	89.0	11.4	12.8	11.3
Limestone	100.0	0.5	0.5	0.4
Total			113.5	100.0

[a]As determined in preceding steps.
[b]Calculated as 63.0/0.882 = 71.4.
[c]Calculated as 71.4/1.135 = 62.9.

back to percentage values, divide the number of parts for each of the ration components by the total number of parts (oat hay: 71.4 parts air dry/113.5 total parts = 62.9% on as as-fed basis). Dry matter intake is converted to an as-fed basis by dividing the daily intake given in Table 8-1 (2.6 kg) by 0.89 (approximate DM in the total ration) to give 2.92 kg (6.4 lb) of the complete diet per day. If the feeds were hand feed separately to the ewes, 62.9% or 1.84 kg (0.629 × 2.92) of the as-fed diet would be oat hay and 1.08 kg (2.92 − 1.84 kg) of the as-fed diet would consist of barley, soybean meal, and limestone.

The total air dry part of the barley–soybean meal–limestone portion of the diet is 42.1. Each feed part is divided by this value and multiplied by 100 to obtain the percent it contributes. The composition of this mixture becomes 68.4% barley, 30.4% soybean meal, and 1.2% limestone on an air dry basis.

The *Pearson square* method may also be used for determining the amount of soybean meal to add to the barley in step 4.

Formulation to provide specific amounts of nutrients per ewe per day (same category of ewe and same feeds)

1. Write down the ewe's daily requirement (from Table 8-1) and the nutrient composition of the three feed ingredients (Table 8-11). Dry matter per se is not a nutrient, but it is an important indication of the amount of feed the ewe can and should consume.

TABLE 8-11 DAILY NUTRIENT REQUIREMENTS AND FEED COMPOSITION FOR 60-KG (132-LB) EWES, FIRST 6 TO 8 WEEKS LACTATION SUCKLING TWINS

	DM (kg)	DE (Mcal)	Crude Protein (kg)	CA (g)	P (g)	Carotene (mg/kg feed)	Vitamin A (IU)
Daily requirements	2.6	7.4	0.403	10.7	7.7		6000
Feed composition, dry matter basis							
Oat hay	88.2%	2.38	9.2%	0.26%	0.24%	101.0	
Barley	89.0%	3.79	13.0%	0.09%	0.47%	—	
Soybean meal	89.0%	3.53	51.5%	0.36%	0.75%	—	

2. Determine the amount of energy that the designated DM intake of the major feed ingredient (oat hay) provides (2.6 kg DM × 2.38 Mcal DE/kg of hay = 6.19 Mcal DE). The ewe requires 7.4 Mcal DE and the oat hay provides 6.19 Mcal; thus the diet is deficient by 1.21 Mcal DE.

3. Assume that 2.6 kg of dry matter is the maximum amount a 60-kg ewe can consume in a day. The shortage of digestible energy in the oat hay is provided by substituting barley for oat hay: 3.79 Mcal DE/kg (barley) − 2.38 Mcal

DE/kg (hay) = 1.41 Mcal DE. Divide the 1.21 Mcal DE in which the hay is deficient (step 2) by the 1.41 Mcal of additional DE that one unit of barley provides to determine the amount of barley required in addition to hay (1.21 ÷ 1.41 = 0.86 kg of barley). Thus the 2.6-kg ration of dry matter should consist of 0.86 kg of barley DM ÷ 1.74 kg (i.e., 2.6 − 0.86 kg) of hay.

4. Determine whether the hay–barley ration provides sufficient protein: 1.74 kg of hay DM × 9.2% protein (from Table 8-6) = 0.16 kg of protein from hay; 0.86 kg of barley DM × 13% protein = 0.112 kg protein from barley; total protein from the hay–barley ration is 0.272 kg. Thus 0.405 kg required − 0.272 kg provided by the hay–barley ration gives a 0.133-kg protein deficiency in the hay–barley ration.

5. To determine the amount of soybean meal to substitute for barley to provide the required amount of protein, calculate the difference in the protein content of soybean meal and that of barley from Table 8-11 (0.515 − 0.13 - 0.385%). To compensate for the 0.133 kg of protein lacking in the hay–barley ration, divide 0.133 by 0.385 to get 0.345 kg DM from soybean meal. The ewe's ration becomes 1.74 kg DM from hay, 0.51 kg DM from barley, and 0.35 kg DM from soybean meal. The amount of soybean meal needed to substitute for part of the barley may also be determined by the Pearson square method.

6. The calcium provided by the hay (1.74 kg × 0.26% = 4.5 g), barley (0.51 kg × 0.09% = 0.5 g0, and soybean meal (0.35 kg × 0.36% = 1.3 g) adds up to 6.3 g, leaving a deficiency of 4.4 g (10.7 − 6.3). To determine the amount of limestone to add to compensate for the calcium deficiency, divide 4.4 by 0.34 (calcium content of limestone) to obtain 13 g or 0.013 kg. A similar comparison for phosphorus shows the diet to be adequate in this mineral. Therefore, the daily diet per ewe on a DM basis becomes 1.74 kg of oat hay, 0.51 kg of barley, 0.35 kg of soybean meal, and 0.01 kg of limestone.

To convert to an as-fed basis, divide each amount of dry matter by the percent of dry matter in that feed. On an as-fed basis, the daily diet per ewe is 1.97 kg of oat hay, 0.57 kg of barley, 0.39 kg of soybean meal, and 0.01 kg of limestone.

These are only two methods of formulating diets. Other methods include the use of simultaneous equations to algebraically arrive at a solution and the use of computers to solve more complex sets of equations. The procedures discussed in this section do not include all nutrients or the effects of feed palatability, economics, and other factors that nutritionists and feed producers should consider.

COMPOSITION OF FEEDS

Feed composition of a wide variety of common sheep feeds is shown in Table 8-12. For a more complete listing of feeds, refer to reference 8. By referring to the previous discussion on balancing rations and formulating diets and to Tables 8-1, 8-2

TABLE 8-12 COMPOSITION OF SOME COMMON SHEEP FEEDS, EXPRESSED ON AN AS-FED AND DRY BASIS (100% DRY MATTER)[a]

Feed	Dry Matter (%)	Digestible Energy (Mcal/kg)	Metabolizable Energy (Mcal/kg)	TDN (%)	Protein Crude (%)	Protein Dig. (%)	Calc (%)	Phos. (%)	Carotene (mg/kg)
Alfalfa, fresh	24	0.62	0.51	14	4.8	3.5	0.48	0.07	45
	100	2.56	2.10	58	19.7	14.6	1.96	0.30	185
Alfalfa hay, early bloom	90	2.22	1.82	51	16.2	12.7	1.27	0.20	126
	100	2.47	2.03	56	18.0	14.1	1.41	0.22	140
Alfalfa hay, mature	91	2.17	1.78	49	11.7	7.7	1.03	0.17	11
	100	2.38	1.95	54	12.9	8.5	1.13	0.18	12
Alfalfa, dehydrated 17% protein	92	2.43	1.99	55	17.3	11.6	1.40	0.23	120
	100	2.65	2.17	60	18.9	12.7	1.52	0.25	131
Alfalfa silage, <30% DM	26	0.73	0.60	17	4.9	3.3	0.37	0.08	
	100	2.28	2.28	63	18.7	12.4	1.40	0.32	
Barley grain	88	3.35	2.74	76	11.9	9.8	0.04	0.34	2
	100	3.79	3.11	86	13.5	11.1	0.05	0.38	2
Barley hay	87	2.16	1.77	49	7.6	4.1	0.20	0.23	46
	100	2.47	2.03	56	8.7	4.7	0.23	0.26	53
Barley straw	91	1.93	1.58	43	4.0	0.7	0.27	0.07	2
	100	2.12	1.74	48	4.3	0.8	0.30	0.07	2
Beans, navy	89	3.43	2.82	78	22.6	19.9	0.16	0.52	
	100	3.84	3.15	87	25.3	22.3	0.18	0.59	
Beets, mangels	11	0.39	0.32	9	1.3	0.9	0.02	0.02	0
	100	3.57	2.93	81	11.8	8.0	0.18	0.22	1
Beet pulp, dehydrated	91	2.96	2.43	67	8.8	4.6	0.63	0.09	
	100	3.26	2.68	74	9.7	5.0	0.69	0.10	
Beet pulp, wet	11	0.37	0.30	8	1.2	0.7	0.10	0.01	
	100	3.35	2.75	76	11.2	6.1	0.87	0.10	
Bermuda grass hay	91	1.97	1.62	45	8.9	4.6	0.43	0.16	53
	100	2.16	1.77	49	9.8	5.0	0.47	0.17	58
Bluegrass, Canada, fresh	26	0.76	0.62	17	4.9	3.7	0.10	0.10	104
	100	2.91	2.39	66	18.7	14.4	0.39	0.39	400
Bluegrass hay Canada	92	2.44	2.00	55	9.6	4.1	0.28	0.25	270
	100	2.65	2.17	60	10.3	4.5	0.30	0.27	293

(*continued*)

TABLE 8-12 (continued)

Feed	Dry Matter (%)	Digestible Energy (Mcal/kg)	Metabolizable Energy (Mcal/kg)	TDN (%)	Protein Crude (%)	Protein Dig. (%)	Calc (%)	Phos. (%)	Carotene (mg/kg)
Bluegrass Ky., fresh	31	0.88	0.72	20	5.4	4.1	0.15	0.14	48
	100	2.87	2.35	65	17.4	13.2	0.50	0.44	156
Bluestem, fresh	31	0.76	0.62	17	3.4	2.7	0.19	0.17	
	100	2.54	2.08	57	11.0	7.2	0.63	0.57	
Brome, fresh	34	1.20	0.98	27	6.1	5.0	0.17	0.10	156
	100	3.53	2.89	80	18.0	14.8	0.50	0.30	459
Brome hay	91	2.20	1.80	49	8.8	4.8	0.31	0.17	31
	100	2.43	1.99	55	9.7	5.3	0.35	0.19	34
Carrot roots	12	0.45	0.38	11	1.2	0.9	0.05	0.04	107
	100	3.84	3.15	87	10.1	7.7	0.42	0.34	890
Citrus pulp	21	0.81	0.67	18	1.5	0.8	0.43	0.03	
	100	3.88	3.18	88	7.3	3.7	2.04	0.15	
Alsike clover hay	88	2.25	1.84	51	13.1	8.7	1.13	0.23	164
	100	2.56	2.10	58	14.9	9.9	1.29	0.26	187
Crimson clover hay	87	2.12	1.73	48	16.1	11.0	1.22	0.19	20
	100	2.43	1.99	55	18.4	12.7	1.40	0.22	23
Ladino clover hay	90	2.61	2.14	59	19.7	15.0	1.21	0.28	75
	100	2.91	2.39	66	22.0	16.7	1.35	0.31	83
Red clover, fresh	20	0.59	0.49	13	3.8	3.0	0.45	0.08	49
	100	3.00	2.46	68	19.4	15.0	2.26	0.38	248
Red clover hay	89	2.34	1.92	53	14.2	9.0	1.35	0.22	18
	100	2.65	2.17	60	16.0	10.1	1.53	0.25	20
Corn fodder	81	2.26	1.85	51	7.2	3.8	0.41	0.20	4
	100	2.78	2.28	63	8.9	4.7	0.50	0.25	4
Corn stover	85	2.21	1.81	50	5.6	2.5	0.49	0.08	4
	100	2.60	2.13	59	6.6	2.9	0.57	0.10	4
Corn silage, well eared	33	1.03	0.85	23	2.7	1.2	0.08	0.07	15
	100	3.09	2.53	70	8.1	3.6	0.23	0.22	45
Corn gluten meal	91	3.54	2.90	80	42.7	36.3	0.15	0.45	31
	100	3.88	3.18	88	46.8	39.8	0.16	0.50	34
Corn grain, yellow	88	3.39	2.78	77	8.9	5.7	0.02	0.31	22
	100	3.84	3.15	87	10.1	6.5	0.02	0.35	25
Cottonseed hulls	91	1.96	1.60	44	3.7	−0.5	0.13	0.09	
	100	2.16	1.77	49	4.1	−0.5	0.15	0.09	

Cottonseed meal	93	2.58	2.12	59	37.9	27.3	0.20	0.90	
	100	2.78	2.28	63	40.8	29.3	0.21	0.97	
Kentucky fescue hay	92	2.43	1.99	55	16.7	12.0	0.41	0.37	
	100	2.65	2.17	60	18.2	13.1	0.44	0.40	
Grama grass, fresh	41	1.09	0.89	25	5.4	3.8	0.22	0.08	
	100	2.65	2.17	60	13.1	9.2	0.53	0.19	
Lespedeza, fresh	28	0.71	0.58	16	4.2	3.1	0.31	0.08	
	100	2.56	2.10	58	15.3	11.3	1.13	0.27	
Lespedeza hay	88	1.95	1.60	44	14.8	6.4	0.78	0.25	44
	100	2.21	1.81	50	16.8	7.2	0.88	0.29	50
Limestone, ground	100						34.00		
Linseed meal	90	3.14	2.58	72	34.6	29.5	0.39	0.80	
	100	3.48	2.86	79	38.3	32.8	0.43	0.89	
Milk, cattle, fresh	12	0.69	0.65	16	3.3	3.1	0.12	0.09	
	100	5.60	5.43	150	26.7	25.4	0.95	0.76	
Milk, sheep, fresh	19	1.15	1.11	31	4.7		0.20	0.15	
	100	6.00	5.82	161	24.7				
Native hay, Intermoutain	93	2.16	1.71	48	8.3	2.6	0.52	0.16	26
	100	2.34	1.92	53	9.1	2.9	0.57	0.17	88
Native hay, Midwest early	91	2.25	1.82	51	7.2	3.6	0.31	0.18	25
	100	2.47	2.02	56	8.1	4.1	0.34	0.21	25
Needle and thread grass, fresh	29	0.75	0.62	17	3.1	2.0	0.27	0.05	
	100	2.60	2.13	59	10.6	6.9	0.93	0.16	
Oat hay	91	2.13	1.75	49	8.5	5.2	0.22	0.20	4
	100	2.34	1.92	53	9.3	5.7	0.24	0.22	4
Oat grain	89	3.02	2.47	68	11.8	9.3	0.07	0.33	14
	100	3.40	2.78	77	13.3	10.4	0.07	0.38	45
Oat straw	92	1.91	1.57	44	4.1	0.3	0.22	0.06	112
	100	2.07	1.70	47	4.4	0.3	0.24	0.07	482
Oat silage	31	0.84	0.69	19	3.0	1.5	0.10	0.24	174
	100	2.73	2.24	62	9.6	4.9	0.34	0.13	191
Orchard grass, fresh	23	0.69	0.57	16	4.3	3.1	0.13	0.54	
	100	2.95	2.42	67	18.4	13.3	0.58	0.32	
Orchard grass hay	91	2.33	1.91	53	10.2	6.5	0.35	0.35	
	100	2.56	2.10	58	11.2	7.1	0.39	0.21	
Prairie hay, midwest early	90	2.50	2.05	57	7.8	3.9	0.44	0.23	
	100	2.78	2.28	63	8.7	4.4	0.49		

TABLE 8-12 (continued)

Feed	Dry Matter (%)	Digestible Energy (Mcal/kg)	Metabolizable Energy (Mcal/kg)	TDN (%)	Protein Crude (%)	Protein Dig. (%)	Calc (%)	Phos. (%)	Carotene (mg/kg)
Phosphate defluorinated rock	100						33.07	18.04	
Phosphate, monosodium	100							22.46	
Redtop, fresh	39	1.02	0.83	23	2.9	1.8	0.13	0.09	
	100	2.60	2.13	59	7.4	4.5	0.33	0.23	
Redtop hay	92	2.19	1.80	50	7.4	3.5	0.39	0.20	
	100	2.38	1.95	54	8.1	3.8	0.43	0.22	
Ryegrass, fresh	25	0.64	0.52	14	3.5	1.5	0.16	0.10	98
	100	2.60	2.13	59	14.5	6.2	0.65	0.41	401
Rye grain	88	3.28	2.69	75	12.1	9.5	0.06	0.32	
	100	3.75	3.07	85	13.8	10.9	0.07	0.37	
Rye straw	90	1.78	1.46	41	2.7	−0.6	0.22	0.08	
	100	1.98	1.63	45	3.0	−0.7	0.24	0.09	
Sagebrush, big, fresh	65	1.43	1.18	33	6.1	3.2	0.46	0.12	10
	100	2.21	1.81	50	9.3	4.9	0.71	0.18	16
Salt-bush, fresh	55	0.87	0.72	20	4.0	1.8	1.78	0.07	14
	100	1.59	1.30	36	7.2	3.4	2.23	0.08	18
Sorghum grain	90	3.48	2.85	79	11.1	8.2	0.03	0.29	1
	100	3.88	3.18	88	12.2	9.1	0.04	0.33	1
Sorghum silage	30	0.74	0.61	17	2.2	0.6	0.10	0.06	5
	100	2.51	2.06	57	7.5	2.2	0.35	0.21	15
Sorghum milo grain	89	3.44	2.82	78	10.0	7.8	0.04	0.30	
	100	3.88	3.18	88	11.3	8.8	0.05	0.34	
Sorghum Sudan grass, fresh	18	0.49	0.41	11	3.0	2.3	0.08	0.07	35
	100	2.78	2.28	63	16.8	12.6	0.43	0.41	108
Soybean meal, mech. extracted	90	3.37	2.77	77	42.9	36.8	0.26	0.61	
	100	3.75	3.07	85	47.7	40.9	0.29	0.68	
Soybean meal, solvent extracted	90	3.48	2.85	79	44.8	41.6	0.30	0.63	
	100	3.88	3.18	88	49.9	46.4	0.34	0.70	
Sugar beet molasses	78	2.64	2.16	60	6.6	3.4	0.13	0.03	
	100	3.40	2.78	77	8.3	4.4	0.17	0.03	
Sugar-cane molasses	75	2.60	2.13	59	4.4	−1.3	0.75	0.08	
	100	3.48	2.86	79	5.8	−1.7	1.00	0.11	

Sunflower meal	90	1.79	1.46	41	23.3	18.9	0.21	0.93	
	100	1.98	1.63	45	25.9	21.0	0.23	1.03	
Sweet clover hay	87	2.04	1.67	46	13.7	10.0	1.11	0.22	86
	100	2.34	1.92	53	15.7	11.8	1.27	0.25	99
Timothy, fresh	26	0.71	0.58	16	4.8	2.6	0.10	0.08	62
	100	2.69	2.21	61	18.0	10.0	0.39	0.32	235
Timothy hay, early	90	2.21	1.81	51	13.4	7.3	0.48	0.22	47
	100	2.47	2.03	56	15.0	8.2	0.53	0.25	53
Turnip roots	9	0.35	0.29	8	1.1	0.8	0.05	0.02	
	100	3.79	3.11	86	11.8	8.9	0.59	0.26	
Wheat grain	89	3.41	2.80	78	14.2	11.4	0.04	0.37	
	100	3.84	3.15	87	16.0	12.8	0.04	0.42	
Wheat hay	88	2.01	1.65	45	7.4	4.0	0.13	0.17	75
	100	2.29	1.88	52	8.5	4.6	0.15	0.20	85
Wheat straw	89	1.60	1.32	36	3.2	−3.1	0.16	0.04	2
	100	1.81	1.48	41	3.6	−3.5	0.18	0.05	2
Wheatgrass, Crested, fresh early	28	0.92	0.76	21	6.0	5.1	0.13	0.10	126
	100	3.31	2.71	75	21.5	18.3	0.46	0.34	451
Wheatgrass Crested, fresh mature	60	1.43	1.17	32	3.3	1.3	0.16	0.09	45
	100	2.38	1.95	54	5.5	2.1	0.27	0.15	75
Wheatgrass, Crested, hay, early	94	3.14	2.57	71	18.0	15.3	0.40	0.25	213
	100	3.35	2.75	76	19.2	16.3	0.43	0.26	238
Wheatgrass, Crested, hay, mature	94	1.99	1.63	45	5.3	3.3	0.18	0.11	29
	100	2.12	1.74	48	5.6	3.5	0.19	0.12	31
Winterfat, fresh, stem cured	80	1.24	1.01	28	8.7	5.4	1.58	0.09	14
	100	1.54	1.27	35	10.8	6.7	1.98	0.12	18

*a*Selected from feeds listed in *Nutrient Requirements of Sheep*, 6th rev. ed., 1985. National Academy Press, Washington, D.C.[8]

TABLE 8-13 NUTRIENTS PROVIDED BY DIFFERENT PASTURE FORAGES COMPARED TO REQUIREMENTS FOR 60-KG (132-LB) EWE, LATE LACTATION SUCKLING TWINS

						Ingredient					
			DM		DE	TDN		Protein		Ca	P
			(lb)	(kg)	Mcal	(lb)	(kg)	(lb)	(kg)	(g)	(g)
Requirement			5.1	(2.3)	6.6	3.30	(1.50)	0.70	(0.32)	9.1	6.6
Feed[a] and amount, lb		(kg)									
Alfalfa	21.2	(9.6)	5.1	(2.3)	5.95	3.0	(1.36)	1.02	(0.46)	46.0	10.0
Bluegrass pasture, Ky.	16.4	(7.4)	5.1	(2.3)	6.54	3.3	(1.50)	0.88	(0.40)	11.3	10.3
Bluestem pasture	16.4	(7.4)	5.1	(2.3)	4.58	2.8	(1.26)	0.56	(0.25)	8.7	12.5
Brome pasture	15.0	(6.8)	5.1	(2.3)	8.16	4.0	(1.83)	0.91	(0.41)	11.6	6.8
Grama pasture	12.4	(5.6)	5.1	(2.3)	6.10	3.1	(1.40)	0.67	(0.30)	12.3	5.6
Lespedeza	18.2	(8.3)	5.1	(2.3)	5.89	2.9	(1.33)	0.76	(0.35)	25.7	6.6
Sagebrush	7.8	(3.5)	5.1	(2.3)	7.58	3.8	(1.75)	0.70	(0.32)	24.4	6.4
Timothy	19.6	(8.9)	5.1	(2.3)	6.32	3.1	(1.40)	0.94	(0.43)	8.9	7.1
Crested wheatgrass	18.2	(8.3)	5.1	(2.3)	7.64	3.8	(1.75)	0.09	(0.50)	10.8	8.3

[a] Amount of feed needed to provide 5.1 lb (2.3 kg) DM.

and 8-12, one can determine how much of any particular feed or combinations of feed is needed to provide adequate nutrition.

PROVIDING FEED AND CONTROLLING COSTS

Once it is clear how to provide proper nutrition for sheep in any situation, the next major consideration is to provide such at the lowest possible cost. It should be obvious from a study of Table 8-12 that there are many different feeds or feed combinations that are suitable for use in sheep diets. Feed costs vary considerably by the type of feed, and seasonal or yearly fluctuations occur in the comparative prices of feeds.

Pastures

Fortunately, sheep are efficient grazers and the easy way to keep feed costs low is to utilize pasture to the fullest extent, and thus avoid any costs related to harvesting, storing, or transporting feed. Sheep can meet their nutritive needs from a wide variety of plants—grasses, legumes, weeds, herbs, and shrubs. Also, they are capable of grazing on almost any type of terrain and in almost any kind of weather. Examples of different pasture diets that meet the dry-matter requirements of sheep are shown in Table 8-13. These include common pastures from various parts of the country and cultivated as well as uncultivated land.

As seen in Table 8-13, some grasses and browse plants in the early growing stage provide nutrients in about the range desired to meet all the major requirements. The only question remaining is whether sheep will eat that much. During the early growing stages of most plants, palatability is high and thus it is easy for sheep to eat enough. Even when grasses and legumes are in mixtures, and a preference for some plants may be seen, pasture use is efficient when the plants are young and growing. Mixtures of this type are often seen in irrigated pastures and sometimes in dryland pastures. Figures 8-3 and 8-4 show examples of highly productive irrigated pasture, with heavy stocking rates. As plants mature, there is a reduction in palatability and a greater difference between plants as to their relative palatability. Figure 8-5 shows an example of a relatively sparse grazing area during the summer. In this type of operation, sheep are allowed to graze over large areas. Some plants become so unpalatable when nearing or reaching maturity that sheep will not touch them unless forced to because they are the only feed available. Crested wheatgrass in dryland pastures is a good example of one that is relatively unpalatable as it reaches maturity and dries out. Proper pasture management is required to obtain maximum use of such feeds. If grazed early the grass is utilized fully and then can be grazed again if sufficient moisture is available to cause regrowth. If not used fully, it gets ahead of the sheep and becomes very fibrous, and at this state it can be better utilized if harvested as hay.

Figure 8-3 Ewes and lambs grazing in lush brome grass and alfalfa pasture. (University of Wyoming photo)

Figure 8-4 Ewes and lambs stocked at heavy rate in brome grass and alfalfa pasture. (University of Wyoming photo)

Not only do grasses become less palatable as they mature, but they also tend to become lower in protein, phosphorus, and carotene. Therefore, when sheep are grazing on pastures comprised of dry grasses, as is common in many areas where sheep are produced, it becomes necessary to add a supplemental source of protein or protein and minerals.[17]

When sheep are grazing, regardless of the type of pasture, the actual intake of feed is not known. Thus, in balancing the diet under these conditions, assumptions as to daily feed intake are necessary. The assumption that sheep will eat the necessary amount of green growing grasses to meet their requirements is a valid assumption when the grass supply is adequate. When sheep graze on sparsely vegetated areas, even if grass or forage crops are green, some adjustment needs to be

Figure 8-5 Ewes and lambs grazing on range during late June. (University of Wyoming photo, Courtesy of LeBar Ranch, Douglas, Wyoming)

made, such as using these pastures for livestock other than lactating ewes or supplementing with a concentrate.

As pastures dry out and become less palatable and less nutritious, it is a good idea to have the feed tested for nutrient content when possible before decisions are made as to when or with which feeds to supplement. If facilities for testing feeds are not available, use of previous observation of the sheep while grazing these pastures can be helpful.

Shrubs make up a large part of sheeps' diets in many areas, particularly for winter grazing in the west. Sagebrush is suitable for winter feed and is reasonably palatable, but it is difficult for sheep to eat enough to meet all requirements when this is the only feed available. Since energy is the most likely to be deficient, supplementation with small amounts of grain is the logical approach. Salt bush more nearly provides a complete diet for sheep than does big sagebrush, partly because of palatability resulting in a high energy intake.

Legumes for sheep pastures are desirable from the standpoint of providing proper nutrition and are used alone or in mixtures with grasses. Mixtures sometimes result in a preference for the legume over grass, and in dryland pastures this may result in loss of the legume. In irrigated pastures, these legumes can be grazed several times, as regrowth is rapid following irrigation; however, adequate time should be allowed for regrowth.

There is some controversy, however, concerning the use of legumes, particularly alfalfa, in pastures for sheep due to the possibility of bloat. Some sheep producers routinely avoid legume pastures, whereas others use alfalfa pastures or alfalfa and grass mixtures with little or no difficulty. Observations would indicate that breed or breed cross seems to be involved in susceptibility to bloat. The ones most often experiencing bloat are black-faced sheep, Suffolks and Hampshires, with western white-faced ewes seemingly less of a problem. Perhaps this is because the white-faced ewes eat less or not as fast as the black-faced ewes. This is not meant to indicate that there is no problem with white-faced sheep; it is wise to be careful when anticipating moving sheep into alfalfa pastures. They should not be introduced into a lush alfalfa pasture when they are hungry. A bit of precaution can help to avoid serious problems.

Pasturing sheep on *by-products* of other crop production is a common method of keeping feed costs low. Wheat pastures are used in some areas, following the emergence of wheat planted in the fall. This provides feed similar to most of the pasture grasses in nutrient content. Its availability is dependent on good growing conditions for the newly seeded wheat crop.

Crop aftermath of various kinds is used in many cases to derive extra benefit from crops. Regrowth of grain fields or of hay fields following harvest and corn stalks, bean straw, or sugar-beet tops are examples of products that are of value to sheep. Weeds along irrigation ditches or fence rows that are hard to reach with harvesting or cultivating machinery are also included as having value to sheep, and sheep can be of help in controlling weeds in such places.

Harvested Forages

In many operations, sheep are dependent on harvested forages—hay, straw, haylage, or silage—for the major part of their diet. Some situations require that sheep be fed harvested feeds during the winter and grazed during the summer. Confinement operations generally have limited pasture and rely heavily on the use of harvested feeds, both roughages and concentrates.

Hay is probably the most common of the harvested forages that are used for feeding sheep. There is great variation in hay, ranging from high-quality legume hay to low-quality grass hay that is discolored by weather or by being harvested too late. Variation occurs in palatability, nutrient content, and digestibility. For sheep to produce at their best when fed on hay only, the hay needs to be both high in nutritive value and palatable. Low-quality forages such as discolored grass hay or straw or corn stalks can be used satisfactorily as part of the diet for sheep, but in many cases these are relatively unpalatable. Hay is fed from stacks or bales or is ground or pelleted before being fed. Grinding or grinding and pelleting tend to increase feed intake and reduce waste of coarse stemmy material in hay. Also, grinding or pelleting a mixture of low-quality roughage and good hay can help to get the sheep to use the poor-quality feeds.

Silages are also common feeds for sheep and are fed usually as part of the diet. Silage is often fed along with hay to provide the forage in the diet. In some phases of production it may be satisfactory to use good silage as the only forage when a concentrate is included as part of the ration. Corn silage needs to be supplemented with a protein concentrate, while grass silage or legume silage needs to be supplemented with a high-energy concentrate. Haylages are also used in sheep's diets in many areas. These are forages that are intermediate between dry forages and silages in dry matter content.

Root crops are often used as feed for sheep, the common ones being stock beets, turnips, and carrots. These are low in dry matter and thus require a high intake to meet nutrient requirements. However, they are highly palatable. Harvesting, storing, and slicing these root crops in some cases make their use economically questionable, although sometimes sheep are turned in and allowed to harvest them.

Concentrates

Concentrates make up much less of a sheep's diet than either pasture or harvested forages. Even so, concentrates can cause more than their proportionate share of feed costs. Concentrates that are commonly included as part of a sheep's diet are cereal grains—barley, corn, milo, oats, rye, and wheat—and high-protein feeds such as soybean meal, cottonseed meal, and linseed meal. A comparison of various cereal grains and other feeds as sources of energy is shown in Table 8-14, and a similar comparison of protein supplements is shown in Table 8-15. In most sheep production operations these feeds are used only as supplements to correct deficiencies oc-

Providing Feed and Controlling Costs

TABLE 8-14 COMPARISON OF SELECTED ENERGY SOURCES

Energy Source (compared with corn with a value of 100)	Relative Energy Value[a] (%)
Corn	100
Milo	85–100
Barley	90
Oats	80
Wheat (up to 50% of grain portion of ration)	100–110
Beet pulp (up to 30% of grain portion of ration)	100
Hominy (up to 20% of grain portion of ration)	105
Molasses (70% dry matter content; up to 10% of ration)	70
Fat (up to 5% of ration)	225

Reprinted with permission from *Sheepman's Production Handbook*.[14]

[a]Processing methods, moisture content, and type of ration may modify these results on a percentage basis.

curring due to stage of production or to the type of pasture or forage being fed. Grains are normally used as supplements in situations where sheep tend to need more energy than provided by the forage, such as sheep grazing on sagebrush or those fed grass or haylage. Also, as energy requirements increase during late gestation and lactation, supplementing the diet with grain is the logical way to meet these needs. Grain is often a major ingredient in diets used for growing and finishing lambs, particularly when lambs are confined to pens without pasture. Grains can be fed to sheep as whole, cracked, ground or rolled, all of which are satisfactory. They should not, however, be ground too finely.

All the cereal grains are high in energy, but they are relatively low in protein and calcium, and all except yellow corn are deficient in carotene (pro-vitamin A). When fed supplementally to pasture or forage diets or when in mixtures with high-quality forages, there is no problem of deficiencies, since the forages are adequate in protein and high in calcium and carotene. In supplementing low-quality forages, in addition to grain it may be necessary to add some high-protein feed and also vitamin A and/or phosphorus.

TABLE 8-15 COMPARISON OF PROTEIN SUPPLEMENTS[a]

Protein Supplement (compared with soybean meal with a value of 100)	Relative Feeding Value
Soybean meal (44%)	100
Linseed meal (38%)	90
Cottonseed meal (43%)	98

Reprinted with permission from *Sheepman's Production Handbook*.[14]

[a]Processing methods, moisture content, and type of ration may modify these results on a percentage basis.

The protein concentrates are high-priced feeds generally, and as such they should be used only when protein-deficient diets are fed and then in proper amounts. In many cases where sheep graze on dry grasses during the winter months, their main problem is a protein deficiency, and this is sometimes accompanied by deficiencies of phosphorus and vitamin A. It is common for these diets to be supplemented with some high-protein feed or a combination of high-protein feed and grain and molasses made into pellets, cubes, or cake. When such combinations are prepared, it is easy to add extra phosphorus or vitamin A during the mixing process. Only small amounts of such supplements are required, 0.25 lb (113 g) per day for example. When fed out in large pastures, these types of supplement are often fed on the ground and scattered over a wide area to allow all the sheep to get a share of them. An example of a device for spreading pellets or cubes is shown in Figure 8-6. Many mixtures of high-protein feeds for sheep contain some nonprotein nitrogen such as urea. Urea concentrations should not exceed 1% of the dietary dry matter or one-third of the total dietary protein.[8]

Water

A source of fresh water is essential to good production. Mature sheep in confinement will drink about 1 gallon or more per day and lambs on feed about half that much. If sheep are in pastures with plenty of green feed, much of their water supply comes from consumption of the plants. The same is true when fed succulent root crops or silages. Water requirements increase markedly during late gestation.[6] Water needs to be available in greater quantities during summer partly because of the weather and partly because usually the ewes are lactating during the summer. When sheep are grazed in semiarid areas during the winter, it is common for them to use snow as their water source part of the time;[2] but snow is not always available, so other sources (streams or tanks) are used when needed. Water facilities are shown in Figures 8-7 and 8-8.

Figure 8-6 Spreader mounted on pickup for distributing cubes to sheep flock. (University of Wyoming photo, Courtesy of Richard Strom, Laramie, Wyoming)

Providing Feed and Controlling Costs

Figure 8-7 Windmill and nearby water tank for providing water in range area. (University of Wyoming photo, Courtesy of Warren Livestock Company, Cheyenne, Wyoming)

Minerals and Vitamins

Salt, or sodium chloride, serves in many functions and is thought to stimulate appetite. Total salt requirement of lambs is about 0.4% of the dietary dry matter.[5,8,9] Sodium requirements are given in Table 8-4, but specific requirements of chlorine are not known. Salt is provided in many ways: as a block in pastures or in corrals, as loose salt either coarse or finely granulated, and incorporated into feeds either as part (0.5%) of complete diets or part (1%) of the concentrate portion of diets. Many range producers provide salt at the rate of 0.5 lb (227 g) per month, while in

Figure 8-8 Electric waterer in pen for ram test. (University of Wyoming photo)

feedlots lambs will consume more nearly 1 lb (453 g) per month. In cases where deficiencies of various trace minerals are evident or suspected, salt can be used as a carrier. For instance, iodized salt is used in some areas, and trace-mineralized salt is used by some producers. Salt can be added at high levels (10% or more) to limit free-choice supplement intake, providing plenty of water is available. In some areas either natural feed or water provide enough salt to meet the requirements of sheep.

Other minerals as well as vitamins are provided by mixing them into complete diets for use in confinement. Also, these can be added to pelleted or ground feeds used as supplements. In some cases several different minerals are offered free choice. However, it has been found advisable to feed a complete diet or, when that is not possible, to offer on a free-choice basis a palatable, complete mineral mixture.[1,11]

REFERENCES CITED

1. Burghardi, S. R., and others. 1982. Free choice consumption of minerals by lambs fed calcium-adequate or calcium-deficient diets. *J. Animal Sci.* 54:510.
2. Butcher, J. E. 1970. Is snow adequate and economical as a water source for sheep? *Natl. Wool Grower* 60(2):28.
3. Coop, I. E. 1962. The energy requirements of sheep for maintenance and gain. I. Pen-fed sheep. *J. Agric. Sci.* 58:179; Coop, I. E. and M. K. Hill. II Grazing sheep. *J. Agric. Sci.* 58:187.
4. Ensminger, M. E. 1970. *Sheep and Wool Science*. Interstate Printers and Publishers, Danville, Ill.

5. Hagsten, L. B., T. W. Perry, and J. B. Outhouse. 1976. Salt requirements of lambs. *J. Animal Sci.* 40:329.
6. Hatfield, E. E. 1968. Nutrient requirements of the ewe during gestation. *Proceedings of Symposium, Sheep Nutrition and Feeding.* Iowa State University, Ames.
7. Hogue, D. E. 1968. Nutritional requirements of lactating ewes. *Proceedings of Symposium, Sheep Nutrition and Feeding.* Iowa State University, Ames.
8. *National Research Council,* 1985. Nutrient requirements of domestic animals. No. 6. Nutrient requirements of sheep, 6th rev. ed. National Academy Press, Washington, D.C.
9. Oh, J. H., W. C. Weir, and W. M. Longhurst. 1971. Feed value for sheep of cornstalks, rice straw and barley straw as compared with alfalfa. *J. Animal Sci.* 32:343.
10. Oxley, J. W. 1968. Nutrition and wool production. *Proceedings of Symposium, Sheep Nutrition and Feeding.* Iowa State University, Ames.
11. Pamp, D. E., D. Goodrich, and J. C. Meiske. 1977. Free choice minerals for lambs fed calcium- or sulfur-deficient diets. *J. Animal Sci.* 45:1458.
12. Pope, A. L. 1971. A review of recent mineral research with sheep. *J. Animal Sci.* 33:1332.
13. Rattray, P. V., and others. 1974. Efficiency of utilization of metabolizable energy during pregnancy and energy requirements for pregnancy in sheep. *J. Animal Sci.* 38:383.
14. Scott, George E. 1975. *Sheepman's Production Handbook,* 2nd rev. ed. Sheep Industry Development Program. Denver, Colo.
15. Shelton, Maurice, P. V. Thompson, and J. E. Huston. 1976. Nutritional efficiency of energy use for wool production. *Texas Agr. Exp. Sta. Progress Report 3403.*
16. Weir, W. C. 1968. Nutrient requirements for maintenance of the dry ewe. *Proceedings of Symposium, Sheep Nutrition and Feeding.* Iowa State University, Ames.
17. Weir, W. C., and D. T. Torell. 1967. Supplemental feeding of sheep grazing on dry range. *California Agr. Exp. Sta. Bulletin 832.*

SUGGESTED READING

Church, D. C. 1984. *Livestock Feeds and Feeding,* 2nd ed. O and B Books, Corvallis. Ore.

Ensminger, M. E., and C. G. Olentine, Jr. 1978. *Feeds and Nutrition—Abridged.* Ensminger Publishing Co., Clovis, Calif.

Morrison, Frank B. 1956. *Feeds and Feeding.* Morrison Publishing Co., Ithaca, N.Y.

9 Prebreeding and Breeding Season

The goal in good management is to have all factors under control to assure a high ovulation rate in ewes and to have a high proportion of these ova fertilized. This is the most important part of a year-round management program, the key to success. A study of ovulation rate as previously depicted in Figure 6-9 indicates that the potential lambing percent, even in common domestic sheep other than Finns, should be 160% or above. The average lambing percent for the United States is just over 100%, so this represents a serious loss of present potential. In addition, the opportunity exists to increase the potential up to 200% or higher through more efficient use of genetic resources, by improving selection effectiveness, by use of new techniques such as embryo transfer, and by acceleration of lambing frequency.

REALIZING PRESENT POTENTIAL

Management aimed at reducing the loss of present potential, or to more nearly achieve the potential production, is the first step. Quite a large portion of this loss in potential occurs at or near breeding time.[2,18] No doubt some ova are shed during the ovulation process which are incapable of being fertilized due to immaturity or to poor health or poor condition of ewes. Also, failure of fertilization can occur because of incorrect timing of mating or failure of mating to occur. Even though fertilization has occurred, some embryonic loss can result from a variety of causes. These losses that occur at or near breeding are "hidden losses" that cannot be observed at the time of their occurrence; they are nevertheless very real.

Breeding time. The direct approach to realizing a high reproductive rate is to have the ewes mated as nearly as possible at their peak of breeding activity or fertility. In actual practice, this is most difficult to achieve, because other considerations outweigh the natural ability of sheep. Thus many sheep are not bred during the peak of the breeding season when they are most fertile. Purebred breeders often turn rams in early, during August or September, and lamb their ewes during the winter in order to have lambs well grown and competitive in size and development with lambs of other breeders. Farm flock operators often breed early so that lambing occurs during winter to use available labor and avoid competition with other farm work. Production of spring lambs (the first crop to market in the spring of the year) is practiced by some sheep producers to realize higher market prices at that time of year.

At the other extreme, many range producers in the northern states, particularly through the mountain area, wait until December to start breeding in order to have lambs born later in the spring when severe weather conditions are less likely. Just as important is that green grass is available before and during the lambing season when the ewes don't start lambing until May. In doing so, breeders are aware that their chosen time of breeding does not coincide with the peak of the breeding season for sheep. Breeding in December presents additional problems quite frequently, as snow cover and blizzard conditions restrict grazing for some time (see Figure 9-1). Producers in these western range areas who turn in rams during October usually have sheds and labor available for lambing time. They also have to provide more harvested feed than do producers who lamb later without using sheds. The resulting increase in lambing rate often more than justifies the extra feed, time, and facilities. However, many producers have difficulty finding enough experienced workers who will do a good job at lambing time. Thus labor availability has a strong influence on decisions as to when breeding begins. In some cases, flock size is limited to what

Figure 9-1 Ewes in a snowstorm during breeding. (Charles J. Belden photo)

Figure 9-2 Example of calendar front.

OCT. 1985

Pasture and range forages in Wyoming are generally high in calcium and low in phosphorus. The recommended mineral supplement for the state is a 1:1 mixture of trace-mineralized salt and sodium tripolyphosphate fed free choice.

SEPTEMBER
1 2 3 4 5 6 7
8 9 10 11 12 13 14
15 16 17 18 19 20 21
22 23 24 25 26 27 28
29 30

NOVEMBER
                1 2
 3 4 5 6 7 8 9
10 11 12 13 14 15 16
17 18 19 20 21 22 23
24 25 26 27 28 29 30

SUNDAY	MONDAY	TUESDAY	WEDNESDAY	THURSDAY	FRIDAY	SATURDAY
		1 274 2-26	**2** 275 2-27	**3** 276 2-28	**4** 277 3-1	**5** 278 3-2
6 279 3-3	**7** 280 3-4	**8** 281 3-5	**9** 282 3-6	**10** 283 3-7	**11** 284 3-8	**12** 285 3-9
13 286 3-10	**14** 287 3-11	**15** 288 3-12	**16** 289 3-13	**17** 290 3-14	**18** 291 3-15	**19** 292 3-16
20 293 3-17	**21** 294 3-18	**22** 295 3-19	**23** 296 3-20	**24** 297 3-21	**25** 298 3-22	**26** 299 3-23
27 300 3-24	**28** 301 3-25	**29** 302 3-26	**30** 303 3-27	**31** Halloween 304 3-28		

Figure 9-3 Page from sheep production calendar, showing gestation table.

one person or one family can handle under shed lambing conditions. Each manager must be aware of the basic nature of sheep and their breeding habits and fit the management system as nearly as is feasible to the abilities of the sheep.

A management calendar can be a useful guide to the best timing. An example of such a calendar, published by the Agricultural Extension Service at the University of Wyoming is shown in Figures 9-2, 9-3, and 9-4. Extension personnel in many states are ready and willing to help in the preparation of such guides.

Accelerated lambing programs, in which ewes are bred more than once per year, also require that breeding occur at times of the year that are not optimum for the sheep. In a twice yearly lambing operation, one of the breeding times may be October or November (the best time of year), but the other time would be April or May, which is about the poorest time of the year, as seen in Figure 6-9. In a program aimed at breeding three times in two years, it may be that none of the breeding times occurs during the peak of the breeding season (October and November). These programs require extra management, such as early weaning and hormone therapy for getting ewes to breed out of season. Sheep of only a few of our domestic breeds are adaptable to acceleration. More labor, feed, and facilities and equipment are required to make such programs work. Thus, the time of year for breeding ewes is only one of the problems in accelerated breeding programs.

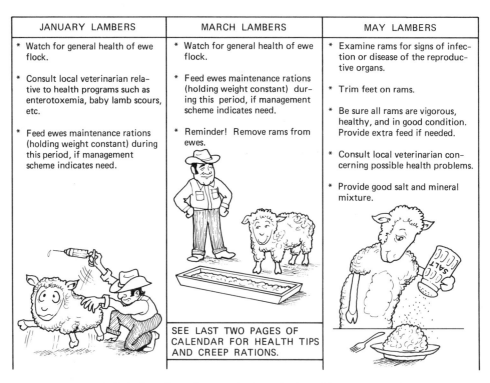

Figure 9-4 Page from sheep production calendar, with management tips.

Age Distribution

The age distribution in a flock and how each age group is handled at breeding time is always a part of management. Ewes are most productive at three or four years of age (see Figure 6-10), but all flocks cannot consist of three- and four-year-old ewes only. It is customary to have all ages represented from one through five or even six or older in some flocks. A quick, easy method of age determination is helpful in making up breeding groups. Permanent ear marks to designate year of birth or different-colored ear tags for each year are common. Also, in flocks where individual identification is used, the first number or the first two numbers on the ear tag can be used to indicate the year born.

Young ewes that have not reached their mature size at the time of breeding often require extra care or feed. Likewise, older ewes are sometimes kept for an extra year even though their teeth may have begun to spread, and these ewes require extra care. All the ewes of a particular age may be daughters of a ram still in the flock. For these reasons, ewes are often kept divided by age or are divided at breeding time. There is probably no advantage to dividing ewes by age that are two, three, or four years of age at breeding time.

Breeding of ewe lambs is more common than in past years, but it is still a practice limited to flocks in which ewe lambs are well grown out and in which ewes lambing at a year of age can be given plenty of care. Ewe lambs that reach puberty and breed during their first year have a greater lifetime production than those not exposed. This offers a special system of breeding and selection in which all ewe lambs are kept and allowed to breed. About two to three months following breeding, they are all tested for pregnancy and all open ewe lambs are marketed. At this time they are still suitable for market lambs with little or no penalty. Recent increases in the number of Finn sheep and their crosses have stimulated more interest in breeding of ewe lambs and increased the probability or degree of success. However, any system of management that involves breeding of ewe lambs requires extra care and more feed than if the ewes are bred as yearlings to produce their first lambs as two-year-olds. Regardless of whether bred as lambs or not, ewe lambs need to be handled as a separate group during the breeding season. Ordinarily, ewe lambs do not continue cycling for as long as mature ewes, so their breeding season is somewhat restricted.

Health and Soundness

Health and soundness are always essential to the success of a sheep operation and should be under constant supervision. Unsound ewes, those with teeth malformations or excess wear on their teeth, those with unsound udders or nonfunctional teats, and those not sound on their feet and legs should have been removed before breeding begins. This is usually done at the time of weaning the previous lamb crop, and thus cull ewes have been disposed of well ahead of breeding time. If any such problems arise during breeding, it is advisable to remove the ewes when they are observed or at least mark them for later disposal. Freedom from disease and parasites is also a necessity for ewes to remain thrifty and capable of gaining in weight. Where parasites are a problem, drenching for internal parasites can be done before breeding begins. Treatments for control of external parasites are most effective after shearing and so are usually not applied near breeding time. However, in cases of reinfestation of either internal or external parasites, continued treatments may be necessary.

In hot weather, shade should be available, and shearing before breeding can be of help. In severely cold weather, protection needs to be available, such as sheds, windbreaks, or natural windbreaks. Pastures with natural windbreaks are sometimes saved just for breeding time. Proper nutrition also contributes to health and thriftiness. Experienced sheep producers learn to spot undernourished or unthrifty ewes, either individuals or groups, no matter what the cause, and to correct the situation.

Feeding before and during Breeding

Nutrient requirements at breeding time are always well above what is needed for maintenance. Young ewes are still growing, and mature ewes are usually recovering from the stress of their previous lactation. In addition, wool growth causes a constant demand for nutrients in order that wool fibers be strong and uniform. Rams, too, are either still growing or are recovering from previous stress and are producing wool. Mature rams have had more time to overcome previous stress than have ewes and thus should approach breeding time in better condition than most ewes.

A list of nutritive requirements for both ewes and rams at various stages of production and at different weights was presented in Tables 8-1 and 8-2. As indicated in the discussion in Chapter 8, these tables are presented as useful guides to providing adequate nutrition, not as rigid standards to be followed exactly. Most sheep at this time of year are grazing on some type of pasture, and in many cases these are dry winter pastures. Thus balancing the diet for sheep at this time often consists of supplementing the pasture intake with a small amount of concentrate. There are many different combinations of feed and many ways can be used to provide proper nutrition. Comparative costs, ease in feeding, and feed availability are factors that determine which feeds or combinations should be used. Feed combinations can be changed if deemed advisable, but a sudden change in feed or feeding practices should be avoided.

Prior to breeding, when mature ewes are recovering from previous stress, there is normally wide variation in condition. These differences result from different performance in their previous lambing and subsequent lactation. Year differences also are sometimes evident in general thrift and condition of ewes. Also, the length of previous lactation or how early lambs were weaned influences the condition of groups of ewes. Both overcondition and undercondition have detrimental effects on fertility,[13] so relative condition should enter into decisions as to how much or which feeds are to be used. The requirements shown in Table 8-1 are by weight with no reference to the sheep's condition. Therefore, judgment is required in applying these recommendations in a practical sheep operation.

For thin ewes, nutritive requirements are higher than indicated in Table 8-1 as based on ewe weight. Ewes in high condition due to early weaning, extra good pasture, poor previous performance, or for any reason, will need less feed than recommended for their weight. It is a serious management problem when both fat ewes and very thin ewes are in the same flock. This can be solved by dividing ewes according to condition and putting the thin ewes in the best pasture or on the best feed, while fat ewes can graze low-quality pastures or be stocked at greater concentrations in pastures. This type of management is relatively easy to accomplish with ewes in a confinement operation or in farm flocks in which pasture management is under the owner's supervision. However, in range operations or in any situation where sheep graze on public lands, this type of management becomes more complicated. It is important that sheep be fed according to their needs. Thus, in many situations of pasture management, a ready solution is to ship the big, fat, dry ewes and the very thin ewes to market beforehand, and to start the breeding season with a relatively even group in weight, as well as condition.

Even though there are some cases in which overcondition results from high-quality pastures, the most common problems in feeding during the breeding season are insufficient feed or specific nutrient deficiencies. Dry grass pastures can provide plenty of energy and usually are adequate in calcium, but they tend to be deficient in protein, phosphorus, and carotene. Generally, protein and phosphorus are the most expensive supplemental nutrients. Care should be taken to provide these deficient ingredients at the lowest cost, but not to try shortcuts that provide cheaper feed but do not correct the deficiencies. Ewes grazing on sagebrush or largely on browse plants of a coarse nature are likely to have difficulty getting enough intake to meet their energy requirements. The most direct approach to correct this is to provide a high-energy supplement, such as cereal grains or high energy cubes or pellets.

Feeding rams before and during breeding often presents as many problems as feeding ewes because of extreme differences in previous care and feeding. Newly purchased rams in high condition need to be fed differently than rams that have been carried through the year in proper condition. Even when those in high condition are held back for a while on their feed intake and then fed to gain before breeding, the combination of breeding and foraging for their own feed is a greater stress than for rams more accustomed to the feed and climatic conditions. As recommended in Table 8-1, it takes more than 6 lb (2.7 kg) per day of pasture grasses or dry harvested forages to meet the requirements of a 250-lb (114-kg) ram. Feed must be both palatable and easily available in order for a ram to eat enough.

Flushing. Flushing results in ewes gaining in condition before breeding. Its purpose is to increase fertility, ovulation rate, and first-service conception rate. Flushing can be accomplished by moving ewes into more palatable or more nutritious pastures or simply by increasing their feed. When ewes are grazing on relatively low

TABLE 9-1 RESULTS OF FLUSHING RANGE EWES FOR DECEMBER BREEDING[a]

Item	Flushing Treatment		
	Control[b,c]	$\frac{1}{2}$ lb/day[c]	$\frac{3}{4}$ lb/day[c]
No. of ewes	127	128	138
Initial weight, lb	128.9	126.5	129.8
Weight change, lb	5.9	6.4	6.4
Lambs born/ewe mated	92.9	89.0	116.6
Lambs raised/ewe mated	81.0	89.0	107.0
Grease fleece weight, lb	11.4	10.4	12.1
Fleece grade, spinning count	62.9	61.6	62.1

[a]In fenced pastures, Eastern Wyoming; Pope, 1982[16]
[b]Control ewes received $\frac{1}{4}$ lb per head per day of high concentrate pellets as normal treatment during flushing and breeding.
[c]Feed allowance above that of controls.

quality pastures during late fall or winter, supplementing their diet with a small amount of concentrates is used for flushing.

There has been some controversy over the benefits from flushing ewes, likely due to differences in condition of ewes before being flushed.[3,4,10,19] Collins[3] points out that flushing tended to improve ovulation rate and lambing rate of thin ewes, but decreased ovulation rate of fat ewes. Even so, the fat ewes ovulated at a higher rate than did the thin ewes. In a study at the U.S. Sheep Experiment Station,[10] a 17-day flushing period immediately prior to breeding, starting November 1, increased lamb production over that of control ewes. Continuing the flushing for another 17 days into the breeding period produced no further response in production, but did increase the cost. Flushing range ewes by supplementing with 0.5 lb (0.23 kg) of concentrate above the normal feeding practice increased lambing rate in a Wyoming study.[16] These results are summarized in Table 9-1.

The effect of flushing ewes before breeding is likely to continue to be somewhat elusive due to differences in feed conditions and is not a universally recommended procedure for every sheep operation. For range producers who use dryland pastures for grazing ewes and whose breeding season is during late fall and winter, flushing definitely is a desirable practice.

Stimulating Estrus

A helpful practice for stimulating and to some extent synchronizing breeding is to put vasectomized rams in with ewes about ten days to two weeks prior to the beginning of breeding. The same effect can result from fence line contact, that is, turning a ram into an adjoining pasture. Be sure that rams have been vasectomized at least three weeks prior to their use as teasers, or if fence line contact is used, see that they are separated by a good fence. These are aimed at getting a large portion of the ewes to ovulate and conceive during the early part of the breeding season, since the early lambers of any flock tend to be more prolific than those that lamb later. Regardless of which months constitute the breeding season, the early lambers seem to be most prolific, as indicated in Table 9-2. However, this sort of stimulus is more likely to be helpful when ewes are bred in August and September, rather than when ewes are bred in October and November (the peak of the breeding season).

Synchronizing Estrus

Synchronization is a management tool that for years has received a lot of attention from researchers and often from the popular press in news stories. Synchronization occurs naturally to some extent simply by weaning lambs all at the same time, a traditional practice in commercial sheep operations. Use of hormone treatments for synchronizing estrus is common in research institutions, but not to any great extent in the commercial sheep business. Materials are not readily available to producers, and even for research purposes, the costs of material and labor are relatively high. Studies have been made using several different hormones, or combinations of hormones, with varying methods of application. Generally synchronization has been achieved.[15] Common results are from 70% to 80% first-service conception, 10% to 20% second service, and a few settled at third service or did not conceive at all. Lambing then occurs in two or three peaks, with a 10- to 12-day time lapse in between, as shown in Figure 9-5.

Complete success, that is, 100% of all treated ewes conceived at first breeding, has not been achieved when large numbers of ewes were included. If complete success could be assured, then there is a fear on the part of many sheep producers that the resulting synchronized lambing would coincide with a severe spring storm. This would not be quite as serious as suspected, since ewes bred on the same day will drop their lambs over a period of several days. However, as seen in Figure 9-5, there

TABLE 9-2 COMPARISON OF LAMBING RATE OF EARLY VERSUS LATE LAMBING EWES

	Lambing Season					
	January and February			April and May		
	Ewes Lambing	Lambs Born	Lambing Rate	Ewes Lambing	Lambs Born	Lambing Rate
Ewe lambing days 1–17	757	1304	1.72	1426	2061	1.45
Ewe lambing days 18–34	638	1017	1.59	402	529	1.32

From lambing records, 1974–1983, University of Wyoming.

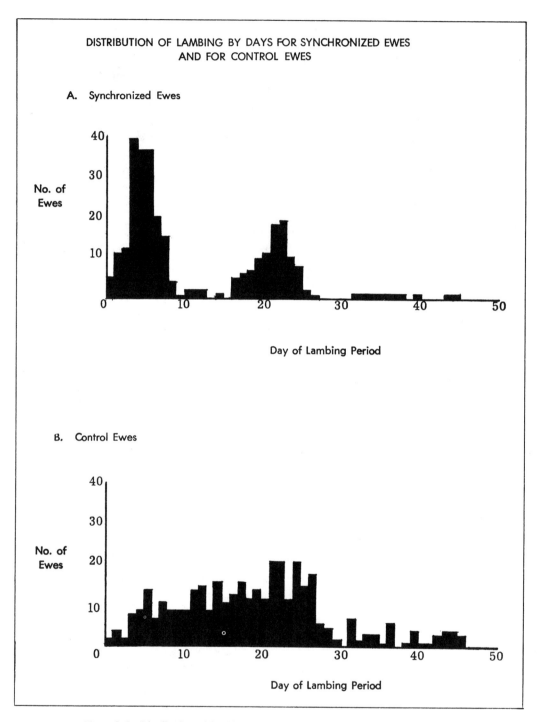

Figure 9-5 Distribution of lambing by days for synchronized and for control ewes. (From Nelms and Botkin[15], with permission)

could be a serious loss if a storm lasting for several days did coincide with lambing time.

Comparing results of synchronized and natural matings, lambing rates are about the same as shown in Table 9-2. Successful synchronization requires more ram power than conventional mating or requires that artificial insemination be used. Because of the foregoing comments, there is very little pressure from the industry to perfect techniques for synchronization.

Realistically, the main accomplishment of synchronized breeding is to allow for three estrous periods to occur in ewes during about the same length of time as two periods would occur with natural mating. For example, synchronized ewes could be bred on days 1, 18, and 35 of the breeding season, which is about the length of two cycles. Also synchronization has a place in any type of accelerated lambing program. Future studies should be helpful in making it more feasible for traditional sheep operations.

Ram Care

Ram care is just as important as is the care and management of ewes. Since rams, too, have limited fertility during the hot summer months and they increase in both fertility and libido to a peak in October and November, season of breeding has the same implications as for ewes. Providing shade or shearing before breeding starts in hot weather is probably more advantageous for rams than for ewes. Acquiring rams 6 to 8 weeks before breeding allows time for adjustment. It is important to provide proper care and environment for at least 6 weeks prior to breeding, as well as during breeding.

Health and soundness are critical for rams. They should be inspected to see that their teeth meet the pad properly and observed for soundness in body structure and of feet and legs. Breeding soundness is just as important as structural soundness, and information is continually being reported on ram fertility and its measurement. Testes size has been found to be highly related to sperm production and fertilizing capacity of rams.[14] Scrotal circumference is an easily measured trait that can help in selection of highly fertile herd sires. Data on normal scrotal size and development as related to body weight have been reported [1,8] and can be used as a basis for comparison.

Testicles of rams can be palpated to be sure that both are present and free of obvious defects. However, some disease and infertility problems in rams are not visible nor recognizable by palpation of the scrotum. Thus more accurate methods of breeding soundness evaluation have been developed. For years an index for evaluating ram semen has been available.[9]

A program for Breeding Soundness Evaluation has been developed by researchers and extension personnel at Colorado State University.[12] This program has been demonstrated in group meetings to aid individual ranchers, veterinarians, and technicians in developing workable programs. This type of approach helps in eliminating rams of low fertility and helping to reduce disease spread.

There are many different views with regard to ram use and ram care, depending on the type of operation and traditional practices. In any case, ram use represents a significant part of production costs. Thus, as stated by Kimberling, "if efficiency of production is considered, a sheep producer must do his best in evaluating ram breeding soundness. He can't afford not to."[12]

It is important to make observations during breeding to be sure that rams are actually breeding the ewes. Plenty of rams should be provided to assure mating of all ewes, and this will vary with age of rams as well as shape and size of pastures. More rams are required for breeding a group of several hundred ewes in a four- or five-section pasture with rough terrain than for breeding a group the same size in a small, level pasture or in a corral. Table 9-3 gives some recommendations for the number of ewes mated per ram in different conditions. The use of breeding soundness evaluation and care in using only rams that are sound and healthy will allow for fewer rams and still result in successful breeding.[12]

In addition to observations, other methods are used to check rams' breeding ability. Marking harnesses are used to indicate which ewes have been bred. When these are used, it is important to check the harnesses frequently, because straps can break or get tangled and the chalk wears out or sometimes breaks off. In cold, wet weather, the chalk will cake and not mark. Colored pigments or dyes mixed with light mineral oil and smeared on the brisket of rams can also be used as a way of detecting whether rams are working and to indicate which ewes have been bred. By changing the color of the chalk or dye after about 15 days, one can tell if ewes are coming back in heat the second time. In using different colors, light colors should be used first and then darker colors used for the second or third period. Also, care should be used to avoid any unscourable materials such as crankcase oil. Not only are these methods used for a check on ram performance, but they also are sometimes used for separating early lambing ewes from those expected to lamb later. An example of a ram harness in place is shown in Figure 9-6 and of a ram's brisket being smeared with dye in Figure 9-7.

Breeding season is a time of stress for rams, and it is important to care for them as well as possible. Rams are truly half the flock, not only in determining inheritance, but also in their contribution to lambing rate. Many systems are used to help maintain rams in good condition. For example, some operators start with 2

TABLE 9-3 AVERAGE NUMBER OF EWES MATED PER RAM

Ram Age and Other Conditions	No. of Ewes per Ram
Ram lamb, well matured	15-30
Rams 1-5 years, small pastures	40-50
Rams 1-5 years, large pastures	25-40
Rams 6 years or older[a]	20-30

Adapted with permission from *Sheepman's Production Handbook.*[17]

[a]Older ram breeding load dependent on good condition.

Figure 9-6 Ram with marking harness. (University of Wyoming photo)

rams per 100 ewes and after two weeks replace them with fresh rams, also at the rate of 2 per 100 ewes. Sometimes in rough terrain and in big pastures, 4 or 5 rams are provided per 100 ewes. In smaller flocks, rams are sometimes separated from the ewes each day for feeding or are left with the ewes during the day and separated for feeding during the night. Special care is particularly needed for rams in high condition that have been purchased just before breeding begins. Extra time and labor are worthwhile when designed to reduce stress on rams.

Artificial Insemination. The use of artificial insemination (AI) in sheep has lagged behind as compared to its use in other species. The potential advantages are just as great with sheep as in any other species. The most obvious advantage is that it will allow for greater use of superior germ plasm or further spread the use of outstanding rams. It can be of benefit to supplement performance and progeny testing programs and could be used when selection is based on carcass traits not easily measured on live rams. It could be of help in overcoming import restrictions

Figure 9-7 Painting ram's brisket before breeding. (University of Wyoming photo)

that prohibit introduction of new genetic material. Its use has been more extensive in other countries, such as Australia, Russia, and some South American countries, than in the United States.

The main limiting factor has been the lack of satisfactory methods for freezing or preserving ram semen.[6,7,11] There has been very little interest in AI on the part of large commercial producers because of extra labor requirements and extra handling of the ewes. Much of the research effort has been devoted to the use of AI in conjunction with controlled or synchronized estrus in order to reduce the length of time during which high labor requirements are necessary. However, trained technicians and special equipment are still necessary for the collection, evaluation, and dilution of semen, as well as for the insemination process itself. Teaser rams are also required for detection of ewes in heat. The genetic gain that could be realized from the use of superior sires in an AI program should be sufficient to justify more effort toward the solution of the problems that limit the practical use of AI today.[10]

INCREASING POTENTIAL PRODUCTION

In an earlier discussion it was pointed out that potential lambing rate can be changed by effective selection or by introduction of new genetic material with the use of newer breeds. In addition to these traditional methods, recent advances in biotechnology and genetic engineering give an indication of tremendous possibilities in increasing potential production above its present level. The combination of superovulation, breeding either naturally or by AI, and subsequent transfer of fertilized embryos allows for extensive use of superior dams or sires. Although these techniques have not reached the point of general application in commercial flocks, there is presently an opportunity to take advantage of such knowledge and at least extend the use of sires or dams that have been proved superior. Continued research and

TABLE 9-4 LAMBING AND REBREEDING PERFORMANCE OF FALL AND SPRING LAMBING EWES

Item	Season of Lambing	
	Fall	Spring
Ewe seasons[a]	537	591
Ewes lambing	188	495
Percent lambing	35	84
Ewes mating after lambing	159	248
Percent mating after lambing	85	50
Ewes conceiving	134	113
Percent conceiving	71	23
Lambing to conception (days)	44	66

Adapted from Whiteman and others.[20] Used with permission.
[a]The number of ewes that could have lambed.

Accelerated Lambing Program

Accelerated lambing programs have been in use for some time, and studies are continually being conducted to find methods of improving their success. This is a management system that can adapt to many types of operations, but it requires extra care and attention to details if it is to be successful. Several different methods of acceleration may be used:

1. Continuous lambing: leaving rams in with ewes year-around.
2. Lamb at regular 8-month intervals: three lamb crops in two years.
3. Lamb half the flock at 4-month intervals by dividing the flock into two groups: three lamb crops in two years from each group, using the same facilities for lambing.
4. Lamb on a regular 7-7-10 month schedule: three lamb crops in two years; normal breeding season can be used more than with 8-month intervals.
5. Breed twice yearly: one breeding time can coincide with peak of breeding season.

TABLE 9-5 TWICE-A-YEAR LAMBING (IDAHO RANGE, 1973)

Age	Item	Selected Control Rambouillet and Targhee	Multiple Lambing Rambouillet	Multiple Lambing Targhee
1	No. lambings/year[a]	0	45	93
	Lambs born/year[b]	0	55	100
	Lambs weaned/year[b]	0	42	65
	Pounds of lamb weaned/year/ewe	0	31	49
2	No. lambings/year[a]	88	137	143
	Lambs born/year[b]	118	163	195
	Lambs weaned/year[b]	76	118	155
	Pounds of lamb weaned/year/ewe	63	86	121
3 and (older)	No. lambings/year[a]	91	122	109
	Lambs born/year[b]	154	193	190
	Lambs weaned/year[b]	113	147	159
	Pounds of lamb weaned/year/ewe	82	106	119

Reproduced with permission from *Sheepman's Production Handbook*.[17]

[a]Per 100 ewes bred.
[b]Per 100 ewes present at lambing.

TABLE 9-6 BREED OR COMBINATION LEAST-SQUARES MEANS FOR NUMBER OF LAMBS BORN PER EWE EXPOSED IN THREE BREEDING SEASONS UNDER ACCELERATED LAMBING

Combinations[a]	n[b]	Winter		n	Fall		n	Late spring		n
$\frac{1}{2}$D, $\frac{1}{2}$R	93	1.46	0.07	69	1.64	0.09	78	0.84	0.09	240
$\frac{1}{4}$D, $\frac{3}{4}$R	105	1.32	0.07	78	1.53	0.09	91	0.66	0.08	274
$\frac{1}{4}$F, $\frac{1}{2}$D, $\frac{1}{4}$R	75	1.62	0.08	54	1.59	0.11	61	0.64	0.10	199
$\frac{1}{4}$F, $\frac{1}{4}$D, $\frac{1}{2}$R	92	1.65	0.07	69	1.65	0.09	78	0.63	0.09	239
$\frac{1}{4}$F, $\frac{3}{4}$R	64	1.59	0.09	50	1.70	0.11	54	0.67	0.11	168
$\frac{1}{4}$F[c]	231	1.63	0.04	173	1.65	0.06	193	0.65	0.05	597
Mean across breeds	429	1.53[d]	0.03	320	1.62[d]	0.04	362	0.69	0.04	1111

Reproduced with permission from Dzakuma, Stritzke, and Whiteman.[(5)]

[a]F = Finn sheep, D = Dorset, and R = Rambouillet.
[b]Number of ewes exposed.
[c]Values listed are averages for the three $\frac{1}{4}$F groups.
[d]Fall and winter versus spring ($p < 0.01$); $\frac{1}{4}$D, $\frac{1}{4}$R versus others ($p < 0.05$).

Any accelerated program can be designed that combines some of the above or can be developed simply to breed ewes as often as possible as determined by the condition of the ewes and their ability to exhibit fertile estrous.

Special requirements. Accelerated programs require the use of sheep with the least restricted breeding seasons, such as Rambouillets, Merinos, Dorsets, Finns, or crosses among these. Very few ewes can be expected to breed while lactating, so early weaning is a part of management. In many experimental attempts at acceleration, hormone therapy to stimulate estrus and synchronize breeding has been included, but some attempts have been made without use of hormones. Since optimum pasture conditions do not prevail throughout the year, ewes and lambs will require some harvested forages and concentrates. Creep feeding of lambs in an accelerated program is recommended so as to make early weaning more simple. Sheep need to be in good condition, well fed, healthy, and free of parasites to ensure successful accelerated programs. Examples of the degree of success are shown in Tables 9-4, 9-5 and 9-6 for different types of accelerated programs. This is also an area of continued research and development, and should become an ever-increasing part of future sheep production.

Record Keeping

Accurate records are a key part of good management and in most cases are easy to obtain. Recording of breeding dates, feed allowances, and details of ram or ewe behavior during breeding is simply part of the sheep business. More detailed records, such as which ewes are mated to each ram, and which requires individual identification, are essential in the purebred sheep business and are useful in flock improvement. Whether a change in production is due to selection, good management, or simply environmental changes is impossible to tell without records. This is the advantage of a small operator over the larger operator, the ability to identify individual performance, but that advantage is easily lost if records are not kept and used properly. Even though we live in a "computer age," the use of computers is not necessary for keeping track of the business. It is likely, however, that continued expansion of computer methods and availability will enable more sheep operators to benefit from their use.

REFERENCES CITED

1. Braun, W. F., J. M. Thompson, and C. V. Ross. 1980. Normal scrotal size of rams. *Sheep Breeder and Sheepman,* Vol. C, May, p. 244.
2. Casida, L. E., C. O. Woody, and A. L. Pope. 1966. Inequality in function of the right and left ovaries and uterine horns of the ewe. *J. Animal Sci.* 25:1169.
3. Collins, Spellman B. 1956. *Profitable Sheep.* MacMillan, New York.

4. Coop, I. E. 1966. Effect of flushing on reproductive performance of ewes. *J. Animal Sci.* 67:305.

5. Dzakuma, J. M., D. J. Stritzke, and J. V. Whiteman. 1982. Fertility and prolificacy of crossbred ewes under two cycles of accelerated lambing. *J. Animal Sci.* 54:213.

6. Dziuk, P. J. and others. 1972. Natural service and AI with fresh or frozen sperm at an appointed time in the ewe. *J. Animal Sci.* 35:572.

7. First, N. L., A. Sevinge, and H. A. Nenneman. 1961. Fertility of frozen and unfrozen ram semen. *J. Animal Sci.* 20:79.

8. Fitch, G. Q., and others. 1985. Analysis of scrotal circumference growth curves in rams. *Proceedings Western Section Amer. Soc. Animal Sci.* 36:57.

9. Hulet, C. V., and S. K. Ercanbrack. 1962. A fertility index for rams. *J. Animal Sci.* 21:489.

10. Hulet, C. V., and others. 1962. Effects of feed and length of flushing period on lamb production in range ewes. *J. Animal Sci.* 21:505.

11. Inskeep, E. K., and C. E. Cooke. 1968. Artificial insemination and preservation of semen. *Proceedings of Symposium, Physiology of Reproduction in Sheep.* Oklahoma State University, Stillwater.

12. Kimberling, C. V., and J. G. Butler. 1985. Colorado's ram breeding soundness program. Colorado Agricultural Extension Service Report.

13. Lamond, D. R., and others. 1973. Influence of nutrition on ovulation and fertilization in the Rambouillet ewe. *J. Animal Sci.* 36:363.

14. Lino, B. F. 1972. The output of spermatozoa in rams. *Australian J. Biol. Sci.* 25:359.

15. Nelms, G. E., and M. P. Botkin. 1966. Controlling estrous cycles in cows and ewes. *Wyoming Agr. Exp. Sta. Research Journal 1.*

16. Pope, Ronald E. 1982. Predicting performance of ram lambs and maximizing production of western ewes. Ph.D. thesis, University of Wyoming, Laramie.

17. Scott, George E. 1975. *Sheepman's Production Handbook,* rev. ed. Sheep Industry Development Program. Denver, Colo.

18. Shelton, Maurice, and Phil Thompson. 1977. Partitioning losses in reproductive efficiency of sheep. *Texas Agr. Exp. Sta. Progress Report 3448.*

19. Torell, D. T., I. D. Hume, and W. C. Weir. 1972. Effect of level of protein and energy during flushing on lambing performance of range ewes. *J. Animal Sci.* 34:479.

20. Whiteman, J. V., and others. 1972. Postpartum mating performance of ewes involved in a twice-yearly lambing program. *J. Animal Sci.* 35:836.

10 Management During Gestation

The gestation period for ewes is one of the easiest parts of the reproductive cycles for management, since labor requirements are lower than during breeding season or later during lambing and subsequent lactation. However, this time during the ewe's life is critical to the overall success of reproductive performance. The ewe provides nourishment as well as waste removal for the fetus and provides a suitable environment in which the fetus can develop. The major concerns of the manager are to provide adequate nutrition, to maintain sanitary conditions, and to avoid injuries due to lack of proper facilities or proper care in handling.

FEEDING DURING GESTATION

Even though early gestation is one of the phases of production during which nutritive requirements are at their lowest and feed can be reduced to a fairly low level, it should not be overdone to the point of undernourishment. Expected weight gains for ewes during gestation range from 15 to 30 lb (6.8 to 13.6 kg), with many cases of gains above 30 lb. This allows for a lamb near full term of 10 to 13 lb (4.5 to 6 kg), plus wool and the placenta and fluids, or two lambs weighing 20 to 24 lb (9 to 11 kg), plus wool and the placenta and fluids. Thus the ewe carrying twins will need to gain considerably more than the ewe carrying a single lamb in order to just maintain body weight and condition. Most of the increase in both the lamb and in placental tissues and fluids occurs during the last 4 to 6 weeks of pregnancy, as shown in Figure 10-1. Changes in weight due to shearing usually occur during the gestation period, even though only about one-third of the fleece weight was produced in that part of gestation prior to shearing. Here, as was the case before and during breed-

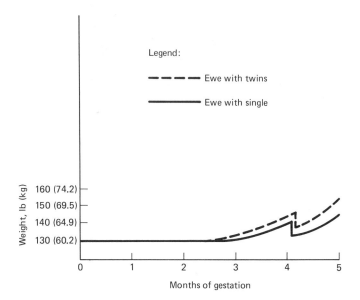

Figure 10-1 Weight change of a ewe during gestation.

ing, relative condition of the ewes has a bearing on weight changes, and thin ewes need to gain more than ewes already in good condition or than fat ewes.

An interpretation of NRC recommendations as previously listed in Table 8-1 can be seen in Figure 10-2. This clearly shows that for average ewes the feed and specific nutrient requirements increase by about 50% in late gestation as compared to early gestation and follow weight changes as shown in Figure 10-1. Further interpretation of the NRC recommended allowances needs to be made in that the change from early to late gestation should be gradual, not a sudden change.[6]

Providing adequate nutrients for early or late gestation is easy as long as plenty of good pasture is available. However, it is sometimes necessary to supplement in cases of low-quality pasture or pasture with only a limited supply of grass. Even in confinement, with the use of low-quality forages or high-moisture feeds it becomes necessary to use concentrates during the last few weeks of gestation. Supplementing normally is done to correct specific deficiencies, but during late gestation all requirements are increased; so quite often it is simply a matter of increasing the amount of whatever feed or feeds are used. During late gestation in March or April in some range areas it is difficult to supplement, because ewes are busily seeking each new spear of green grass. If they start to lose weight or condition, it may be necessary to confine them long enough each day to feed some concentrate. Fortunately, at that time of year in well-managed pastures, ewes get some dry grass along with the fresh green grass, and the latter soon becomes plentiful.

With ewes in confinement that receive diets made up entirely of forages during early gestation, it is often necessary to add some concentrate during late gestation. In balancing diets as described in Chapter 8, it becomes obvious that most forages

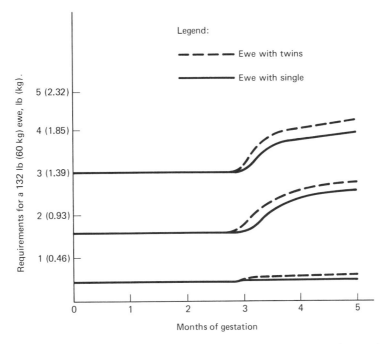

Figure 10-2 Changes in requirements for total feed, total digestible nutrients, and digestible protein during gestation.

do not provide enough energy for meeting requirements during late gestation, and some forages are also deficient in protein or vitamins or minerals. Special care is needed for ewes carrying twins to be sure that sufficient nutrients are provided during late gestation, for pregnancy disease or lambing paralysis occurs primarily in twin-bearing ewes. Thus, a need is readily seen for some practical method of determining fetal numbers in ewes before late gestation approaches. Providing extra feed and care for twin-bearing ewes not only helps to avoid immediate problems before lambing, but also helps to prepare ewes for the subsequent stress of lactation.

ROUTINE MANAGEMENT CHORES

Pregnancy diagnosis. In most cases, it is of value to know whether or not ewes are pregnant, and the earlier such information is available, the better. For the well-managed flock in which there are 5% or less of open or dry ewes, there is little to be gained by the pregnancy test. However, in some situations the percentage of dry ewes exceeds 5%, particularly in flocks made up of young ewes. Other factors such as poor health or poor feed conditions or a short exposure to rams can result in relatively high numbers of ewes that failed to breed or that are bred but in which pregnancy was not maintained. Several methods have been used for detecting pregnancy, with varying degrees of practicality and varying degrees of success.

Use of the ram. When the breeding season is over, a ram or rams can be equipped with marking harnesses or with a marking pigment and turned in with ewes to detect those which are open. If for any reason late lambs are not wanted, rams that have been sterilized can be used for detection. This method is fairly successful when used during the time of year during which ewes are cycling,[7] even though there will be some errors due to pregnant ewes being marked. Many ewes in commercial flocks are not bred until near the end of the normal season, so this method is not expected to be of use in such cases.

Sorting by udder development. Late in the gestation period, ewes can be palpated for udder development in chutes and marked to be sorted. The same approach can be used for ewes on the shearing floor simply by observing and marking the ewes with no sign of udder development. By this method both dry ewes and late lambers are marked and can be sorted from those which are expected to lamb early. The degree of success is dependent on several factors: how late in gestation, length of exposure to rams, and somewhat on age of ewes. Udder development does not begin as early as in young ewes lambing for their first time nor does the udder develop as much size as in mature ewes. Usually only a very slight error occurs using this method in identifying ewes as pregnant.[7] The majority of errors occur in ewes designated as open, so the manager needs to be aware that some of these ewes will be pregnant. Sometimes, instead of udder development, observations as to probable pregnancy from body fill or size of middle are used as a basis for sorting pregnant ewes from nonpregnant or late-lambing ewes. With these two methods experience is helpful and accuracy is expected to improve with continued practice. The biggest disadvantage of these methods is that they are effective only late in gestation and, in many cases, ewes have already been fed through the winter.

Rectal-abdominal palpation. A technique for detecting pregnancy by the use of rectal-abdominal palpation has been developed by Hulet.[1] This consists of placing a ewe on her back in a cradle constructed for surgical use,[4] and then passing a rod 1/2 in. (1.27 cm) in diameter and about 18 in. (45.7 cm) long into the rectum and palpating for the presence of a fetus between the rod and the operator's hand placed on the ewe's abdominal wall. This method has been reported as effective in identification of pregnant ewes as early as 43 days postbreeding and nearly 100% accurate at 65 to 70 days following breeding. When performed by experienced operators, this method is relatively harmless to ewes, but rectal damage can result from use by inexperienced operators.[3] This method has been used with a high degree of accuracy for determining the fetal number in pregnant ewes.[2] In its application special equipment is needed for catching ewes and for restraining them, as well as extra labor. With proper help and equipment, it can be a rapid method for diagnosing pregnancy.

Intrarectal Doppler technique. Results from several studies have indicated a high degree of success with the use of a Doppler-shift principle in which an ultrasonic beam is directed through maternal tissue by a probe placed just anterior to the udder in an area devoid of wool. Positive diagnosis of pregnancy is based on a pulsating sound created by arterial blood flow. Further developments have resulted in the use of a button-type Doppler transducer inserted into the rectum. Fetal heartbeat, fetal pulse, or fetal movement are taken as positive criteria of pregnancy. It has not been possible to distinguish between multiple and single fetuses with this method.[5]

Ultrasonic scanning instruments. The results of studies with these instruments indicate greater success (Tables 10-1 and 10-2) than with those previously mentioned in that accuracy of pregnancy diagnosis is above 90% at 70 to 100 days, and it is a safe method.[8] The devices that have been tested and for which reports are available are the Scanoprobe[R] and Scanopreg[R] (for sheep) manufactured by Ithaca, Inc., Ithaca, New York. Costs of determining pregnancy are usually not included in research reports, but generally the main cost involved is in the time and labor required for constraining the sheep. The cost of the testing device can be kept at a minimum by forming pools for joint purchase of a device; likewise, labor costs can be reduced by pooling labor forces.

The accuracy with which pregnancy is diagnosed is only of minor importance in many operations. The ability to detect pregnancy early and particularly the ability to determine fetal numbers are more useful in determining proper management than

TABLE 10-1 ACCURACY OF PREGNANCY DIAGNOSIS

Method	Number of Ewes Tested	Stage of Gestation when Tested, days	Percentage Accurate Diagnosis[a]		
			Pregnant	Nonpregnant	Combined
Rectal-abdominal palpation[b]	498	60–96	$\frac{173}{262} = 66.0$	$\frac{139}{236} = 58.9$	$\frac{312}{498} = 62.7$
Intrarectal Doppler device	498	60–96	$\frac{250}{366} = 68.3$	$\frac{112}{132} = 84.8$	$\frac{362}{498} = 72.7$
Scanoprobe	212	69–103	$\frac{130}{152} = 85.5$	$\frac{59}{60} = 98.3$	$\frac{189}{212} = 89.1$
Scanopreg	212	78–112	$\frac{124}{128} = 98.9$	$\frac{77}{84} = 91.7$	$\frac{201}{212} = 94.8$
Scanopreg (retest)[c]	84	79–113	$\frac{7}{8} = 87.5$	$\frac{76}{76} = 100.0$	$\frac{83}{84} = 98.8$

Reproduced with permission from Trapp and Slyter.[8]

[a] $\frac{\text{No. of ewes diagnosed correctly within the classification}}{\text{No. of ewes diagnosed in each classification}} \times 100$ = percentage accurately diagnosed.

[b] Rectal-abdominal palpation as described by Hulet.[1]

[c] Ewes diagnosed nonpregnant initially were retested the following day.

TABLE 10-2 EFFECT OF LENGTH OF GESTATION ON ACCURACY OF PREGNANCY DIAGNOSIS

Length of Gestation When Tested, days	Percentage of Ewes Lambing Considered Pregnant			
	Rectal–Abdominal Palpation	Intrarectal Doppler	Scanoprobe	Scanopreg
60–69	$\frac{32}{45} = 71.1$	$\frac{34}{36} = 94.4$	$\frac{2}{2} = 100.0$	
70–79	$\frac{63}{113} = 55.8$	$\frac{93}{105} = 88.6$	$\frac{42}{42} = 100.0$	$\frac{7}{7} = 100.0$
80–89	$\frac{61}{82} = 74.4$	$\frac{93}{96} = 96.9$	$\frac{46}{47} = 97.8$	$\frac{55}{56} = 98.2$
90–99	$\frac{17}{30} = 56.7$	$\frac{30}{33} = 90.9$	$\frac{22}{22} = 100.0$	$\frac{27}{31} = 87.1$
100–109			$\frac{18}{18} = 100.0$	$\frac{30}{32} = 93.7$
110–113				$\frac{5}{5} = 100.0$

Reproduced with permission from Trapp and Slyter.[8]

just finding out whether or not ewes are pregnant. Too often, by the time pregnancy information is available, ewes have already been fed through most of the winter, and market prices for cull ewes are poor; so ewes are kept at least until shearing, whether pregnant or not. Thus the cost of diagnosing ewes for pregnancy is more than potential return.

The real value lies in the ability of a pregnancy test to differentiate between single-bearing ewes and those with two or more lambs. This would allow for managing and feeding according to producing ability. Twin-bearing ewes could be separated, given more care and better feed, and then kept separate during and following lambing. Such a group of ewes could well be the basis of a "super herd" to be used for raising replacements. Even with large flocks in which no replacements are raised or in which replacements come from the entire flock, dividing ewes on the basis of fetal numbers and giving better care to the most productive ewes is a good management practice. Unfortunately, these ewes are not the best wool producers, but giving them extra care and feed will improve their fleece weights. Likewise, the extra attention given these ewes will reduce the stress of producing wool, lamb, and milk all at the same time and should help to avoid a serious drop in production in their future years.

The system in which ewe lambs are bred and then pregnancy tested, with open ewes being marketed, has been mentioned in the previous chapter. This represents another important practical use of pregnancy testing and avoids the problem of low salvage value when they are more than a year of age. With ewe lambs in many situations, the twinning rate is not very high, so determining whether or not ewe lambs are pregnant is about as important as determining fetal numbers.

SHEARING

Shearing of sheep is the harvesting of a crop and should be approached with this in mind. It should be accomplished with two goals in mind: removal of wool in the best possible condition and removal with as little stress as possible. In spite of the potential danger of injury due to careless or rough handling during shearing and subsequent abortion, sheep are generally sheared during late gestation. The management decision to shear before lambing is prompted by the fact that there are no young lambs in the way and thus there is no danger of lambs being trampled or "bummed" due to separation from their mothers while being penned. Shorn ewes are more likely to seek shelter in inclement weather than are unshorn ewes. Also, shorn ewes are cleaner at lambing time, resulting in less chance for infection or "fly blow," and it is easier for newborn lambs to nurse if ewes have been shorn beforehand. For this method to prove successful, protection must be available (either artificial or natural shelter) following shearing, and in some cases extra feed may be required. Also, shearing crews need to be closely supervised to avoid excess injury to ewes. Shearing facilities not only need to be constructed where protection can be provided before, during, and after shearing, but should also be constructed for ease in handling and to avoid crowding.

For a complete description of the shearing process and preparation of wool following shearing, see Chapter 19.

REFERENCES CITED

1. Hulet, C. V. 1972. A rectal-abdominal palpation technique for diagnosing pregnancy in the ewe. *J. Animal Sci.* 35:814.
2. Hulet, C. V. 1973. Fetal numbers in pregnant ewes. *J. Animal Sci.* 36:325.
3. Lamond, D. R. 1963. Diagnosis of early pregnancy in the ewe. *Australian Vet. J.* 39:192.
4. Lang, D. R. 1969. A mobile surgical cradle for sheep. *New Zealand J. Agr. Res.* 12:611.
5. Lindahl, I. L. 1971. Pregnancy diagnosis in the ewe and intrarectal Doppler. *J. Animal Sci.* 32:922.
6. National Research Council. 1985. Nutrient requirements of domestic animals, No. 6. Nutrient requirements of sheep, 5th rev. ed. Academy of Sciences, Washington, D.C.
7. Shelton, Maurice. 1974. Pregnancy diagnosis methods in sheep. *Texas Agr. Exp. Sta. Progress Report 3287.*
8. Trapp, M. J., and A. L. Slyter. 1983. Pregnancy diagnosis in the ewe. *J. Animal Sci.* 57:1.

11 Lambing Time

For most producers, lambing time is the busiest time of the year. In spite of this, it can and should be the most satisfying and rewarding time due to the feeling of accomplishment. Saving of lambs that would otherwise have been lost not only is a satisfaction to the operator, but also an important boost to income. Lambs raised per ewe is the most important trait in any flock and thus should be everyone's primary consideration. It is not enough to have achieved a high lambing rate, and thus everything at one's command should be used to assure a high survival rate.

GETTING READY FOR LAMBING

Before ewes are due to start lambing, it is important to have all details under control. Records of breeding dates should be sufficient to give an approximate date for the first lambs. To begin with, ewes need to be healthy and fed so that they are gaining in weight. Plenty of palatable feed and adequate water should be readily available. The more productive the ewes, the more essential that their feed contain plenty of available energy, since lambing paralysis occurs most commonly in ewes carrying two or more lambs.[4] Any handling or moving of ewes should be done carefully to avoid crowding or injury. A protected area should be available where ewes can be comfortable and kept under observation. Buildings and equipment need to be arranged for convenience, feed storage readily accessible for the drop band, and lambing pens or "jugs" located for easy access from the drop band, (Figures 11-1 through 11-4). The drop band may be the entire flock in many cases or it may be only those picked out by the manager as the first ewes to lamb. Sorting ewes that are expected to lamb first from ewes expected to lamb later can be done by observa-

Figure 11-1 Area near lambing shed. (University of Wyoming photo, Courtesy of Warren Livestock Company, Cheyenne, Wyoming)

tion of udder development and general appearance of ewes. It requires some extra time and effort, but tends to simplify the lambing operation. In some cases a drop band is designated at breeding time either by breeding part of the flock two to three weeks ahead of the rest or by use of marking devices on the rams and separating the ewes that were marked during the first 14 to 17 days of breeding. Separation on the basis of either udder development or marks by the ram at breeding time can help simplify lambing but may result in some error. Thus, the group or groups designated for later lambing should not be neglected.

Wherever the drop band is held, whether in pastures or under sheds, it is necessary that the area be clean, sanitary, and free of disease organisms. If in sheds, good bedding should be used and yet not overdone so as to create difficulty in movement. A plentiful supply of fresh water, soap, and towels along with commonly used drugs and medicines should be on hand. Drugs or medicine should be in a handy location,

Figure 11-2 Idaho-type sheds with "drop band" nearby. (Photo by U.S. Forest Service)

Figure 11-3 Lambing jugs in small shed with heat provided by overhead heater. (University of Wyoming photo)

but one that provides protection from weather damage or stray sheep, and well away from the reach of children.

Adequate competent help needs to be available to ensure the success of any lambing operation. Although many range operations are managed without sheds or without any care at lambing time, it is easy to justify the extra expense of providing shelter and labor when viewed from the aspect of increased lambing percent.[2] In some operations, lambs are born in open areas and then transported to sheds. The interior of a large shed is shown in Figure 11-5 and a "lambulance" for transport of ewes with their newborn lambs is shown in Figure 11-6. The availability of competent help is sometimes a problem, and in many cases entire families get involved and put in extra time during lambing. Unavailability of suitable help for lambing is the main reason that some sheep producers have switched to range lambing.

Probably the most important attribute of the shepherd or lambing attendant is the ability to be observant. Just by looking, it should be obvious when a ewe is

Figure 11-4 Ewe with newborn lamb in individual jug. (University of Wyoming photo, Courtesy of Warren Livestock Company, Cheyenne, Wyoming)

Getting Ready for Lambing

Figure 11-5 Interior of large lambing shed. (University of Wyoming photo, Courtesy of Warren Livestock Company, Cheyenne, Wyoming)

getting ready to lamb. For a novice, this is the first task, to recognize the symptoms that indicate a ewe is about ready to lamb. By being aware, it is then relatively easy to identify which ewes are having difficulty. This ability is not difficult to learn, provided the person is willing.

The majority of ewes will give birth to their lambs normally without any assistance. Likewise, in most cases the ewe will provide proper care and nourishment for her lamb or lambs. These are obvious statements since we know many producers that allow ewes to lamb by themselves in large pastures. However, many ewes have some difficulty, either in the birth process or in nursing their lambs.

Figure 11-6 Lambulance for transporting ewes and lambs to the lambing shed. (Courtesy of Warren Livestock Company, Cheyenne, Wyoming)

CAUSES OF DEATH IN LAMBS

There are many different causes of death; the main categories are as follows:

- Abnormalities and deformities
- Diseases
- Dystocia, or difficult birth
- Chilling or inclement weather
- Injury by ewes
- Predation
- Starvation
- Stillbirth
- Unknown, no visible lesions

The extent of loss due to any particular cause is dependent on the type of lambing operation. For instance, losses due to predators and inclement weather are the major categories in many range sheep operations, but are not very serious in farm flocks where shed lambing is practiced. Total loss of lambs also is variable, with reported losses ranging from 10% to over 30% born.[1,2,5,6,7,8,9,11,12,13] Tables 11-1 through 11-3 show examples of extent of loss and causes of loss under different conditions.

Abnormalities and Deformities

This category usually makes up a small portion of the death loss. Common hereditary defects, or lethals, have previously been listed in Chapter 7, but in addition to those there are many different kinds of abnormalities, both hereditary and environmental, that occur in rare cases. An example of an environmental abnormality is the "monkey-faced" lamb shown in Figure 11-7. Although the lamb was alive at

TABLE 11-1 LOSSES OF LAMBS BY CAUSE, WYOMING, 1980-1984

Year	Total Loss, All Causes[a]	Causes of Loss, Number (% of total)			
		Predation	Weather	Other Causes	Unknown[c]
1980	157,000	71,500 (55.0)	18,500 (14.2)	28,700 (22.1)	11,300 (8.7)
1981	130,000	71,900 (56.6)	16,400 (12.9)	22,600 (17.8)	16,100 (12.7)
1982	127,000	52,800 (49.6)	18,200 (17.1)	19,300 (18.1)	16,200 (15.2)
1983	106,500	55,300 (47.8)	33,200 (28.7)	20,200 (17.5)	6,900 (6.0)
1984	190,300	67,900 (35.7)	96,000 (50.4)	16,600 (8.7)	9,800 (5.2)

Adapted from *Wyoming Agricultural Statistics*.[13]

[a]Lamb loss from birth.

[b]Includes losses due to identifiable causes.

[c]Unidentified causes.

Causes of Death in Lambs

TABLE 11-2 PREWEANING MORTALITY IN LAMBS BY BREED, SEX, AND MULTIPLICITY[a]

Breed	Hampshire	Rambouillet	Shropshire	Southdown	Total
Ram					
Number	206	210	329	264	1009
Preweaning losses, %	19.4	10.5	14.6	22.3	16.7
Ewe					
Number	171	234	364	251	1020
Prewaenning losses, %	18.1	9.4	12.6	13.1	12.9
Ram co-twin of ram					
Number	118	136	164	116	534
Preweaning losses, %	35.6	21.3	17.7	31.9	25.7
Ram co-twin of ewe					
Number	113	102	164	95	474
Preweaning losses, %	23.9	17.6	22.6	18.9	21.1
Ewe co-twin of ram					
Number	113	102	164	95	474
Preweaning losses, %	23.0	17.6	20.7	27.4	21.9
Ewe co-twin of ewe					
Number	120	128	212	96	556
Preweaning losses, %	23.3	21.9	22.2	18.8	21.8
Singles					
Total	377	444	693	515	2029
Percent	18.8	9.9	13.6	17.9	14.8
Twins					
Total	464	468	704	402	2038
Percent	26.5	19.9	20.9	24.6	22.7
Total	841	912	1397	917	4067
Preweaning losses, %	23.1	15.0	17.3	20.8	18.8
Lambs dropped, %[b]	139	132	135	127	133
Lambs weaned, %[b]	103	110	106	95	104

Reproduced with permission from Vetter, Norton, and Garrigus.[(9)]

[a]Excluding data on 96 triplets, 47 lambs for which sex was not recorded, and 21 lambs for which sex of co-twin was not recorded.

[b]Only ewes lambing figured in these calculated values. Triplets were excluded, but would not alter these percentages over five percentage points.

Figure 11-7 Monkey-faced lamb. (University of Wyoming photo)

TABLE 11-3 AUTOPSY FINDINGS ON 1051 LAMBS FROM THE MONTANA AGRICULTURAL EXPERIMENT STATION RANGE FLOCK

Autopsy Findings	Number of Lambs	Percent of Lambs Born	Autopsy Findings	Number of Lambs	Percent of Lambs Born
Pneumonia	169	2.35	Miscellaneous (total)	62	0.86
No visible lesions	167	2.32	Trauma	5	
Atelectasis (stillborn)	151	2.09	Docking infection	5	
Starvation	146	2.03	Fractured ribs	5	
Dysentery	125	1.73	Fractured tibias	4	
Delayed parturition	53	0.73	Castration infection	3	
Enterotoxemia	48	0.66	Liver infarcts	3	
Prepartum death	39	0.54	Fractured skulls	2	
Liver rupture	29	0.40	Fractured metatarsus	2	
Anomalies (total)	27	0.37	Fractured femurs	2	
Diaphragmatocele	4		Fractured pelvis	2	
Megacolon	2		Blackleg	2	
Hydronephrosis	2		Arthritis	2	
Cardiac anomaly	2		Mechanical choke	2	
Leg anomaly	2		Coccidiosis	2	
Omphalocele	2		Hemorrhage, pyloric valve	2	
Contracted flexor tendons	1		Anasarca	1	
Megalo-ureters	1		Gastroenteritis	1	
Schistosomus reflexus	1		Cirrhosis, liver	1	
Mandibular anomaly	1		Pleuritis	1	
Hepatomegaly	1		Icterus	1	
Cranial anomaly	1		Intussusception, intestine	1	
Agnathia	1		Gastritis	1	
Apoctia	1		Myocardial abscesses	1	
Cervical anomaly	1		Duodenal torsion	1	
Lung anomaly	1		Cretin	1	
Epigastric hernia	1		Engorgement	1	
Alopecia	1		Adrenal neurofibroma	1	
Intestinal atresia	1		Pyemia	1	
Umbilical hemorrhage	12	0.16	Suffocation	1	
Navel infection	12	0.16	Tetanus	1	
Hemorrhagic enteritis	10	0.13	Pericarditis	1	
			Abdominal hemorrhage	1	
			Abomasitis	1	
			Ruptured cecum	1	
			Total	1051	14.7

Reproduced with permission from Safford and Hoversland.[7]

birth, it survived for only a few days. These types of abnormalities are often beyond the control of the shepherd and occur in spite of good breeding and management practices. In some cases a temporary deformity, such as crooked legs, is a result of difficulties at birth or of improper position of two or more lambs in the uterus before birth. By being present and providing proper care, this kind of problem can be corrected.

Diseases

Many diseases cause losses in newborn lambs, some of which can be cured or at least their effect kept to a minimum by an alert manager. In many cases it may be necessary to have the help of a veterinarian to recognize or diagnose the cause and then to attempt to correct the problem. Diagnostic laboratories are available in many states to help in identifying specific causes of death and should be used in cases of doubt.

The most important role of good management is the prevention of disease as much as possible by sanitation. Clean lambing grounds and lambing pens should be provided. Where possible, pens should be disinfected and allowed to dry between lambing seasons. Some producers make a practice of dipping the navel of all newborn lambs in a tincture of iodine solution to avoid the occurrence of navel ill. Some clean out lambing jugs regularly and use slacked lime as a drying agent and then use new bedding. Others let bedding build up during lambing, adding some dry bedding as needed to avoid messy pens, but clean and disinfect only when lambing has been completed. Decisions are based on previous history of success or failure or on advice from others who are experienced in the lambing operation. The disease of most concern that sanitary conditions are aimed at preventing is *lamb scours,* or dysentery, or diarrhea, which can be the scourge of shed lambing operations. Various kinds or degrees of scours are involved depending on causative organisms. Some types are relatively mild and of little concern, but some cause severe dehydration and are often fatal and extremely difficult to bring under control.

Pneumonia is a cause of death loss or unthriftiness, but it often is predisposed by a lack of resistance resulting from some other disease or problem. Several common problems are related in young lambs. Weak lambs, chilled lambs, or hungry lambs are more susceptible to any stress, and these are the lambs most likely to be affected by pneumonia. Treatment with antibiotics is common in such cases and often helps lambs to survive, but it does not help to solve the original problem.

Several other diseases can be involved in death loss of young lambs. *Mastitis* in ewes results in starvation of lambs, and often has caused permanent udder damage before being observed. Treatment of the udder or treatment with antibiotics, to be effective, must be administered early, and thus early observation is necessary. *Sore mouth* virus, when allowed to spread, sometimes gets on the udder and teats and can result in starvation of lambs. This can be prevented by routine vaccination or can be corrected by the same vaccine, but it does take time to clear up the sores. *Enterotoxemia* can be the cause of early loss of some of the biggest and best lambs.

Immunization at a few days of age helps, but does not provide permanent immunity, so lambs may need to be treated again. Some use immunization of ewes ahead of lambing to prevent enterotoxemia in their lambs. *Pink eye* or eye irritations from entropion (inverted eyelids) or eye problems caused by contaminations in forages such as awns of foxtail barley or cheat grass can cause blindness and resulting starvation. Treatment with an antibiotic ointment is effective, but the causative problem needs to be corrected, the impediment removed, or eyelid clipped.

Several diseases are merely the result of specific nutritive deficiencies and need to be corrected or avoided by proper feeding. Poisonous plants, too, can take their toll of ewes or lambs, requiring care in selection of grazing areas. Infections occur at the site of injuries, or in warm weather fly blow can be a problem; these can be corrected by disinfectants at the site. Some lambs develop abscesses in critical areas such as the mouth or throat and require care by draining or disinfecting.

Unlike several other causes of baby lamb losses, disease can continue to be a problem for some time after birth, even though the major part of disease loss occurs early in life. Discussion here centers mostly around the various diseases involved and routine observation and means of correction. Our discussion is not meant as a guide to disease treatment, but more as a help to prevent disease. For a detailed account of sheep diseases, see references 3 and 5.

Dystocia, or Difficult Birth

The following discussion and illustrations were prepared by Dr. Norman Gates, Washington State University. The authors are grateful for his helpful contribution. It is reprinted with permission from *"A Practical Guide to Sheep Disease Management."*[3]

SHEEP OBSTETRICS

Introduction

Lambing is one of the most critical activities in a sheep operation. The highest level of mortality in the flock usually occurs during the lambing period. There are various causes of death in ewes and lambs during lambing, one of which is difficult birth, or dystocia. The practice of breeding ewe lambs has certainly increased the frequency of dystocia. When a ewe begins labor, the moment of truth has arrived. If that ewe doesn't give birth to one or more live lambs, the entire investment in that ewe is lost for one year. If the flock has been managed properly to prevent lamb losses due to pregnancy toxemia and infections that cause abortion in pregnant ewes, normal birth of live lambs is generally accomplished by the ewe and mother nature. Unfortunately, however, most producers observe a number of ewes that mother nature has overlooked in her busy schedule and the shepherd is forced to assume the role of obstetrician, or midwife.

Dystocia is a problem that should be expected since it occurs rather frequently. Therefore, the producer should be properly equipped to handle the problem effectively when it does occur. Most dystocia cases can be corrected by the shepherd if he or she has a reasonable understanding of why there is a problem and what must be accomplished to correct the problem. People have sometimes attempted to "assist" ewes without adequate understanding of the problem or its solution. The theory seems to be "if I can get hold of anything and pull hard enough, something has to give." That theory is sometimes correct but also results in dead lambs or disabled ewes. The art of obstetrics is the combined application of knowledge, cleanliness, gentleness, patience and interest.

Equipment

A minimum of equipment is required for delivering difficult lambs. A clean bucket, preferably stainless steel, should be readily available and used only for lambing. An iodine-base surgical scrub liquid soap should be added to the bucket of warm water for cleansing and disinfecting the hands and arms of the operator and the hindquarters of the ewe. Obstetrical lubricants are essential to minimize trauma to the ewe's birth canal while manipulating lambs into deliverable positions. Frequent application of lubricant to the hands and arms of the operator is essential. Grease or oil are not recommended as obstetrical lubricants. A lamb puller can be a useful and effective instrument, particularly when the operator's hands and arms are too large to permit easy passage through the birth canal of the ewe. Once the puller is properly secured to the lamb, the operator's hand is withdrawn from the birth canal and traction applied to the lamb puller. Intrauterine antimicrobial boluses should always be inserted deeply into the uterus after the delivery is completed if the operator entered the uterus to assist the ewe. This practice is intended to lessen the chance of subsequent uterine infections and is quite effective.

Assisting the Ewe

Generally, dystocia is the result of abnormal position of the unborn lamb in the uterus or birth canal. Normal positioning is the head and two front feet forward (Figure 11-8). Although the backward presentation (Figure 11-9) is considered an abnormal position, either presentation shown in Figure 11-8 or 11-9 is considered normal from the standpoint of delivering the lamb by assisting the ewe. Regardless of the initial malpresentation, one of the foregoing positions must be established before traction is applied to the lamb. The following discussion of abnormal fetal positions and their correction is intended to familiarize the reader with the most common problems.

Normal presentation: A ewe can experience dystocia when the lamb is in a normal position (Figure 11-8). The problem is usually the result of an unusually large single lamb, overly fat ewe, or small pelvic diameter, as may be the case in ewe lambs bred to lamb as yearlings. In these cases, exert traction on the lamb outward and slightly downward. A lamb in the normal position is best delivered with two hands. Simultaneous traction is applied to the lamb with one hand inserted into the ewe's vagina and cupped over and slightly behind the lamb's head while the front legs of the lamb are grasped by the other hand. The two-handed delivery will greatly facilitate exteriorizing the lamb's head through the vulva.

Backward presentation: The backward lamb (Figure 11-9) is not an uncommon cause of dystocia. The shepherd diagnoses the problem by observing that only feet are protruding from the vulva and that the soles of the feet are turned upward. Upon

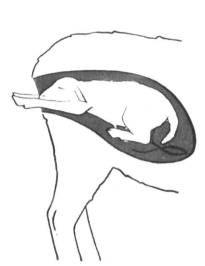
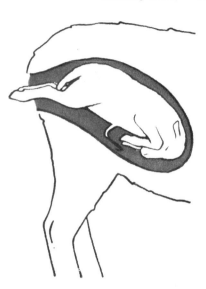

Figure 11-8 Normal presentation. (Courtesy of Dr. Norman Gates, Pullman, Washington)

Figure 11-9 Backward presentation. (Courtesy of Dr. Norman Gates, Pullman, Washington)

further examination the tail of the lamb is palpated and pulled down between the back-legs. Traction is then applied to both hind legs. The lamb should be delivered fairly rapidly without pausing during the delivery. There is considerable danger of the backward lamb drowning midway through delivery; thus the need for a forthright delivery. Once delivered, the lamb's nose and mouth should be wiped to remove as much mucus as possible. Gently swinging the lamb by its hind legs is fairly effective in clearing mucus from the upper respiratory passage. Gentle artificial respiration applied to the rib cage of the lamb is sometimes beneficial in stimulating respiration.

Locked elbows: When there is not enough room in the birth canal (pelvic canal) to allow passage of normally positioned head and front feet, a common form of dystocia results wherein the head is forced into the canal and the elbows are forced downward (Figure 11-10). In effect, the elbows are essentially locked behind the pelvis and the more forceful the ewe's labor, the tighter the lamb becomes locked. This situation occurs more frequently in ewe lambs bred to lamb as yearlings, but also in mature ewes giving birth to a large single lamb. The condition is diagnosed simply. After half an hour, only the tip of the lamb's nose and its toes are seen protruding through the birth canal. Prolonged, excessive pressure from a laboring ewe can cause death in this type of lamb, so don't hesitate to assist her. With one hand, gently push the lamb's head back into the birth canal (just enough to relieve the pressure against the front legs) and, at the same time and with the other hand, pull on one front foot until you feel the elbow pop out of its locked position and slide smoothly into the canal. Grasp the other foot and pull gently, unlocking the other elbow, and the lamb is in the normal position, ready for delivery. This is accomplished by sliding the fingers of one hand over the back of the head of the lamb, grasping the two front legs with the other hand, and exerting a gentle pull with both hands simultaneously.

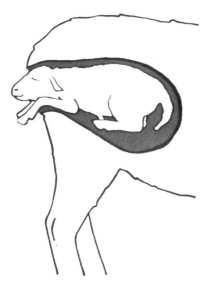 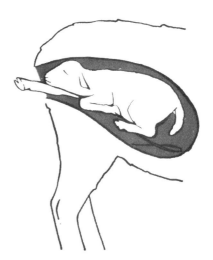

Figure 11-10 Locked elbows. (Courtesy of Dr. Norman Gates, Pullman, Washington)

Figure 11-11 One leg back. (Courtesy of Dr. Norman Gates, Pullman, Washington)

One leg back: Lambs are frequently positioned with one front leg back (Figure 11-11). Although many of these lambs can be delivered in this position, particularly if the lamb is small, it is preferable to get the malpositioned leg out and in normal position. The manipulation is easily accomplished by entering the birth canal with one hand, proceeding back along the shoulder until the foot of the turned back front leg is located. Grasp the foot with cupped hand to avoid lacerating the uterus as the foot is drawn up and into the birth canal until it is in a normal position (Figure 11-8). Occasionally the lamb will be jammed into the pelvic canal, which may require pushing the entire lamb back into the uterus to allow room to accomplish the manipulation.

Head back: Another common position causing dystocia is when the lamb begins its way through the birth canal, both front feet and nose together, but because of either a small birth canal or the large size of the lamb, the head is forced back and to one side (Figure 11-12). This is one situation when the novice is tempted to grab the two front feet and start pulling. However, gently insert one hand through the birth canal, proceed back along the lamb until the head is located, grasp the head and withdraw it until the lamb is in the normal lambing position (Figure 11-8). Now, with one hand cupped over the top of the lamb's head and the other hand pulling gently on the two front legs, the lamb is delivered.

Two legs back: This is another very common form of dystocia but generally one that is easily corrected. By referring to Figure 11-13, correction of the problem is simply a matter of entering the birth canal with one hand, sliding down the lamb's neck and across the shoulder to the foot. The foot is grasped in the cupped hand, flexed, and drawn through the birth canal. Repeat this procedure with the other foot and deliver the lamb. Note: In attempting to withdraw either foot, the lamb may have to be pushed back into the uterus (slightly) to allow room to withdraw the foot.

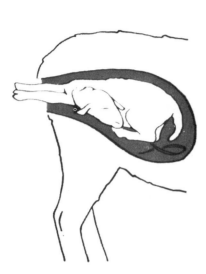
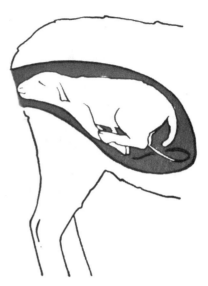

Figure 11-12 Head back. (Courtesy of Dr. Norman Gates, Pullman, Washington)

Figure 11-13 Two legs back. (Courtesy of Dr. Norman Gates, Pullman, Washington)

Inverted lamb with back presented: There are two alternative manipulations that can be accomplished with this type of lamb (Figure 11-14): (1) position the lamb for a backward delivery, or (2) position the lamb for a normal two front legs and head delivery. The initial challenge, however, will be to diagnose the problem. After entering the birth canal and locating the lamb, the operator's hand is moved up, down, and across the lamb until some identifiable anatomy of the lamb can be found. A determination can then be made as to the logical type of delivery.

Breech presentation: The breech lamb (Figure 11-15) is easily diagnosed after entering the birth canal. Slide one hand, palm up, under and forward until a hind foot can be grasped in cupped hand. Flex the foot and withdraw it into the birth canal. Repeat the procedure with the other rear leg. With both hind feet in the birth canal, simply grasp them both and deliver the lamb, forthwith.

Inverted lamb with four feet presented: The lamb shown in Figure 11-16 can be difficult to straighten out and requires identification of front legs from rear legs. If the lamb's head can be grasped and withdrawn toward the birth canal, the two front legs are easily located and manipulated for a normal position delivery (Figure 11-8). If the head cannot be located, identification and withdrawal of the two rear legs into the birth canal will position the lamb for a backward delivery.

Twin lambs, normal and backward presentation: In most cases when lambs are positioned as illustrated in Figure 11-17, the normally positioned lamb will be delivered by the ewe without assistance. After cleaning up the first lamb, the ewe will begin labor again but is likely to have difficulty with the second lamb. The rear feet of the backward lamb may be observed protruding through the vulva, soles of feet turned upward

Causes of Death in Lambs

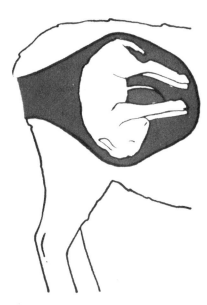

Figure 11-14 Inverted lamb with back presented. (Courtesy of Dr. Norman Gates, Pullman, Washington)

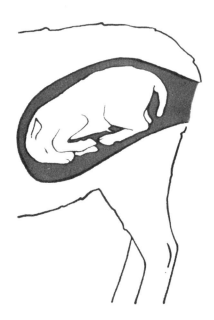

Figure 11-15 Breech presentation. (Courtesy of Dr. Norman Gates, Pullman, Washington)

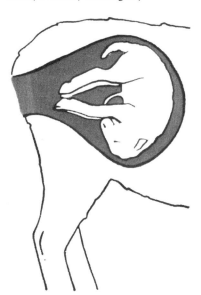

Figure 11-16 Inverted lamb with feet presented. (Courtesy of Dr. Norman Gates, Pullman, Washington)

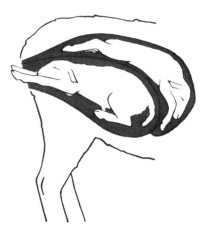

Figure 11-17 Two lambs, normal and backward. (Courtesy of Dr. Norman Gates, Pullman, Washington)

or the lamb may not be visible at all. If the rear feet are observed, the lamb should be delivered as soon as possible by the shepherd. If the lamb is not visible after one half-hour of labor, the ewe should be examined to determine the nature of the problem, and the lamb delivered.

Four legs, one head presented: This type of presentation (Figure 11-18) is fairly common and should be corrected by the shepherd as soon as the situation is observed. Both lambs must be gently pushed back into the uterus to allow room to deliver one lamb in the normal position. The second lamb is positioned as shown in Figure 11-12, and should be manipulated into the normal position as described under the "head back" discussion.

Partially dilated cervix: A completely different type of dystocia results from partial dilation of the ewe's cervix. In such cases the ewe may labor normally but cannot deliver a lamb. The operator will be able to get one to several fingers through the cervix, into the uterus. The partially dilated cervix can be opened up either by gentle, gradual pressure with the fingers or of use of injectable hormones. If the cervical opening will permit passage of all the fingers when brought together to shape a cone, then normal pressure with the fingers to open the cervix is usually the preferred method of correction. Care must be exercised to avoid tearing the cervix by attempting manual stretching too rapidly. If only one or two fingers can be inserted through the cervix, the operator is well advised to attempt the use of hormones: (1) inject 3 mg of estradiol, intramuscularly, and (2) inject 20 to 40 IU (depending on size of ewe) oxytocin, intramuscularly, one hour after the estradiol injection. This sequence will usually cause the ewe to begin labor. (3) Reexamine the cervix one hour after administering oxytocin, and (4) oxytocin may be repeated at hourly intervals for a maximum of three injections. If the ewe has not lambed after three oxytocin injections or dilated enough to permit manual dilation of the cervix by the operator, a caesarean section should be performed.

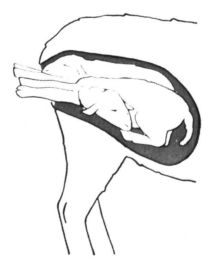

Figure 11-18 Four legs and one head. (Courtesy of Dr. Norman Gates, Pullman, Washington)

Causes of Death in Lambs

Conclusion: Several considerations should be reiterated with regard to obstetrical procedures because of their importance:

1. Cleanliness of the shepherd's hands and arms is necessary to prevent secondary infection of the uterus.
2. Gentle manipulation of the lamb is necessary to prevent death of the lamb and damage to the ewe.
3. Traction is never applied to the lamb until the lamb is properly positioned (Figure 11-8 or 11-9).
4. Intrauterine antimicrobial boluses are always placed deeply in the uterus if the operator has entered the uterus with hand and arm.
5. Caesarean section (surgery) is sometimes necessary to deliver lambs that cannot be delivered normally.
6. Seek competent assistance when you are unable to deliver the lamb after a reasonable amount of time and effort.

Chilling and Inclement Weather

In the northern states, lambing often occurs during cold weather, and cold rains in any area can be a cause of chilling and subsequent death. Therefore, some precautions are needed, first, to avoid chilling as much as possible and second, to correct the problem when lambs do get chilled. Chilled lambs tend to be lethargic, inactive, and don't get up to nurse; thus they sometimes starve. For pasture lambing, protected areas with natural windbreaks are preferred. For shed lambing, buildings need to be free of drafts, and usually there is some special arrangement to help warm up lambs that have been chilled. Heat lamps or incubators can be of help, and sometimes lambs are moved to warm buildings, even into homes, to assure their getting warm (Figure 11-19). Using a stomach tube for infusing milk or colostrum into the lambs' stomach may help to speed up their recovery and reduce death loss.

Figure 11-19 Heat lamp to provide warmth during cold lambing weather. (University of Wyoming photo)

Injury by Ewes

Most ewes are careful with their lambs, but they may not be so careful with lambs other than their own. Also, some ewes tend to be easily excited or nervous and can cause injury to any lambs, including their own. Ewes with flighty dispositions are good candidates for market at the first opportunity. Being startled by people or dogs can cause ewes or groups of ewes to crowd into corners or step on their lambs. Overcrowding can be a cause of lambs being stepped on or crushed. This kind of problem is more prevalent in flocks with many twins and triplets than in flocks with predominantly single lambs. Not only are there more lambs in a limited amount of space, but also the lambs tend to be weaker and less able to get out of the way or to withstand stress. Thus more space and more care are justified for highly prolific ewes.

Predation

Most sheep producers have problems from predation. This varies from a minor problem involving occasional losses to one of serious proportions in many large flocks. In flocks located near urban areas, stray dogs often present a serious problem. Several different kinds of predators are involved in creating problems, but the most serious problem is caused by coyotes.[10,13] In many large sheep operations, the loss due to predation by coyotes is greater than that from any other cause. The problem can be solved in some cases by complete protection, such as predator-proof fencing, close herding or corralling sheep at night, but such practices are prohibitive in cost in most flocks. There are various agencies involved in control of predation that help in many situations. The Fish and Wildlife Service of the U.S. Department of Interior, state departments of agriculture, and country predator control districts are useful in helping to control losses. However, in spite of efforts by these agencies and efforts by sheep producers themselves, losses to predators are still one of the most serious problems.

The inability to control predation adequately is a source of frustration and has caused many good operators to quit the sheep business. The whole matter is highly political and is usually a highly emotional issue. Thus decisions by government agencies are often influenced more by emotion than by factual information. Research has been of help in providing information, but cannot overcome the emotional and political aspects of the situation. For further reading, see Wade[10] and Wade and Bowns.[11]

Starvation

Many lambs starve because of neglect or desertion by their mother, or because of insufficient milk due to improper nutrition or diseases in ewes. The problem is much more common in multiple births than in singles. This cause of baby lamb loss can be reduced markedly by skilled operators in the lambing shed. In cases where milk is insufficient, one of a pair, or even singles when necessary, can be transferred to another ewe with plenty of milk. There are many successful methods of encouraging adoption. Smearing milk from the ewe on the lamb she is expected to adopt, or smearing her afterbirth, or using perfume or commercially prepared substances will work. Dipping the lamb in warm salty water also works. Generally, transfers are easy in cases of ewes lambing at or near the same time. In pasture lambing, a useful method of preventing a ewe from deserting one of a pair of twins is to tie the lambs together with a 3- to 4 ft length of cord attached to one front leg of each lamb. Within a day, or if lambs are "mothered up" earlier, the cord is removed. Sometimes it becomes necessary to confine a ewe to a stanchion (Figure 11-20) to get her to allow her lamb or lambs to nurse. As a last resort, artificial rearing can be a method of reducing losses due to starvation; it is discussed more completely in Chapter 13.

Stillbirth

Everyone in the livestock business is faced with and perplexed by the problem of apparently normal lambs being dead at birth. Dystocia (difficulty in lambing or delayed lambing) cannot help but be of major concern when assigning causes for stillbirth. This, as much as for any reason, is the purpose of having experienced help in lambing sheds. Some of the difficulties that are most commonly encountered have been previously discussed in the section dealing with dystocia and sheep obstetrics.

Unknown

In every study or survey of lamb losses, there are always some lamb deaths for which the cause is unknown. This varies due to experience and to availability of

Figure 11-20 Stanchion for restraining ewe. (University of Wyoming photo)

veterinary or diagnostic laboratory services. It is important to do the best possible job of diagnosing causes of loss in order to be successful in reducing losses.

IDENTIFICATION

Marking lambs for identification is an essential part of purebred operations and is encouraged for everyone wherever possible. Without proper identification, improvement by selection is impossible. Many lambs are placed in "lambing jugs" with their mother merely until the identification process is completed. Particularly for healthy single lambs, there is no need for them to stay confined to small pens for more than a few hours. If there is a scarcity of pens, they need to be used for ewes or lambs with difficulties or for twin pairs. It is helpful for any future observations to mark the ewe and her lambs with the same number (Figure 11-21). Some put these numbers on the shoulder for a ewe with a single lamb and on the rump for ewes with twin lambs. Sometimes the number is branded on the side of both ewes and lambs, and in cases of twins, a 2 is branded on the other side. The ability to identify and keep records and use them is one of the advantages of lambing in confinement as compared to range lambing.

FEEDING EWES AT LAMBING TIME

For ewes that have their lambs in sheds, it is a good idea to provide them with good quality hay and plenty of fresh water, and to start feeding grain after about 12 hours following lambing. Ewes require a little time to recover from the stress of parturition, and it normally will take several days for them to reach their maximum milk production or for their lamb or lambs to consume large quantities of milk. Therefore, the increase in nutrient requirements as recommended for lactating does not

Figure 11-21 Ewe and lamb branded alike for quick identification. (University of Wyoming photo)

occur suddenly at lambing time, but is more gradual, and feed allowances need to follow the same pattern. Full feeding can be resumed after about 1 or 2 days following parturition.

For ewes lambing on fresh grass pastures, this type of feeding regimen is readily accomplished. As ewes recover from the stress of parturition and milk flow begins to increase, they tend to increase their own feed intake and can provide for high nutrient requirements due to the high palatability of grass. This requires good range management and proper timing of lambing to assure that plenty of grass is available.

FEEDING NEWBORN LAMBS

When care is being provided for ewes at lambing time, the attendant will strip out both teats when a ewe has her lamb to be sure milk is easily available for the lambs. Each lamb needs to get colostrum. Most lambs will find the teats and start nursing without assistance, but in any flock there are likely to be some that require assistance. Quite often these are lambs born somewhat prematurely and weak or are lambs that have been stressed during the birth process due to dystocia. Lambs should be checked twice daily while in jugs to make sure they are nursing. Whether lambs are able to get started on their own or are assisted by an attendant, it is of help to have ewes recently shorn and free of tags around their flanks or udders.

REFERENCES CITED

1. Dennis, S. M. 1970. Prenatal lamb mortality in a purebred Southdown flock. *J. Animal Sci.* 31:76.
2. Faulkner, E. K., and James R. Tigner. 1977. Birth rates of sheep from range operations in Carbon County, Wyoming. *University of Wyoming Agr. Extension Bulletin B-643.*
3. Gates, N. L. 1984. *A Practical Guide to Sheep Disease Management.* Washington State University, Pullman.
4. Groverman, Frederick A. 1985. Pregnancy toxemia or twin lamb disease. *Natl. Wool Grower,* April, p. 13.
5. Jensen, Rue, and Brinton L. Swift. 1982. *Diseases of Sheep.* Lea and Febiger, Philadelphia.
6. Kirk, John H., and Bruce C. Anderson. 1981. Reducing lamb losses. *Natl. Wool Grower.* October 1981, p. 14.
7. Safford, J. W., and A. S. Hoversland. 1960. A study of lamb mortality in a western range flock. Autopsy findings on 1051 lambs. *J. Animal Sci.* 19:265.
8. Tigner, James R., and Gary E. Larson. 1977. Sheep losses on selected ranches in Southern Wyoming. *J. Range Management* 30:244.

9. Vetter, R. L., H. W. Norton, and U. S. Garrigus. 1960. A study of pre-weaning death losses in lambs. *J. Animal Sci.* 19:616.
10. Wade, Dale A. 1982. *Impacts, Incidence and Control of Predation on Livestock in the U.S. with Particular Reference to Predation by Coyotes.* Council for Agricultural Science and Technology. Special Publication No. 10.
11. Wade, Dale, and James E. Bowns. 1983. Procedures for evaluating predation on livestock and wildlife. *Texas Agr. Exp. Sta. B-1429.*
12. Welch, James A. 1960. Lamb deaths. *West Virginia Agr. Exp. Sta. Bulletin 452.*
13. *Wyoming Agricultural Statistics.* 1985. Wyoming Crop and Livestock Reporting Service, Cheyenne.

SUGGESTED READING

Collins, Spellman B. 1956. *Profitable Sheep.* Macmillan, New York.

Ensminger, M. E. 1970. *Sheep and Wool Science.* Interstate Printers and Publishers, Danville, Ill.

12 Management During Lactation

MANAGEMENT AFTER LAMBING

The first few days following lambing are the most critical as far as lamb survival is concerned, but the entire lactation period is critical in terms of growth rate and efficiency of lambs. Also, lactation is the period of greatest stress to ewes. For operators who use sheds for lambing, it is customary to be constantly moving sheep into and out of the shed lambing area. Ewes with their lambs are usually confined to individual "jugs" for only a day or two, and often are then combined with other ewes and lambs into mixing pens (see Figures 12-1 and 12-2). These groups are of various sizes, for instance, 10 to 20 ewes with their lambs, but this depends on the size of flock, size of pens, and sometimes on weather. In small flocks all the ewes

Figure 12-1 A handy method of moving lambs from lambing shed to mixing pens. (University of Wyoming photo, Courtesy of Warren Livestock Company, Cheyenne, Wyoming)

Figure 12-2 Ewes and lambs in group pen, their first move from lambing shed. (University of Wyoming photo, Courtesy of Warren Livestock Company, Cheyenne, Wyoming)

and lambs may be in the same mixing pens. Keeping ewes and their lambs apart in small groups after they leave the lambing shed allows for gradual adjustment to group conditions. Ewes and their lambs can learn to find each other, and observations can be made of any lambs that are not getting along as well as expected. It is much easier to notice if lambs have problems if they are in relatively small groups. If plenty of pens are available, any ewes and lambs that are problems can be kept in a separate group for a longer time. These would include weak lambs, cripples, triplets, and ewes with big teats or those that have undergone extreme stress or are poor milkers (see Figure 12-3) or those that do not claim their lamb.

Sometimes it is necessary to confine the ewe even further by tying or putting her in a stanchion if she does not allow her lamb or lambs to nurse. Also when trying to "graft" a lamb not her own onto a ewe with plenty of milk, it becomes easier if a stanchion is available (see Figure 11-20).

After a day or two in small mixing pens, groups can be combined into larger groups and moved to small pastures, such as seen in Figure 12-4, until they are

Figure 12-3 Ewe with distended udder and little milk. (University of Wyoming photo)

Management After Lambing

Figure 12-4 Ewes with lambs in pasture before being moved into range area. (Courtesy of Warren Livestock Company, Cheyenne, Wyoming)

combined into the final grazing flocks. These, too, are variable in size according to size of operation and grazing conditions. An example of a flock of ewes and lambs grazing on a good range area during late spring is shown in Figure 12-5.

When facilities allow, ewes nursing twins or triplets can be kept separate during the entire lactation period. These ewes can be allowed the best pastures or given extra feed and care, since they are the most productive. For ewes raising triplets or yearling ewes raising twins, stress on ewes can be reduced by weaning one lamb (4

Figure 12-5 Ewes and lambs grazing on summer range. (Courtesy of U.S. Forest Service)

to 5 weeks). Ewes raising more than one lamb produce more milk than do ewes raising single lambs, so feed requirements are higher. It is not only for the immediate nutrient needs for production that extra feed and care are required, but also to avoid letting ewes get too thin from the stress of lactation and thus depress their future production.

MILK PRODUCTION OF EWES

The lactation curve for ewes is shown in Figure 12-6. This represents an average and varies considerably by particular individuals or groups. Some ewes are persistent in their lactation pattern, maintaining a high level of milk production for as much as 4 months, but most ewes have a rapid decline in milk yield.

Ewe's milk is higher in nutrient content than cow's milk (see Table 12-1). The table is a comparison of averages. In both species there is a great variation in composition as well as volume due to breed, age, size, and condition. When substituting cow's milk for that of ewes, it is advisable to supplement it with added fat and protein. This is important in the preparation of milk or milk substitutes for use in artificial rearing of lambs.

FEEDING EWES DURING LACTATION

The trend in milk production, as shown in Figure 12-6, is the basis for the recommended allowances previously listed in Chapter 8 differing between early and late lactation. Another interpretation of nutrient requirements for ewes during early lactation is shown in Table 12-2. A study of this or of Table 8-1 reveals that a lot of

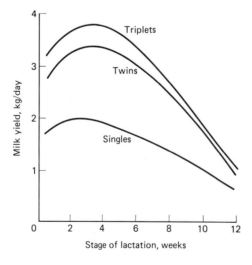

Figure 12-6 Smoothed milk yield curves for 70-kg ewes rearing single, twin and triplet lambs. (Courtesy of John Robinson, from Church[1], with permission)

Feeding Ewes During Lactation

TABLE 12-1 COMPARISON OF EWE'S MILK TO COW'S MILK

	Composition					
Species	DM (%)	DE (Mcal/kg)	TDN (%)	Crude Protein (%)	CA (%)	P (%)
Cattle	12	0.69	16[a]	3.3	0.12	0.09
Sheep	19	1.15	31	4.7	0.20	0.15

Adapted from National Academy of Sciences, *Nutrient Requirements of Sheep*.[3]

[a]National Academy of Sciences, *U.S. and Canadian Tables of Feed Composition,* 3rd ed.[7]

feed is required at this time. A ewe weighing 150 lb (68 kg) must eat about 7 lb (3.2 kg) per day of ordinary feed such as hay and grain to meet these requirements. For feeds low in dry matter, such as silage, this means about 20 lb (9.1 kg) of feed, which is virtually impossible for a ewe to eat. Thus, it is necessary to have some high-energy feed in the diet of lactating ewes. Since ewes are normally losing weight during lactation, care should be taken to feed according to the needs based on their normal weight as measured at breeding or during early gestation, rather than actual weight during the lactation period.

The length of the lactation period in ewes is variable. An often-heard comment is that greater efficiency should result from direct feeding of lambs than indirectly feeding ewes so they can feed their lambs. Such a comment ordinarily is meant to apply during late lactation or in dry lot, because of the trend seen previously in Figure 12-6. This leads to greater emphasis on early weaning of lambs, which is discussed further in Chapter 13.

TABLE 12-2 DAILY REQUIREMENTS FOR EWES OF DIFFERENT WEIGHTS DURING EARLY LACTATION, FOR EWES NURSING TWIN LAMBS

	Requirements					
Ewe Wt., lb (kg)[a]	DM, lb (kg)	DE, Mcal	TDN, lb (kg)	Crude Protein, lb (kg)	CA, g	P, g
110 (50.0)	5.3 (2.4)	6.9	3.40 (1.56)	0.86 (0.39)	10.5	7.3
120 (54.4)	5.5 (2.5)	7.1	3.54 (1.61)	0.87 (0.39)	10.6	7.5
130 (59.0)	5.7 (2.6)	7.3	3.67 (1.66)	0.89 (0.40)	10.7	7.7
140 (63.5)	5.9 (2.7)	7.6	3.81 (1.73)	0.90 (0.41)	10.8	7.9
150 (68.0)	6.1 (2.8)	7.9	3.95 (1.79)	0.92 (0.42)	11.0	8.1
160 (72.6)	6.3 (2.9)	8.2	4.08 (1.85)	0.93 (0.42)	11.1	8.3
170 (77.1)	6.5 (2.9)	8.4	4.22 (1.91)	0.95 (0.43)	11.2	8.5
180 (81.6)	6.7 (3.0)	8.7	4.35 (1.97)	0.97 (0.44)	11.3	8.7

Interpolated from "NRC Recommended Allowances," National Academy of Sciences, *Nutrient Requirements of Sheep*.[3]

[a]Weight of ewe following parturition.

ARTIFICIAL REARING

Nearly every operator is faced with the problem of "bum lambs" due to many causes. Some of these are high incidence of multiple birth, poor milking ability of ewes (no milk in some cases), death of ewes during or following parturition, and lambs being lured away from their mothers by "granny ewes." For years it has been common to raise lambs by bottle feeding, and this is still practiced, but artificial rearing has progressed beyond that and is now easier and more successful. Equipment is available for nurseries that enable the feeding of a few or of many lambs (see Figures 12-7 and 12-8). Also, milk replacers are available commercially or can be prepared that provide an adequate diet for lambs. The first step, according to Frederickson, Jordan, and Terrill,[2] is to understand that it can be done.

Milk replacer is the most important part of an artificial rearing program, and a high-quality milk replacer has the following properties:

1. It results in a minimum of digestive disturbances.
2. It has a high level of digestibility and utilization.
3. It is easy to mix.
4. The ingredients stay in suspension.

To help ensure success, a product with a good reputation should be used, even if it is expensive. Controlling costs can be more effective by limiting the time lambs are fed the high-priced feed. Before newborn lambs are started on any artificial diet, it is essential that they receive colostrum, either by nursing their mother or by use of a supply on hand. Either cow's or ewe's colostrum can be frozen to provide a source of supply. As soon as possible, lambs are started on solid feed such as a complete pellet so that they can be weaned from the milk replacer at an early age. For a complete discussion of artificial rearing and description of equipment, see references[2] and.[4]

Figure 12-7 Nursette used for artificial rearing of lambs. (University of Wyoming photo)

Figure 12-8 Large facility for artificial rearing of lambs. (Courtesy of U.S. Sheep Experiment Station, Dubois, Idaho)

CREEP FEEDING LAMBS

Providing supplemental feed to lambs during the nursing period, known as creep feeding, is desirable in many situations; its value depends on the production and management program. The use of creep feeders is common in farm flock production in which maximum early growth is desired. It is often a part of purebred operations and of spring lamb production, and is usually part of any operation that includes early weaning.

A creep feeder can be easily constructed and placed in a pen or pasture (see Figure 12-9). Lambs will start eating solid feed at about 10 days to 2 weeks of age and often are seen eating alongside their mothers when feed is provided. They do not eat very much at first, but will increase their feed intake rapidly as long as the feed available in the creep area is palatable. Examples of mixtures for creep feeding are shown in Table 12-3. These relatively simple diets are fairly high in protein and are palatable. Grinding the feeds helps to ensure good mixtures that provide

Figure 12-9 Creep feeder for young lambs. (University of Wyoming photo)

TABLE 12-3 EXAMPLES OF DIETS FOR CREEP FEEDING LAMBS

Ingredient	Diet			
	1 (%)	2 (%)	3[a] (%)	4 (%)
Corn	80	75	55	55
Oats	10	10	5	25
Oil meal	10	10		15
Molasses		5		5
Alfalfa pellets			40	
Alfalfa hay[a]	Free choice	Free choice		
Antibiotic, 40 g/t	+	+	+	+

Adapted with permission from *Sheepman's Production Handbook*[(5)]

[a] A complete ration can also be made by grinding and mixing alfalfa hay instead of alfalfa pellets as suggested in diet 3.

uniform consumption of all ingredients. Milo grain or barley may be substituted for all or part of the corn, and wheat can be used to replace up to half of the corn. Lambs will eat an average of 1.5 lb (0.7 kg) per day from 10 days to 120 days of age, ranging from only a little at first up to about 3 lb (1.4 kg) when near 120 days of age. Lambs will eat more and gain faster if the feed is pelleted, but pelleting is expensive and not essential for good results. Mixing an antibiotic or coccidiostat into creep diets is an option that can provide protection against disease, and this is an easy method of administration. Care needs to be taken to assure that the proper amount is used and that it is thoroughly mixed.

DOCKING AND CASTRATING LAMBS

Docking (removing lambs tails) is done mainly for purposes of cleanliness and to facilitate breeding of ewes. Castrating (removal of testicles in male lambs) is done mostly for a marketing advantage, but also is helpful in making management simpler. The process of castration does have a detrimental effect on growth, as intact male lambs grow 5% to 7% faster and are more efficient than are wether lambs (males that are castrated when young). Keeping ram lambs intact requires extra care in that the lambs must be weaned before reaching sexual maturity, and after weaning they must be kept separate from ewe lambs. Although some producers have been successful in marketing ram lambs, it is customary to receive price reductions for rams. For further discussion as to the effects of castration on carcass quality, see Chapters 16 and 17.

Both docking of tails and castration of male lambs are normally done at the same time, along with vaccination and branding, and the whole process is referred to simply as docking. The best age for docking is 4 to 7 days, but this does not fit into many management systems. Several methods have been used for both removal

Docking and Castrating Lambs

of tails and castration. A tool that is commonly used is an "all in one" instrument (see Figure 12-10) for cutting off the lower part of the scrotum, for clamping the testicles and pulling them out, for cutting off the tail, and in some cases for ear notching or ear marking. Any type of cutting instrument can be used to cut off the scrotum, but if they do not have the clamp for pulling out testicles, the docker may use disinfected pliers or fingers or teeth. When removing the testicles, it is wise to clamp the fingers of the other hand over the spermatic cord and press against the abdominal wall while pulling the testicles, to avoid any chance of rupture. Also, sanitation should be used at all times. If a knife is used, it is not advisable to slit the scrotum and remove testicles with the scrotum left intact as is often done with calves, because the wool causes matting and prevents proper drainage of the wound. Depending on the management system, it may be necessary to use insect repellent on the open wounds, and sometimes a commercial "blood stop" is used, particularly on the tail stub to prevent excess bleeding.

Elastrator bands applied with an elastrator (see Figure 12-10) are used as a bloodless method of both docking and castration. This has an additional advantage in that one person can do both without any additional help to hold the lamb. However, another instrument is needed if lambs are to be earmarked for purposes of sex or age identification. Also, some producers avoid the use of elastrator bands because they cause more danger of tetanus and pain to lambs than does cutting, but studies of different methods do not indicate any permanent effect.[6]

An effective tool for use in docking tails is an emasculator (see Figure 12-10). However, when this is used, another tool needs to be used for castration, if both are to be done at the same time. This is relatively bloodless if care is used to keep the crushing edge of the emasculator toward the lamb. The use of burdizzos is a method for crushing the spermatic cord without cutting the skin. It requires care in its use to be effective and may result in incomplete castration. Hot irons or hot pincers are sometimes used to sear off the tail without blood loss. This leaves a wound that is slow to heal and again would require the use of two tools and two people when castration is done at the same time.

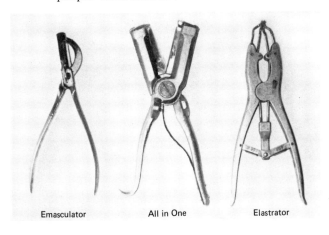

Emasculator   All in One   Elastrator

Figure 12-10 Tools used for docking and castration. (University of Wyoming photo)

Vaccinations are commonly included as part of the docking operation to avoid extra handling of the lambs. Sometimes lambs are vaccinated against enterotoxemia at this time to avoid early losses even though this does not provide permanent immunity. Some are also routinely vaccinated against soremouth at this time.

REFERENCES CITED

1. Church, D. C. 1984. *Livestock Feeds and Feeding,* 2nd ed. O & B Books, Corvallis, Ore.
2. Frederickson, K. R., R. M. Jordan, and C. E. Terrill, 1980. Rearing lambs on milk replacer diets. *USDA Farmers' Bulletin 2270.*
3. National Research Council. 1985. *Nutrient Requirements of Sheep,* 6th rev. ed. National Academy Press, Washington, D.C.
4. Nicholson, J. W. G. 1982. *Artificial Rearing of Young Lambs.* Publication 1307, Minister of Supply and Services, Ottawa, Canada.
5. Scott, George E. 1975. *Sheepman's Production Handbook,* 2nd ed. Sheep Industry Development Program, Denver. Colo.
6. Terrill, Clair E., and John A. Stoehr. 1950. Comparison of rubber rings with cutting for docking and castrating. *Natl. Wool Grower* 40 (3):23.
7. National Academy of Sciences. *U.S. and Canadian Tables of Feed Composition,* 3rd ed. 1982. National Academy Press, Washington, D. C.

13 Weaning and Postweaning Management

WEANING TIME

With traditional pasture management systems, it is common to wean lambs when they are 4 to 5 months of age, all lambs within a band or pasture group being weaned at the same time. For operators that lamb early, weaning time is during the spring and early summer, and for those involved in spring lambing operations, lambs are weaned in the fall.

The process of weaning is simple; it is simply the separation of lambs from their mothers. However, there are often other things that go along with weaning that make it one of the major chores in year-round management. It is usually the other things going on at or near weaning that dictate when it is done.

First, this is the end of lactation for ewes and some care in management is required to avoid udder problems in ewes. Their feed supply should be reduced for a few days to help stop the flow of milk. Some producers practice a gradual weaning of lambs, allowing lambs to nurse once or twice a day for a few days, to help reduce stress from distended udders. This requires a lot of labor and is not justified in large-scale operations. If possible, ewes should be moved to relatively low quality pastures that are beyond hearing distance of their lambs.

Weaning time is also the usual time to sort and cull ewes for various reasons. It is common to check ewes for age, general appearance, teeth soundness, and udder defects. Old ewes or ewes with defects can be sorted out and disposed of. Also, any ewes previously designated because of some defect in their offspring or for being dry can be sorted off at the same time. This is the best time to get a measure of ewes' producing ability, and weighing of lambs at the time they are weaned is a routine practice for many sheep producers. Weaning weight is used as a tool for selection in many instances. Total weight of lamb weaned is a true measure of production and can be used for culling of ewes as well as selection of replacement lambs. Care should be taken to avoid discrimination against twins in cases where sheep are not identified. Also, care should be used in observations of the general appearance of ewes that dry ewes or low producers are not favored over ewes that are most productive and tend to be in relatively poor condition. Ewes retained for further breeding will be somewhat synchronized for their next breeding as a result of their lambs having been weaned at the same time.

Weaning time is the beginning of a new way of life for lambs and generally results in some stress. Leaving the lambs in the pen or pasture where they have been, with plenty of palatable feed and water available, and removing the ewes is helpful to reduce stress from weaning. Usually, additional stress results from extra handling, since this is a common time to weigh and to decide which lambs are to be kept and which are to be sold or finished for market. Lambs shipped to market or to other owners undergo still further stress. Careful handling of lambs and weaning when weather is pleasant can help in reduction of stress, but, unfortunately, weaning dates are often predetermined.

EARLY WEANING

Observing the lactation curve shown in Figure 12-6 suggests that weaning at a younger age than the traditional 4 to 5 months might be of benefit for improving efficiency. During the latter part of lactation, it is likely to be more efficient to use high-quality feeds directly for lambs than to feed them to ewes with low levels of lactation. There are several ways that operators can gain by early weaning of lambs.[4] Pastures of poor quality can be utilized without any worry about maintaining a high level of lactation in ewes. Ewes can be combined into larger groups if pastures are arranged to warrant such a change. There is more flexibility to management of ewes after lambs have been weaned. Benefits for the lambs result from greater efficiency at younger ages. Lambs that are weaned and kept in confinement are safer, with less danger from predation, poisonous plants, or inclement weather. Early weaned lambs can utilize good pasture and perform about as well as if in confinement on dry feed, but the availability of good pasture is dependent on time of year and land holdings. Some good pasture needs to be available for the next breeding season.

The purpose in early weaning is to increase efficiency of the operation by more direct use of harvested feeds or by taking advantage of early efficiency of lambs. It is not to improve performance.

Lambs can be weaned as early as 1 month, but cannot be expected to adapt as readily as when weaning occurs at 2 or 3 months. Likewise, special care is needed to provide a suitable diet for those weaned early. Weaning at 30 days or less is a logical part of accelerated programs, particularly if ewes are bred to lamb twice yearly. In ordinary situations, early weaning age is more likely to be in the range of 60 to 90 days, and apparently anywhere within this range is satisfactory. Lambs will perform as well as if on the ewes.

FEEDING LAMBS FOR REPLACEMENTS

Ewe lambs that are to be bred to lamb first at about 1 year of age require special treatment if success is to be achieved. Following weaning, the level of nutrition must remain high to assure adequate growth in order that the lambs will reach puberty at an early age. Lambs that are not grown out properly are not likely to exhibit estrus at the proper age.[5,9] This method is used to increase the lifetime productivity of ewes, but it requires more effort in management and feeding than if ewes are handled more traditionally.

Ewe lambs that are to be bred first at about 18 to 19 months of age to produce their first lambs when they are 2 years of age also need good care and nutrition, even though requirements are not as high as for the former case. It is not unusual in range flocks for ewe lambs to weigh about the same at shearing time just before they are a year of age as they previously did at weaning time. This means they have actually lost weight, since fleece weight has increased. Such ewes have a difficult time reaching mature size and also likely will not be at their peak of ovulation rate by breeding time. Special care can be given when ewe lambs are wintered on dryland pastures, such as supplemental feeding, for example, feeding 0.5 lb (0.22 kg) per day of a high-energy feed or emergency feeding when pastures are snow covered. Some ewe lambs are moved to farm areas and grazed on crop aftermath or pastures or even fed forage diets in confinement. Requirements as recommended by the NRC (Table 8-1) are easy to meet except when these ewe lambs are grazing on mature and weathered grasses during the winter. Ewes that do not reach sexual maturity by breeding time, no matter at what age first breeding occurs, will have reduced productivity during their first lambing as well as during their lifetime.

Ram lambs being kept as replacements or raised by breeders for future sale should have special care to assure adequate growth and sexual development before being used for breeding. NRC recommendations[6] are somewhat incomplete for growth above 220 lb (100 kg). Performance test information from Texas and Wyoming tests indicates that rams will continue to gain at the rate of 0.7 lb (0.33 kg) per day until they are over a year of age and weigh in excess of 250 lb (119 kg). Generally, the rams on these tests are of breeds known for large size, Rambouillets being by far the most numerous. To do this requires about 6 to 7 lb (2.7 to 3.3 kg) of feed per day, and when feeding a complete pelleted diet, this is readily achieved. It has not been established just how long such a feeding program should continue for the best breeding performance of rams.

FINISHING LAMBS

The majority of range lambs are still in the feeder category when weaned. Likewise, many farm flock lambs on pasture that have not been creep fed or that are weaned early are feeder lambs. The phase of bringing lambs from feeder to slaughter weight or slaughter grade is the easiest phase of the entire sheep operation. Whether lambs are placed directly in feedlots or are on a deferred feeding program before going into the feedlot for rapid finishing, less labor and time are involved than in the care and management of breeding flocks. With proper equipment, thousands of lambs can be fed and handled by one person (see Figures 13-1 and 13-2).

Feedlots vary from open lots with no cover to those in complete confinement on slotted floors, with many types of partial confinement in between. Sheds, windbreaks, or shades are provided in some cases for protection depending on climate or time of year. The majority of lamb feeding is in the West, though lamb feeding is at a low level as compared to past years. In Figures 14-2 and 14-3, a general view of numbers and location of lamb feeding can be seen. Equipment used for feeding is also extremely variable, ranging from automated feeders to fence line feeders to the use of grain and hay troughs for hand feeding. In most cases, lamb feeding is seasonal simply because of the seasonal nature of lamb production. In some cases, though, lambs are fed nearly year-round.

Traditionally, lamb feeding is a more risky operation than is the production operation, but both suffer from fluctuations in market price as well as in feed prices. From the viewpoint of several large operators that are totally integrated, the long-term trend is no different for the finishing phase than for the production phase.

Rations for Feeder Lambs

There is no best diet for finishing lambs. The diet that produces the fastest gain may not be as economical as another diet due to extra processing costs or due to equipment used for feeding. Deferred programs that use pastures for grazing or utilize various types of crop aftermath are often more economical in cost per unit of gain than are fast finishing programs, even though rate of gain may be slower.[2]

Figure 13-1 Fence line of large lamb feedlot. (University of Wyoming photo, Courtesy of Harper Livestock, Eaton, Colorado)

Finishing Lambs

Figure 13-2 Two pens of lambs, shorn and unshorn, in large feedlot. (University of Wyoming photo, Courtesy of Harper Livestock, Eaton, Colorado)

Balancing or preparing diets to meet NRC recommended allowances (Table 13-1) for feeder lambs is a relatively simple process. Many feeds or combinations of feeds can be used successfully to meet the requirements. For example, the recommended levels of the various nutrients for a 75-lb (34-kg) lamb as shown in Table 13-2 can be provided by any combination of alfalfa hay and corn ranging from 20% hay and 80% corn to 80% hay and 20% corn. Likewise, other forage, such as different kinds of hay, silages, or haylages, can take the place of alfalfa hay. Other grains can take the place of corn either completely or in part and not change the levels of various nutrients provided by the diet (see Table 8-14). Either milo or barley can be substituted for all or a portion of the corn and will provide about 95% of the feed value. Wheat is equal to corn in feed value and can be used to replace up to half of the corn or other grain. Dried beet pulp (up to 30% of the total grain) is equal to corn in feed value.[3]

Lamb performance, both rate of gain and feed per unit of gain, tends to improve as the ratio of concentrate to roughage increases.[10,11] All-concentrate diets, such as 90% grain and 10% protein concentrate, with the addition of minerals and a lasalocid are being used in some feedlot operations.

Feed preparation also is quite flexible for lamb feeding, with almost any method acceptable.[1] Most rapid gains can be obtained by self-feeding a complete pelleted diet, but costs of preparation may outweigh the advantage of fast gain. Daily feeding of hay and grain separately has the advantages of low preparation cost and allows the operator to more easily observe progress of lambs and pick out problems, but it requires a lot of labor and achieves less than maximum gain. The decision must be on an individual basis, since all methods of feeding and preparation are satisfactory in some situations. If pelleted diets are fed for extended periods of time, lambs tend to develop a depraved appetite and will eat fences or wooden

TABLE 13-1 RECOMMENDED ALLOWANCES FOR FEEDING LAMBS

Body Weight		Weight Change/Day		Dry Matter per Animal[a]			Energy[b]				Nutrients per Animal					
							TDN		DE	ME	Crude Protein		Ca	P	Vitamin A Activity	Vitamin E Activity
(kg)	(lb)	(kg)	(lb)	(kg)	(lb)	(% body weight)	(kg)	(lb)	(Mcal)	(Mcal)	(g)	(lb)	(g)	(g)	(IU)	(IU)
30	66	295	0.65	1.3	2.9	4.3	0.94	2.1	4.4	3.4	191	0.42	6.6	3.2	1410	20
40	88	275	0.60	1.6	3.5	4.0	1.22	2.7	5.4	4.4	185	0.41	6.6	3.3	1880	24
50	110	205	0.45	1.6	3.5	3.2	1.24	2.7	5.4	4.4	160	0.35	5.6	3.0	2350	24
Early weaned lambs: moderate growth potential[c]																
10	22	200	0.44	0.5	1.1	5.0	0.40	0.9	1.8	1.4	127	0.38	4.0	1.9	470	10
20	44	250	0.55	1.0	2.2	5.0	0.80	1.8	3.5	2.9	167	0.37	5.4	2.5	940	20
30	66	300	0.66	1.3	2.9	4.3	1.00	2.2	4.4	3.6	191	0.42	6.7	3.2	1410	20
40	88	345	0.76	1.5	3.3	3.8	1.16	2.6	5.1	4.2	202	0.44	7.7	3.9	1880	22
50	110	300	0.66	1.5	3.3	3.0	1.16	2.6	5.1	4.2	181	0.40	7.0	3.8	2350	22
Early weaned lambs: rapid growth potential[c]																
10	22	250	0.55	0.6	1.3	6.0	0.48	1.1	2.1	1.7	157	0.35	4.9	2.2	470	12
20	44	300	0.66	1.2	2.6	6.0	0.92	2.0	4.0	3.3	205	0.45	6.5	2.9	940	24
30	66	325	0.72	1.4	3.1	4.7	1.10	2.1	4.8	4.0	216	0.48	7.2	3.4	1410	21
40	88	400	0.88	1.5	3.3	3.8	1.14	2.5	5.0	4.1	234	0.51	8.6	4.3	1880	22
50	140	425	0.94	1.7	3.7	3.4	1.29	2.8	5.7	4.7	240	0.53	9.4	4.8	2350	25
60	132	350	0.77	1.7	3.7	2.8	1.29	2.8	5.7	4.7	240	0.53	8.2	4.5	2820	25

Reprinted with permission from National Academy of Science, Washington, D.C.[6]

[a] To convert dry matter to an as-fed basis, divide dry matter values by the percentage of dry matter in the particular feed.
[b] One kilogram TDN (total digestible nutrients) = 4.4 Mcal DE (digestible energy); ME (metabolizable energy); = 82% of DE.
[c] Maximum weight gains expected.

TABLE 13-2 EXAMPLES OF MIXTURES FOR LAMB FINISHING RATIONS

Ingredient	Amount, lb (kg)	
1. Shelled corn	1175	(533)
Good-quality legume hay	800	(363)
Soybean meal	25	(11)
Plus antibiotic, 20 g	2000	(907)
2. Shelled corn	1000	(454)
Low-quality hay	800	(363)
Soybean meal	100	(45)
Molasses	100	(45)
Plus antibiotic, 20 g	2000	(907)
3. Shelled corn	540	(245)
Corn silage	1350	(612)
Soybean meal	100	(45)
Limestone	10	(4.5)
Plus vitamin A, 1,000,000 IU	2000	
Plus vitamin D, 100,000 IU		
Plus antibiotic, 20 g		

Adapted from *Sheepman's Production Handbook*.[8]

feeders or even pick wool in extreme cases. For short-term feeding, this usually is not a serious problem. For lambs fed pellets over a long time period, 2 months or more, this can be prevented by simply providing some kind of forage, such as 0.25 lb (0.11 kg) per day of good hay.

Examples of lamb feeding diets are shown in Table 13-2. Regardless of feed, feed preparation, or method of feeding, some precautions need to be taken to prevent problems such as enterotoxemia, urinary calculi, and parasites. Vaccination and drenching are customary management operations at the beginning of feeding. Lambs should be handled carefully to avoid stress as much as possible when being started on feed. A gradual change from a high-forage or maintenance diet to full feed is advisable, particularly when high-energy diets are fed.

Growth stimulants in the form of implants or feed additives[7,11] are effective for improving lamb performance in feedlots. The difference between lambs that are treated and control lambs is often not very large, but generally there is some benefit from the use of growth stimulants.

REFERENCES CITED

1. Botkin, M. P., C. J. Kercher, and Leon Paules. 1965. Preparation of hay and grain for lamb fattening rations. *Wyoming Agr. Exp. Sta. Bulletin 429.*
2. Botkin, M. P., and others. 1976. Use of crop aftermath or harvested forages for "warming up" lambs prior to finishing in feedlots. *Wyoming Agr. Exp. Sta. Research Journal 3.*

3. Church, D. C. 1984. Livestock Feeds and Feeding, 2nd ed. O and B Books, Corvallis, Ore.
4. Dahmen, Joe, and Ed Duren. 1977. Early weaning of lambs. *Idaho Current Information Series 418.*
5. Hulet, C. V., E. L. Wiggins, and S. K. Ercanbrack. 1969. Estrus in range lambs and its relationship to lifetime reproductive performance. *J. Animal Sci.* 28:246.
6. National Research Council. 1985. *Nutrient Requirements of Sheep,* 6th rev. ed. National Academy Press, Washington, D.C.
7. Nockels, C. F., D. W. Jackson, and B. W. Berry. 1978. Optimum level of monensin for fattening lambs. *J. Animal Sci.* 47:788.
8. Scott, George E. 1975. *Sheepman's Production Handbook,* revised ed. Sheep Industry Development Program. Denver, Colo.
9. Southam, E. R., C. V. Hulet, and M. P. Botkin. 1971. Factors influencing reproduction in ewe lambs. *J. Animal Sci.* 33:1282.
10. Thomas, V. M., and J. J. Dahmen. 1986. Influence of roughage to concentrate rations, Bovatec and feed processing on lamb feedlot performances and carcass characteristics. *S.I.D. Research Digest* Vol. 2, No. 2, p. 11.
11. Thompson, W. R., K. K. Bolsen, and H. J. Ilg. 1982. Mixtures of corn grain and corn silage, nitrogen source and Zeranol for feeder lambs. *J. Animal Sci.* 55:211.

14 Marketing

A crystal ball is not required to predict the fate of an industry left without a market for its product. A reliable and consistent market is basic to any industry. The survival of the sheep industry must stem from the profitable sale of lamb and wool. Howard Wyman of the National Lamb Feeders Association makes the following comments about lamb marketing.[6] These comments which follow will serve as an introduction to this chapter and chapters 15, 16, and 17:

> What will lamb marketing be like in the future? To start with, many lamb feeders will no longer be in business. High prices for feed grain and high interest rates will force many from the lamb feeder ranks. But these and other factors will also change the way all lambs are produced and marketed.
>
> World demand for grains will increase in the coming years and as demand climbs, so will grain prices. Oh, we hear of surpluses now, but we also hear of people starving. I suspect if the world-wide grain distribution system were better, we'd find the demand just about matches supply, right? And the world's population is still climbing. This, along with rising energy, money and equipment costs, will force grain production costs up despite improved technology. As a result, we'll see more marginal land currently producing feed grains converted to forage production, which means lambs will be fed on grass longer.
>
> Because of predators and the pressures of environmentalists, traditional ranching operations will be abandoned. Instead, farm flock production will increase, and more of our lamb supply will come from part-time producers and operators diversifying into sheep production.
>
> In the past few years, we've seen packers going out of business, moving their plants, closing some of their plants and just plain teetering on the verge of bankruptcy. [See declining supplies in Figure 14-1.] To overcome the problem, we need smaller,

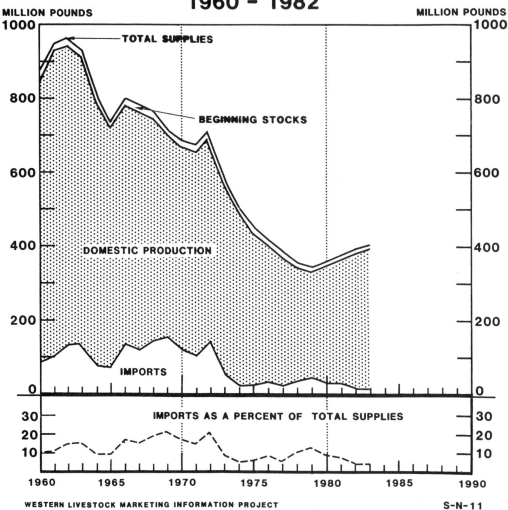

Figure 14-1 Total lamb and mutton supplies, 1960–1982. (Courtesy of Western Livestock Marketing Information Project)

more efficient, cooperatively owned packing plants. These plants will have to be geographically located to move lambs to the plant with minimal transportation cost. They could process consumer-ready meat, and with adequate storage facilities, sell lamb at the most profitable times. Cooperatively owned plants would also encourage producers to promote their own product, something that is needed even today.

Geographics have always played an important role in determining where agricultural production takes place. But it will be even more important for the sheep industry in the next few years. Mother Nature has taken her toll on range operators and feedlot owners who can provide little protection from the elements for their animals. This means not only high death losses, but problems marketing muddy lambs and wool. Many cornbelt farms now have idle lots, barns, silos and facilities that will be used by lamb feeders in the future. So we will see the trend towards more lamb production in the Midwest continuing.

Electronic marketing will become prominent. And we will see the computers used more for inventory control, cash flow and budgeting.

The consumer will see a wider variety of products available. Sausages, lunch meats and even restructured lamb products will be readily available. Most of these products will be prepared at the point of slaughter. New packaging techniques that extend shelf life will also be a part of the retail scene.

Many of these changes are already happening. The old method of pushing carcasses out the back door hoping there is a market for them—will go by the wayside.

FEEDER LAMBS

Sheep and lamb marketing is seasonal and this is particularly true of feeder lambs.[3] The majority of feeder lambs are produced in Western range states. Lambs destined to become feeders from the West are usually born on open range. Therefore, lambing occurs late enough in the spring to avoid much of the cold weather, and it allows for sufficient green forage to ensure good milk production from the ewes. In the Southwest, where the temperature is warmer, range lambs are born in late winter and early spring, while in some parts of the Rocky Mountain region, lambs are born in late spring.

Sixty-one percent of the lambs in the West sell as feeders because the growing season is too short or the feed supply is depleted too soon to produce the weight of lambs the packer demands. Packers encourage feeding because a healthy feeding industry means a more uniform supply of slaughter lambs. Lighter lambs coming from ranges often consume vast amounts of crop residues before entering feedlots at 6 to 10 months of age. The preferred feeder lamb weight has increased from 66 lb (30 kg) 20 years ago to approximately 88 lb (40 kg) at present. The increase in feeder lamb weights allows for production of 110- to 121-lb (50- to 55-kg) slaughter lambs without increasing time in the feedlot. Many Western feeder lambs enter feedlots close to slaughter facilities in states like Colorado, California, and Texas, while others are trucked to Midwestern feedlots (Figure 14–2). It is common for lambs coming from some management systems to be sorted into fats and feeders when they are weaned. Fat lambs usually weigh 100 lb (45 kg) or more and are sent directly to slaughter. The average feeder lamb producer usually ships all lambs at the same time when the feed runs out. However, a more orderly marketing system and higher prices can often be achieved by marketing the heaviest one-third of the feeder lambs a month earlier and the last one-third a month later than average. This system allows

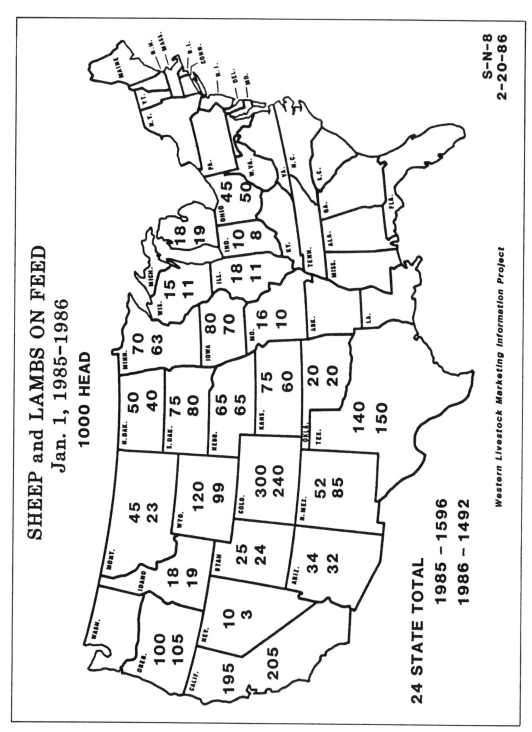

Figure 14-2 Sheep and lambs on feed, January 1, 1985–1986. (Courtesy of Western Livestock Marketing Information Project)

all lambs to be sold at more uniform weights, and it enhances the feedlot operators' chances of producing a more salable product.

Feeder lambs produced in the Midwest and in other parts of the country are often produced under strikingly different conditions than their Western counterparts. They are usually born into small farm flocks in late winter and are housed in heated buildings until they can withstand the cold. More and more of these lambs are early weaned at 60 to 90 days of age and fed high levels of concentrate in confinement until slaughter at 100 to 110 lb (45 to 50 kg). The majority of the feeder lambs produced in confinement are fed by the farmer who owns the ewes so the number of Midwestern-born feeder lambs that are marketed as feeders is minimal and the price is lower than for Western lambs with fewer parasite and disease problems. The early-weaned and confinement-fed lambs reach slaughter weight at 5 to 6 months of age in late spring or early summer before the glut of Western lambs reach the market, thereby smoothing the marketing pattern and easing the price fluctuation that would occur if all lambs were marketed in the fall.

Classes of Feeder Lambs

Feeder lambs are described according to management system, breed, and geographical location. Management systems include lambs that come straight from open ranges, mountains, crop residues, and improved legume or grass pastures. The lambs may have been weaned for several months or recently separated from the ewes.

Individual breeds may be specified or the lambs may be described simply as Western white faced, black × white-faced crosses, or meat breeds. A high percentage of the feeder lambs are shorn before entering the feedlot. Some feeder lambs are preconditioned before they leave the producer's property. This involves starting on feed, vaccination, and under some conditions drenching and/or shearing. Preconditioning helps minimize the considerable stress some lambs undergo when they are gathered, sorted, weaned, weighed, and shipped.

Both the total lamb crop (Figure 1–9) and lambs on feed (Figure 14–3) have decreased over the last 20 years, but lambs on feed have decreased at a slower rate than the total numbers. This a reflection of the greater demand for heavier slaughter lambs. Lambs often need to be fed to reach the weights packers prefer. Packers prefer 110- to 121-lb (50- to 55-kg) lambs because they have higher carcass yields and are more economical per unit weight to process.

Marketing Alternatives for Feeder Lambs

Feeder lamb marketing has become increasingly more difficult for producers not fortunate enough to be located near lamb feeding areas. As a result, producers often retain ownership of their lambs and contract with a lamb feeder to feed the lambs

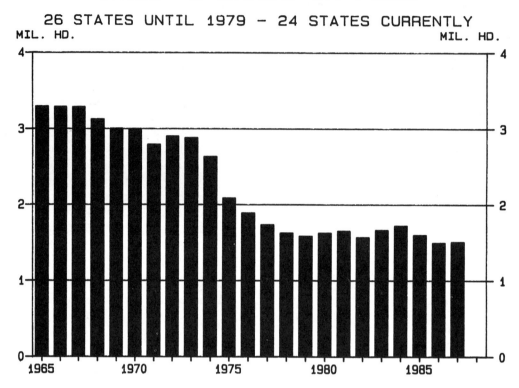

Figure 14-3 Sheep and lambs on feed, January 1, 1963–1985. (Courtesy of Western Livestock Marketing Information Project)

to slaughter weights. This eliminates the need to market feeder lambs and allows marketing of slaughter lambs where demand and profit in some geographic regions are better. Alternatives for marketing feeder lambs include direct from producer to feeder and indirect through order buyers, dealers, terminal markets, auction sales, special sales, and electronic marketing systems.

Producer to Feeder

Direct negotiation between feeder lamb producer and feedlot management is the most common method used to market feeder lambs where relatively large numbers are involved. The feedlot operator is often aware of producers who will market lambs as feeders. In turn, producers contact feedlot operators. Direct negotiation may result in immediate delivery of lambs or in forward contracting. Forward contracting of lambs 30 to 60 days ahead of the delivery date is utilized to ensure a supply for the feedlot and to reduce selling risks. The price may be established at the time of the contract or may be based on some specified market at the time of delivery.

Direct shipment of feeder lambs from producer to feeder reduces the stress of handling, shrink, and exposure to other lambs. To minimize trucking costs, it is best to sell at least a semiload. Small producers can sell to feedlots directly by pooling their lambs after sorting to ensure uniformity. The lambs are usually purchased at an agreed-on price based on an overnight stand in a dry lot without feed and water or on weight at the nearest inspected scale less a pencil shrink of 3% to 4%. The price is negotiated based on the producers' and feeders' knowledge of the market. In this situation the feedlot operator usually has the advantage because of his greater experience and knowledge of the market. Producers seldom take advantage of all the sources of price information that are available to them. In addition to the local paper, farm publications, and farm reports on radio, producers should make greater use of the Sheep Industry Development Program reports and of USDA reporters in their area. Other sources of price information are feedlot operators, auction markets, agents, commission men, dealers, order buyers, and university extension specialists. The producer should be aware that weighing conditions and shrink in addition to price are important when price quotations are compared.

Producer to Feeder through Order Buyers and Dealers

It is common for many feedlots, particularly those in the Midwest, to obtain many of their lambs through order buyers who purchase lambs for feedlot operators based on their specifications. An order buyer acts as an agent for the lamb feeder by locating and purchasing lambs for a specified type within a designated price range. The feedlot operator pays the producer for the lambs directly and pays the order buyer a commission.

A dealer buys lambs and then resells them to a feedlot operator. The dealer assumes the risk of price decline, shrink, trucking, and death loss. Reputable order buyers and dealers can benefit both producer and feeder. Bonded and licensed dealers as well as order buyers often advertise their services in sheep magazines and in other farm publications.

Other Methods of Marketing Feeder Lambs

Feeder lambs are also marketed through terminal markets, auction sales, special sales, and electronic marketing systems. These systems are described in more detail in the section on slaughter lambs. Private agents, marketing firms, and producers have not used electronic marketing systems extensively.

SLAUGHTER LAMBS

Seasonal slaughter by month is shown in Figure 14-4, and area of the country where lambs are slaughtered is shown in Figure 14-5. The bulk of the spring lambs slaughtered in March and April are from areas in California, Arizona, Texas, and the South where spring comes early. Spring slaughter lambs from the Cornbelt, the San Luis Valley of Colorado, Washington, Oregon, and Idaho follow these lambs to market. In August and September, lambs from several areas of the country are marketed, with the bulk coming from the Rocky Mountain region. October is traditionally the start of the fed lamb marketing, and the bulk of the slaughter lambs marketed from November through April are from feedlots in Colorado, California, Texas, the Cornbelt, and the Rocky Mountain region, including the Northwest and the Dakotas. Over 20% of all feedlot lambs come from Colorado. More than one-half of the lambs slaughtered in the United States are from large commercial feedlots.

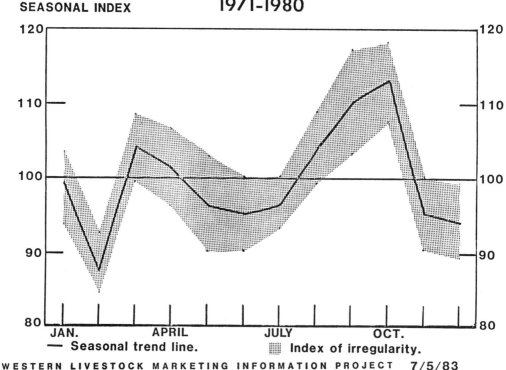

Figure 14-4 United States commercial sheep and lamb slaughter, 1971-1980. (Courtesy of Western Livestock Marketing Information Project)

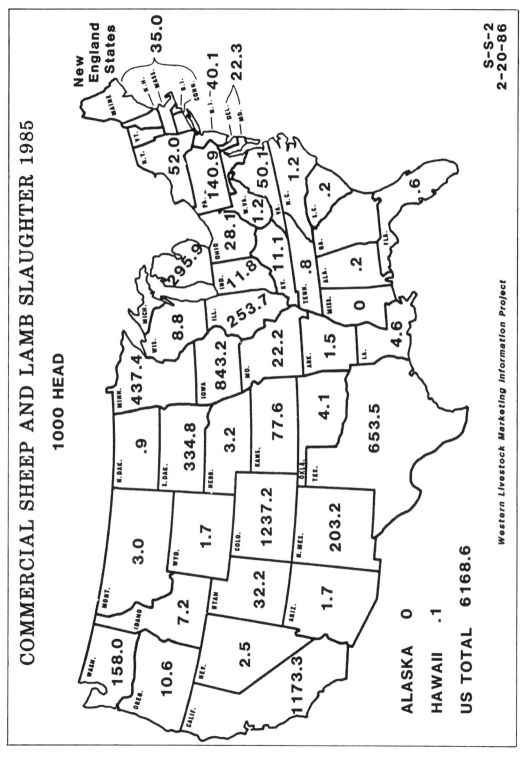

Figure 14-5 Commercial sheep and lamb slaughter, 1985. (Courtesy of Western Livestock Marketing Information Project)

Slaughter weights have increased at a rate of about 1.1 lb (0.5 kg) per year for the last 30 years. Lamb weights of 110 to 115 lb (50 to 52 kg) for the heavier domestic breeds are now readily accepted by the packing industry except in periods of abundant supply. Slaughter weights could continue to increase another 15% to 30% with further selection and management.

Price discounts for heavy lambs (those over 110 to 115 lb or 50 to 52 kg liveweight) reach a peak in February and March when many extremely heavy lambs are coming from the feedlots and the lighter, more attractive spring lambs begin to replace feedlot lambs. During the summer, when very few heavy feedlot lambs are marketed, price discounts for a smaller supply of heavy lambs are practically nonexistent.

Producers who receive the highest prices for lambs are those who produce the kind of lambs the packer wants. Producers should keep in mind that the packer wants the slaughter chain to run smoothly. Anything that slows down or stops the line is a problem. Overall, the packer wants to get the lamb from the front door to the back door as fast as possible, and any lamb that is likely to slow down this process will be discounted.

Packers object to lambs that have full GI tracts because they are harder to dress and the carcass yield is lower. On the other hand, lambs that have been fasted too long have pelts difficult to remove and they undergo tissue shrinkage. Ram lambs are objectionable because of lower dressing percentages, pelts that are harder to pull, additional time spent with testicle removal, and, in the case of older ram lambs, slightly lower quality grades. Shorn lambs, lambs with clean fleeces, and lambs that are free from tags are preferred because there is less chance of carcass contamination during dressing and the dressing percentage is increased. Lambs with tails and/or horns are a problem because they slow down dressing, they decrease the dressing percentage, and the tails collect manure.

The problems of difficult pelt removal will be partially taken care of as mechanical pelt pullers replace hand labor. Objections to lower dressing percentages can be met by selling lambs on a hot carcass weight basis and ignoring live weight. Producers should make every effort to market lambs that are free from mud, manure, and tags to increase profits to themselves and wholesomeness of lamb to the consumer. Excessively fat lambs are a problem to sell to the consumer as retail cuts. Fat lambs can be prevented by shorter feeding periods or by marketing the top end of the lambs at the proper finish. The consumer-preferred lamb discussed in Chapter 15 is not always preferred by producers and packers because under the present marketing system it does not make money for them. The producers' consumer is the packer and the packer wants fat lambs that have high dressing percentages. This could change as more lambs are cut and trimmed at the packer level.

Heavy lambs sometimes create marketing problems. The advantages and disadvantages of producing and marketing heavy lambs and ram lambs are discussed in Chapter 16.

Marketing Alternatives for Slaughter Lambs

Direct sales. Direct sales from producer or feeder to the packer is the most common method of marketing slaughter lambs, and forward contracting of lambs is common.[5] About 80% of all slaughter lambs are marketed direct, and practically all slaughter lambs in major marketing areas are sold direct to the packer. Other considerations regarding price information and weighing conditions discussed under direct sales of feeder lambs also apply to slaughter lambs.

In negotiating price with a packer buyer, the producer should make sure that each of the following points is covered:

1. Number of lambs to be sold
2. Weight ranges and grade of lambs to be delivered
3. Time, place, and method of weighing
4. When and where lambs will be delivered
5. Time of payment
6. Responsibility for transportation and insurance costs
7. Health guarantees
8. Age and sex of lambs
9. Uniformity of lambs

Discussing each of these points before a price for the lambs is agreed on will avoid surprises when the check from the packer is received.

A suggestion for improving procurement practices. The primary determinants of the animal's value are the estimated proportion of lean meat and the estimated quality thereof. In the sheep industry it is generally accepted that fatness, carcass weight, and quality grade are all positively associated with each other and that they are all negatively associated with cutability. Furthermore, carcass yield as a percentage of live weight (dressing percentage) is positively associated with the first three criteria and is negatively associated with cutability. Therefore, when the differences in value between yield grades of lamb are balanced against the contradictory effect of quality grade, carcass weight, and dressed yield, the price advantage for a higher cutability lamb with a more desirable yield grade disappears.

The example given in Table 14-1 explains why, when the packer buys on a live basis, he cannot afford to buy the type of lamb the consumer wants. In the example, the packer pays the same for yield grade 2 and 4 lambs weighing 110 lb (50 kg), but the yield grade 4 lamb has 2.2 lb (1 kg) more carcass and returns the packer $2.79 more. Cutability of the yield grade 2 lamb is higher when retail cuts from both carcasses are trimmed to contain the same amount of fat. Therefore, the yield grade

TABLE 14-1 EXAMPLE SHOWING WHY PACKERS PAY MORE FOR FATTER LAMBS

	Yield Grade	
	2 (leaner)	4 (fatter)
Producer		
Live weight	110 lb (50 kg)	110 lb (50 kg)
Live value:		
0.65¢/lb	$71.50	$71.50
Packer		
Dressing, %	50	52
Carcass weight	55.1 lb (25 kg)	57.3 lb (26 kg)
Carcass value:		
$1.27/lb	$69.98	$72.77
Retailer		
Cutability, %	76	68
Retail cut weight[a]	41.9 lb (19 kg)	37.5 lb (17 kg)
Retail cut value:		
$1.76/lb	$73.74	$66.00

[a]Bone-in retail cut weight if all cuts are trimmed to contain 3 mm of fat.

2 carcass should be worth more to the retailer. However, retailers seldom trim fat lambs of all excess fat, and they have not insisted that the packer supply carcasses that have been yield graded. As a result, yield grade 2 and 4 carcasses continue to sell at the same price, and excessively fat cuts from yield grade 4 lambs at retail counters and in white tablecloth restaurants continue to discourage consumers from making further lamb purchases.

The incentive to increase dressing percentage by producing excessively fat lambs would end if producers were paid on a carcass-weight basis instead of a live basis. Some other reasons for selling lamb on a hot-carcass-weight basis in place of selling live lambs are as follows:

1. Fat cover would decrease, making lamb more acceptable to the consumer.
2. Feed per kilogram of gain would decrease because lambs would go to market with less fat.
3. Arguments about weighing conditions and pencil shrink for live lambs would end. Distances that lambs are shipped before they are weighed, amount of moisture in the wool, hot or cold weather, and time off feed have very little influence on hot carcass weight. However, hot carcass weight does decrease after lambs have been off feed for more than 48 hours, and this is particularly true for younger milk-fed lambs.
4. Price fluctuations due to differences in dressing percentage would end, and quoted prices for hot carcasses would be more comparable than live prices.

5. Producers in leading lamb-producing countries like New Zealand have sold their lambs and cattle on a carcass-weight basis for years without ever being concerned about what the live weight was.

The reasons should be enough to make producers insist on being paid on a hot-carcass-weight basis. Producers need to understand that they are producing meat, not fill, mud, moisture, and other things that affect live weight. It is a simple procedure to sell carcasses. The hot carcass weight of each lamb is recorded as it comes off the kill floor. This scale is just as accurate as the one that weighs live lambs. It is the scale that determines dressing percentage under the live system of selling and the one that corrects errors in estimated yield figures.

When a producer or feeder sells carcasses, he or she can continue to bargain with the buyer on price but does it on a hot-carcass-weight basis. Producers or feeders can keep the lambs and get other bids just as they do now on a live-weight basis (Figure 14-6). When the producer gets the best bid on a hot carcass basis, the lambs are sold. Bargaining is done on a carcass basis because the carcass is the primary determinant of an animal's value. Almost all of the other components of an animal also have value, and the value often varies independently of the carcass value. Pelts can vary widely in value because of mud, seeds, cockle, needles, and color. Therefore, pelt credits should be specified when carcasses are sold. Pelt credit and other by-products often do not cover the packer's cost of slaughter.

It should be apparent that values of lamb and mutton can be more accurately determined on a carcass basis than on a live basis. One might then ask, why have

Figure 14-6 When carcasses are sold, lambs remain in the feedlot until the seller gets the best bid.

all lambs and mutton not been sold on a carcass basis in the past? Some packers have not encouraged carcass selling and those who have often continued to use a complicated system of pricing that includes dressing percentage to calculate live weight and payment is on a calculated live-weight basis. In addition, there is some concern among producers and feeders that control of the final price received is shifted, at least in part to the packer when selling on a carcass-weight basis. Another consideration is that some producers and feeders prefer to accept an overall average penalty for bruised, needle-grass contaminated or condemned carcasses rather than risk a higher discount if their sheep and lambs are actually below average. Of concern for some is the dressing procedure, which can affect carcass weight. For example, will kidney fat and spleens be included with the carcass weight that is the basis for payment? Conditions of sale such as including kidney fat as part of the carcass weight will continue to be a matter of negotiation between buyer and seller. Because livestock sellers are often inexperienced in carcass selling, the USDA has issued the following regulations (Title 9, Chapter 2, Part 201.99 of the Packers and Stockyards Act) governing purchase of livestock by packers on a carcass grade, carcass weight, or carcass grade and weight basis:

(a) Each packer purchasing livestock on a carcass grade, carcass weight, or carcass grade and weight basis shall, prior to such purchase, make known to the seller, or to his duly authorized agent, the details of the purchase contract. Such details shall include, when applicable, expected date and place of slaughter, carcass price, condemnation terms, description of the carcass trim, grading to be used, accounting, and any special conditions.

(b) Each packer purchasing livestock on a carcass grade, carcass weight or carcass grade and weight basis shall maintain the identity of each seller's livestock and the carcasses therefrom and shall, after determination of the amount of the purchase price, transmit or deliver to the seller, or his duly authorized agent, a true written account of such purchase showing the number, weight, and price of the carcasses of each grade (identifying the grade) and of the ungraded carcasses, an explanation of any condemnations, and any other information affecting final accounting. Packers purchasing livestock on such a basis shall maintain sufficient records to substantiate the settlement of each transaction.

(c) When livestock are purchased by a packer on a carcass weight or carcass grade and weight basis, purchase and settlement therefore shall be on the basis of carcass price. This paragraph does not apply to purchases of livestock by a packer on a guaranteed yield basis.

(d) Settlement and final payment for livestock purchased by a packer on a carcass weight or carcass grade and weight basis shall be on actual (hot) carcass weights. The hooks, rollers, and gambrels or other similar equipment used by a packing establishment in connection with the weighing of carcasses of the same species of livestock shall be uniform in weight. The tare weight shall include only the weight of such equipment.

(e) Settlement and final payment for livestock purchased by a packer on a USDA carcass grade shall be on an official (final—not preliminary) grade. If settlement and final payment are based upon any grades other than official USDA grades, such other grades shall be set forth in detailed written specifications which shall be made available to the seller or his duly authorized agent. For purposes of settlement and final payment for livestock purchased on a grade or grade and weight basis, carcasses shall be final graded before the close of the second business day following the day the livestock are slaughtered.

Order buyers and dealers.   Functions of these agents are similar to those described under feeder lambs. These individuals or companies are important marketing alternatives for all segments of the sheep industry.

Terminal markets.   A stockyard performs the services of receiving, buying or selling on a commission basis or otherwise, marketing, feeding, watering, holding, delivery, shipment, weighing, or handling livestock. Commission personnel in terminal markets negotiate a sale to a buyer and collect the purchase price. For conducting these services, they collect a fee. The advantage of terminal markets is that large numbers of sheep and lambs can be brought together where a buyer can purchase the kind and number needed. In the past the producer or feeder who "topped" the market in any one of the large terminal markets could indeed be proud of the lambs he or she produced. The day of the cattle or sheep drives has ended and, in many areas, terminal markets have given way to other methods of marketing sheep and lambs. Prices of livestock sold at the major remaining terminal markets are often still used for reporting market price of lambs. Because the number of lambs handled is small, it is important to have additional price quotes before negotiating price based on the terminal market price.

Auction.   Auction selling of lambs is effective as long as several packer buyers are present to compete with each other. Auctions make it possible for small producers to have a ready market for a few lambs that can be bought by packer buyers and pooled into truckload lots.

Special auctions organized by producers, where several lots of lambs are brought to a central location, have been successful in some areas. Producer pooling of lambs into truckload lots of uniform size for sale in auctions or terminal markets, directly to packers, or through other marketing alternatives has often been successful. In many areas the smaller auction markets have increased as terminal markets decreased.

Producer to consumer.   Most producers who sell directly to consumers receive a premium price for their lambs. The lambs are often delivered to a locker plant by the producer after they have been purchased by consumers. The consumer then pays for slaughter and cutting before taking the wrapped packages home to the freezer. In other cases, properly licensed USDA inspected locker plants or slaughter plants custom kill lambs for the producer, who then sells fresh or frozen cut and wrapped lamb carcasses to individuals or restaurants. The producer-to-consumer transactions have been popular in the eastern portion of the United States and Canada, but some producers have operated successfully in other areas, including the range states. Sales are often made through ads placed in newspapers and through word of mouth. Producers should make sure that they check local laws on sales of meat, as well as regulations on inspection, advertising, and taxes before selling carcasses or cuts.

Consumers who purchase lambs directly from producers are often educated, middle-class people who earn a good salary and are of central European or Mediterranean ethnic backgrounds and live in a city suburb. Therefore, producers with small flocks that live next to metropolitan areas are most likely to be successful marketing lambs in this manner. Those who market all their lambs in this manner often produce ram lambs because they grow faster and can be marketed at younger ages. Consumers prefer meat from young rams because it is leaner and the chops are larger.

ELECTRONIC MARKETING

The three categories of electronic marketing are telephone auctions, computerized exchanges, and video auctions.[1,4]

Telephone auctions. These are the oldest and most popular of the three systems and they are used more by small producers than by large producers. A telephone auction is a sale conducted in the normal auction manner, except no buyer, seller, or animals are present before the auctioneer. Buyers are linked to the auctioneer via conference telephone calls. The auctioneer describes the lambs to potential bidders as to weight, sex, breed background, grade, and location where they will be assembled for pickup. Buyers are given a bidding number in advance of the sale to prevent personal identification. Lambs then go to the highest bidder, who, once he or she has bought them, arranges for pickup.

Prior to the telephone auctions, lambs are graded. Only those of a specified weight range that will grade choice or prime are selected for sale. Sales are held once a week or whenever a truckload (about 400 to 450 lambs) can be assembled. At times two or three assembly points in the same general area will combine to supply a truckload. In some areas a producer may never have more than 10 or 12 lambs ready for market at any one time, but these few lambs, packaged in a load lot, may be attractive to buyers hundreds of miles away. Almost all telephone auctions have resulted in a net return to producers over some of the older more traditional methods of selling lambs, even after fees assessed by the marketing co-op for graders, weighing, handling, telephones, and operation are deducted. The auctions have resulted in several buyers bidding on lambs by telephone where previously only one or two buyers in an area were typical.

Computerized exchanges. Three types of computerized systems are being used for marketing purposes. One system provides information, but no trading occurs. The second exchanges products at administered prices. The third system exchanges products by negotiated trading and is being used for several agricultural products, including lambs. Computer sales are similar to telephone auction sales. In one case the sale is conducted over a confirmed conference telephone call, and in the other the sale is conducted over computer terminals through the main computer, all of which are tied together by a network of telephone lines. Each buyer has a terminal, and by dialing a number, he/she can get a listing of the lambs available for sale, or they can participate in the auction of those lambs.

A computer auction proceeds after the computer terminal in each buyer's office describes the lambs and tells where the lambs will be assembled. Lambs need to be graded for finish and weighed and classified as to breed and sex just as for telephone auctions. The listing shows all information except the reservation price. The computer will not sell the lambs unless the lambs bring at least the reservation price. The buyer can study all lots of lambs listed on the computer terminal and tell which lots he or she is interested in before the auction starts.

When the auction starts, the computer operator conducting the auction lists a price to start the bidding. If there is no bid the price drops until a bid is received. The computer tells all the buyers that there is a bid and what price was bid. Bidding then continues with other buyers using their computer terminals. If there is no bid within a specified time, for example, 20 seconds, and if the last price bid is above the reservation price, the computer sells the lambs. It tells the successful bidder that he or she has bought the lambs, and it tells everyone else on the sale that they are sold to another buyer. The computer then proceeds to the next lot.

At the end of the sale the computer tells each buyer what they have bought and where they are located. The buyer then calls and states when he or she wants the lambs and arrangements are made to have them delivered by producers to the assembly area. As is the case with the telephone auction, producers with any number of lambs from 1 to 1000 or more can participate. Each lot of lambs is priced after being weighed and graded at the assembly point. The computer figures the commission for selling and subtracts that from the producer's check.

Computerized exchanges permit sellers to accurately describe and disseminate information regarding their products to prospective buyers. They permit buyers to place bids using marketing agencies to perform their functions in a least-cost and efficient manner. It is claimed that the computer can conduct sales cheaper than telephone auctions because it takes less time and cost per minute. Computerized exchanges also provide accurate reporting of prices and collection and analysis of data regarding supply, but to date they have not been used as extensively as telephone auctions.

Video auctions. Television video systems are being used for everything from x-rated movies to selling livestock. Selling of livestock by video takes different forms. The first step for any form of video sales is filming of stock in color. Companies sometimes provide this service. The tape is "voiced over" by the company giving a full description of the sheep.

In one system, potential buyers are contacted by the video company and asked to view tapes with sheep of interest to them. Thus, it is basically a service to bring producers and buyers together. Price negotiation is directly between buyer and seller. The advantage is that the buyer has the ability to see the sheep before buying them. Furnishing tapes for viewing by individual buyers seems to be more adapted to sale of high-priced breeding stock. Purebred breeders may also be able to take advantage of video auctions in future years to avoid hauling livestock long distances and then introducing them to concentrations of other animals where disease transmission is always a possibility.

In video auctions the livestock are filmed in color at the producer's farm or feedlot. Potential buyers are then brought together in a central location and competitive bidding proceeds using a large screen. The auction system is being used for feeder cattle and to some extent for feeder lambs.

The problem with all systems of electronic marketing is that many producers are hesitant to use them. Others argue that there are already too many marketing systems for the small number of sheep and lambs being marketed and that the new systems if used will reduce the number of traditional marketing alternatives. When it comes to the newer marketing systems, producers must decide if they want to use tractors or stay with the horse.

SLAUGHTER EWES AND RAMS

Marketing of stock sheep that are no longer of value for breeding is one of the most serious problems facing the industry today.[2] Old ewes sell for one-quarter to one-third as much as cows or sows that are being marketed at the end of their productivity. The drastically reduced price for slaughter ewes and rams in comparison with other species is because demand for mutton is lacking. Seasonality, small volume, high death losses during transportation, high condemnation rates on the kill floor, and high costs of hand boning for the quantity and quality of meat obtained are other factors involved in the low prices.

Producers can help by marketing ewes in the best condition possible. This includes (1) holding shorn ewes a week or two after shearing to allow them to better withstand the rigors of being transported long distances, (2) holding sheep off water overnight before they are loaded to avoid slick truck beds that result in bruised or dead ewes on arrival, (3) avoiding overloading trucks, and (4) avoiding shipment of animals that are too weak to make the trip.

Carcass and carcass part condemnations result from caseous lymphadenitis, pneumonia, emaciation, and parasitic conditions, particularly sarcosporidiosis and worms. Pneumonia can be controlled by proper handling of ewes before and during transport, emaciation by sending the lower end of the culled ewes directly to the renderer, and worms by following recommended management practices during the productive life of the ewe. There is no known control for sarcosporidiosis, and some feel that it is not an important parasitic disease because it seldom affects the live animal. Nevertheless, thousands of mutton carcasses are condemned because small white parasites that are often barely visible to the eye continue to be found embedded in muscles of mutton.

Unstable markets, large up and down movement of prices, and excessive shrink are other problems involved in marketing mutton. The processor of slaughter ewes and rams is also faced with high condemnation rates of livers, hearts, and kidneys. These are usually the only edible mutton by-products saved.

Marketing Alternatives for Slaughter Ewes and Rams

Packers who slaughter old ewes and rams usually obtain their supply from auctions, order buyers, dealers, or directly from producers. Some producers are also able to market slaughter ewes directly to consumers. An example would be sale of mutton for barbecues, which are popular in the South and in some other areas. Other marketing alternatives, including electronic marketing discussed under slaughter lambs, should be explored by those who have slaughter ewes and rams for sale.

In New Zealand the cost of slaughter and boning mutton has exceeded the value obtained for the meat. This example helps to emphasize the need for the development of new products from mutton that can be sold at a profit. These products are discussed in Chapter 16.

BREEDING SHEEP

Breeding sheep come in all ages, prices, and even colors, but the largest numbers of breeding sheep that are marketed are those marketed as replacements for the ewe flock.

Commercial Ewe and Ewe Lamb Replacements

Replacement ewe lambs, yearling ewes, and aged ewes are marketed direct through dealers or order buyers and to a lesser extent through other means. Producers in the Midwest and South often look to range operators in the Western states, including Texas, for replacements because the range operator is often the most efficient producer of breeding stock. In years of drought, high-quality yearling ewes, young ewes, and even older ewes are often sold at bargain prices.

Even in normal years some producers in areas where feed is scarce have available high-quality ewe lambs that make excellent breeding stock but are too old to be sent to the feedlot for lamb production. Wether lambs from these same flocks are suitable for fat lamb production because breakjoints of wether lambs fuse at older ages than those of ewe lambs. One-year-old ewe lambs often produce yearling carcasses, while one-year-old wether lambs from the same flocks produce lamb carcasses. Therefore, some range producers specialize in production of ewe lambs suitable for replacement purposes.

The buyer can often purchase thrifty lightweight ewe lambs at around 66 lb

(30 kg) live weight that will make good producing ewes at reasonable prices. Sometimes higher-quality ewe lambs can be purchased than can yearling ewes because only a limited number of operators specialize in wintering ewe lambs for sale as yearlings. Those who do sell yearlings want to sell them in the spring out of the shearing pen. The greatest demand is for Western white-faced ewes, but the market for crossbred black-faced sheep may return with the increased interest in sheep production in the Southeastern states and in some other farm flock regions. There may also be some Finn crosses wanted in these areas in the future.

The best source of high-quality old ewes is directly from producers who earmark their sheep for sale at a certain age. For example, many range producers plan to sell all their six-year-old ewes every fall. Included are many solid-mouthed ewes that will produce one or more additional years under farm flock conditions. Reputable dealers who purchase these ewes and sort out those with broken mouths, bad bags, and other problems are also good sources for older breeding ewes.

In addition to age, weight, and problems that would prevent ewes from raising healthy lambs, buyers should be concerned about the cycle the ewes have been lambing in, because the cycle may not coincide with the time of year the farm flock operator plans to lamb.

Prices quoted on yearling ewes or cull ewes through the auction ring or terminal markets are often lower than the replacement ewe market because these sheep are of lower quality or they are sold in smaller bunches.

There are reports of buyers being cheated on old ewes, and this occurs especially when buyers do not know much about the ewes they are buying. The best way to avoid being cheated is to buy through producers and reputable dealers who make a business of selling replacement sheep. A group of small producers who agree to take a truckload of ewes delivered directly from the West can almost always obtain high-quality replacement ewes when buying from larger range producers or reputable dealers. County extension agents, sheep specialists at land grant universities, and marketing agencies, including electronic marketing systems, can often help locate truckloads of replacement sheep. Farm flock operators should consider contracting for delivery of ewe lambs or other replacements a year ahead to guarantee the age, weight, and breed desired.

PUREBRED SHEEP

Purebred rams and ewes can be purchased through private treaty, auction houses, purebred auctions, dispersal sales, and order buyers and may soon be available through some electronic marketing systems. Traditional methods of advertising breeding sheep include breeders listing themselves in breeders' directories, classified ads, or block ads. Consigning sheep to annual sales is also a method used by purebred breeders to build their reputation. While a few breeders still show at the "right shows" and display ribbons they have won, this questionable method of advertising breeding stock is discouraged. Instead of spending time making sure that sheep are neatly trimmed and in exactly the right condition for show, more time should be spent on performance testing. Ram performance trials sponsored by some universities and by some breed or wool grower associations make it possible to compare rate of gain, wool growth, and other economically important traits among breeders. High-indexing twin-born rams at well-planned and properly conducted ram trials are usually worth more to the buyer who is interested in efficient lamb and wool production than are rams with blue ribbons at the top shows. Because they have not been pampered by the breeder, they will also be more likely to stand up to a variety of grazing conditions.

Breeding soundness is an important consideration and some method of judging this characteristic before the sale is essential. Suggested minimum standards for yearling rams on a semen examination are outlined in Chapter 9. Most buyers also give a lot of attention to the type of ram they need to correct weak points in their flock. Faults like inverted eyelids, incisor teeth that do not fit the upper jaw, weak pasterns, weak topline, and excessive skin folds need to be eliminated. Equally important is the absence of disease conditions like epididymitis, sheath rot, and foot rot.

A few breeders specialize in unusual types of sheep, such as those with colored wool. Their market is smaller, but advertising breeding stock must be done in the same way as for other sheep. Keeping records of past buyers, making sure that they are satisfied, and making repeat sales in addition to a few new sales each year is the way every purebred breeder survives. Electronic marketing systems and video tapes of breeding stock will be used more widely in the future.

REFERENCES CITED

1. McCoy, J. H. 1979. *Livestock and Meat Marketing,* 2nd ed. AVI Publishing Company, Westport, Conn.
2. Pand, S. A. 1973. *The Lamb Industry: An Economic Study of Marketing Structure, Practices and Problems.* Packers and Stockyards Administration Research Report No. 2 U.S. Government Printing Office, Washington, D.C.
3. Sheep Industry Development Program. 1981. *National Lamb Marketing Symposium.* American Sheep Producers Council, Denver, Colo.
4. Sheep Industry Development Program. 1986 marketing. In *Sheepman's Production Handbook.* American Sheep Producers Council, Denver, Colo.
5. Ward, C. E., M. T. Detten, and F. M. Epplin. 1985. Marketing tool for slaughter lambs and sheep. *Sheep Industry Development Program Research Digest,* Vol. 1, No. 2.
6. Wyman, Howard. 1984. Marketing in 1990. *Natl. Wool Grower,* August, p. 6.

15 Meat from Sheep

Performance, live appraisal, and carcass appraisal are all important determinants of the economic value of a lamb, but they do not necessarily define the kind of lamb the consumer wants. The consumer, in addition to other segments of the industry, must be satisfied if lamb is to survive. The purpose of this chapter is to share with the reader information about value, grades, and identity of lamb, because these factors, in addition to composition of the cuts discussed in Chapter 16 and palatability of the meat discussed in Chapter 17, influence lamb consumption.

CARCASS VALUE

Costs for slaughtering and processing a 115-lb (52-kg) lamb at $61 cwt are found in Table 15-1. The difference between what the packer pays the producer and trucker ($70.15) and what the packer receives for the carcass and by-products ($75.93) must cover all expenses and return some profit in order for the packer to remain in business. If the packer breaks the carcass into trimmed cuts and sells it as boxed lamb, he must obtain more for the cuts ($81.62) than he does for the intact carcass because additional expenses for labor, equipment, and packaging materials are involved.

Retail cut out value can vary considerably depending upon closeness of trim, cutting method (bone-in or boneless cuts), and yield grade of lambs. It is always lower than yield of boxed lamb cuts sold by the packer because of additional trimming. Retail cut yield or a yield grade 2 carcass cut by the bone-in method often averages 76% of the carcass weight. Therefore, retail cuts from the 59.23-lb (27-kg) carcass in Table 15-1, which sells FOB plant for $1.22/lb and has a total value of $72.26, must sell for $1.60/lb to recover the $72.26. If the carcass is fatter or if

Carcass Value

TABLE 15-1 COSTS FOR SLAUGHTERING AND PROCESSING A LAMB[a]

	Packers Income When Selling the Carcass			
	Percent of Live Weight	Weight	FOB Plant per Lb	Value per Head
Carcass	51.50%	59.23 lb	$ 1.22	$72.26
Heart	0.40	0.46	0.28	0.13
Liver	1.35	1.55	0.16	0.25
Tongue	0.21	0.24	0.35	0.08
Casing, per set				0.45
Pelt, each				2.00
Rendering items:	Raw Wt.	Yield	Value	
Blood	4.0 lb	18.0%	$390.00/ton	0.14
Bone/meat meal	9.0 lb	1.50 lb	650.00/ton	0.49
Tallow		0.75 lb	0.17 lb	0.13
				$75.93
	Packers Income When Selling Boxed Lamb (Yield Grade 2)			
	Percent of Carcass Weight	Weight	FOB Plant per Lb	Value per Head
CPC leg[b]	27.90%	16.53 lb	$1.75	$28.93
Loin	10.50	6.22	3.00	18.66
Rack	10.50	6.22	2.60	16.17
Shoulder	22.60	13.39	1.10	14.73
Flank	4.70	2.78	0.15	0.42
Breast	8.90	5.27	0.15	0.79
Shank	3.10	1.84	0.65	1.20
Neck	2.50	1.48	0.10	0.15
Kidney	0.50	0.30	0.45	0.14
Fat	3.50	2.07	0.19[b]	0.39
Bone	3.70	2.19	0.02[b]	0.04
Cutting loss	1.60	0.94		
	100.00%	59.23 lb		$81.62

[a]Live cost to packer: 115-lb live weight at $61 cwt (delivered to plant) = $70.15
[b]Adjusted to reflect equivalent rendered product value.

more boneless cuts are made, the retail cut yield can easily be reduced to 68%. If the carcass in Table 15-1 yielded 68% retail cuts, the average cut would need to average $1.79 to recover the $72.26 paid for the carcass.

When a carcass that yields 68% retail cuts is marked up 30% on sales to recover the retailer's expenses plus a small 1% to 5% profit, the price per pound for the average retail cut becomes $2.56. Retailers vary this average price based on demand for the cuts. Loin and rib chops are in the greatest demand and sell for the highest price per pound, while neck slices and ground lamb sell for the lowest price.

Calculations for the price of boneless mutton based on a live price of 15¢/lb

TABLE 15-2 CALCULATIONS FOR PRICE OF BONELESS MUTTON

	Price/lb (¢)
Live price of boning ewes	0.15
Carcass price assuming a 39% yield (yield is based on chilled carcass weight after an average condemnation rate of 5% to 10% is subtracted)	38.46
Carcass price after a $7.50/ewe slaughter and boning cost is included (assumes carcass weighs 56 lb): $21.54 + 7.50 = $29.04	51.86
Price of boneless meat based on a 63% yield from a carcass	82.31

are found in Table 15-2. When carcass yield after carcass condemnations plus slaughter and boning costs are accounted for, boneless mutton in this example must average 82¢/lb for the packer to break even. Condemnation rates as well as slaughter and boning costs vary, but the 82¢/lb is a realistic cost for boneless mutton based on a 15¢/lb live weight. At times the live price of boning ewes has risen to 25¢/lb. When the price reaches this level, many of the boning ewes in the Western United States are sold live to Mexico and slaughtered where processing costs are lower.

Calculations for the price of a barbecued restructured mutton steak sandwich using $82.31/cwt boneless mutton are found in Table 15-3. Prices would vary depending on the product, but the required price of $1.48 for a sandwich containing 0.21 lb (95 g) of boneless mutton is typical of the costs and markups involved for many fast-food items that could utilize boneless mutton.

Efforts should be made to help more producers and consumers understand these pricing examples. Information on how retail cut value relates to carcass value and live value can help producers make better marketing decisions, and it can reduce suspicions that a middleman is "ripping off" both the producer and consumer. Lamb moves from packers through retail supermarkets to consumers, or it moves through breakers and purveyors to hotels, restaurants and institutions, and then to

TABLE 15-3 CALCULATIONS FOR PRICE OF BARBECUED MUTTON STEAK SANDWICHES

	Price/Lb
Frozen steak	
Price of boneless mutton	82.31¢
Price assuming 140% of meat cost to process (spice, casings, equipment)	$1.15
Price assuming 180% of meat cost to process, advertise, and sell	$1.48
Price of total assuming that meat is 60% of total food cost	$2.46
Price of total assuming that 35% of total food cost is sales price of sandwich	$7.04
Amount of meat/sandwich is 0.21 lb; therefore, 0.21 × $7.04 equals total cost for the sandwich	$1.48

consumers. Brokers are often involved in the transactions, but there is no middleman reaping large profits in the sheep and lamb industry.

INSPECTION

Inspection of lamb and mutton is the responsibility of the Food Safety and Inspection Service of the U.S. Department of Agriculture. In addition, some states have meat inspection programs that are supervised by the USDA. The Food and Drug Administration has jurisdiction over certain animal products. These agencies are charged with continuous inspection and monitoring of the wholesomeness of America's meat products.[1] If meat passes inspection, an inspection stamp (Figure 15-1)

This is the stamp used on meat carcasses. It is used only on the major cuts of the carcass, and so it may not appear on the roast or steak the consumer buys.

This mark is found on every prepackaged processed meat product--soups to spreads--that has been federally inspected.

This is the mark used on federally inspected fresh or frozen poultry or processed poultry products.

Cottage cheese and pasteurized process cheese may bear the USDA "Quality Approved" shield if they are of good quality and are made under USDA supervision.

Figure 15-1 Some of the federal inspection marks used on animal products.

GRADING

There are two different USDA systems for grading lamb and mutton: quality grading and yield grading.[3] Quality grading is designed to reflect age, conformation, and quality differences, while yield grading reflects differences in amount of fat. Grading of lamb and mutton is voluntary, and packers pay USDA graders for the service. In contrast, inspection is mandatory because it involves wholesomeness, sanitary preparation and handling, and freedom from disease, adulteration, and misbranding.

Quality Grades

The quality grades for lamb and yearling mutton carcasses are Prime, Choice, Good, and Utility. For mutton the grades are Choice, Good, Utility, and Cull. Evaluation of carcasses to assign quality grades is based on flank streakings in relation to maturity (Figure 15-2). Older lambs and mutton require more fat streaking in the flank than do younger animals to qualify for a grade. A minimum level of lean and fat firmness is specified for each grade. In assigning quality grades, conformation is just as important as flank streaking and maturity. The rate of compensation of conformation for quality and quality for conformation is on an equal basis. For example, a carcass that has evidence of quality equivalent to the midpoint of the Choice grade may have conformation equivalent to the midpoint of the Good grade and remain eligible for Choice. Also, a carcass that has conformation at least one-third grade superior to that specified as minimum for the Choice grade may qualify for Choice with a development of quality equivalent to the upper third of the Good grade.

Quality grading of lamb and mutton carcasses began in 1931. The standards were amended in 1940, 1951, 1957, 1960, 1980, and 1982. Most of the revisions allowed carcasses with less finish to grade higher. Today over 98% of quality graded lamb carcasses grade Choice or Prime. Choice and Prime lambs of similar carcass weights almost always sell for the same price, indicating that the trade does not distinguish differences in value between quality grades. In addition, there is very little evidence that cuts from Prime lambs are more palatable than cuts from Choice lambs.

Because quality grading lambs is expensive in terms of time and labor and because almost all lambs are graded Choice or Prime and then sold for the same price per pound consideration should be given to dropping the quality grade standards. Yield grading standards should be retained because real differences in retail

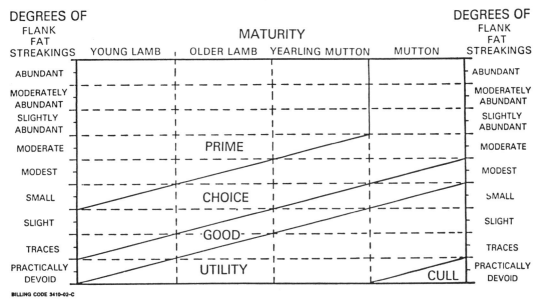

Figure 15-2 Relationship between flank fat streakings, maturity, and quality as outlined by the USDA.[3]

value exist and because consumers prefer lean lamb and mutton, which the yield grades identify.

Yield Grades

There are five yield grades denoted by the numbers 1 through 5, with grade 1 representing the leanest carcasses and 5 the fattest. Major closely trimmed boneless cuts from yield grade 1 carcasses account for 47.3% or more of the carcass weight, while the same cuts from yield grade 5 carcasses account for 41.9% or less of the carcass (Table 15-4). The yield grades of lamb, yearling mutton, and mutton are determined

TABLE 15-4 LAMB YIELD GRADES

Yield Grade	% Boneless, Trimmed Retail Cuts (leg, loin, rack, and shoulder)
1	47.3+
2	45.4–47.2
3	43.7–45.4
4	41.9–43.6
5	−41.9

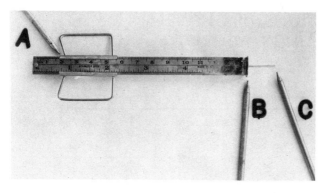

Figure 15-3 A probe for measuring external fat thickness on the unribbed carcass. Point A shows the actual fat depth. Point B is a plate that presses against the fat surface, while point C is the probe that passes through the fat and stops on the epimysial tissue covering the rib-eye.

by considering three characteristics: amount of external fat, amount of kidney and pelvic fat, and conformation grade of the legs.

External fat thickness over the rib-eye muscle between the twelfth and thirteenth ribs is estimated by the grader because lamb carcasses are seldom ribbed. Nevertheless, an accurate measurement can be obtained using a probe developed at the University of Wyoming (Figure 15-3). On carcasses that do not have a normal distribution of external fat, the measurement may be adjusted to reflect unusual amounts of fat on other parts of the carcass. In many carcasses no such adjustment is necessary; however, an adjustment in thickness of fat of 0.05 to 0.10 in. is not uncommon. As a guide in making these adjustments, the standards for each yield grade include an additional related measurement, body wall thickness, which is measured 5 in. laterally from the middle of the backbone between the twelfth and thirteenth ribs. This measurement can be easily obtained with the probe developed at Wyoming (Figure 15-4) or with the Hennesey and Chong grading probe developed in New Zealand (Figure 15-5).

As the amount of external fat increases, the percentage of retail cuts decreases; each 0.05-in. change in adjusted fat thickness over the rib-eye changes the yield grade by one-third of a grade. The amount of kidney and pelvic fat considered in determining the yield grade is evaluated subjectively and is expressed as a percentage of the carcass weight. As the amount of kidney and pelvic fat increases, the percent-

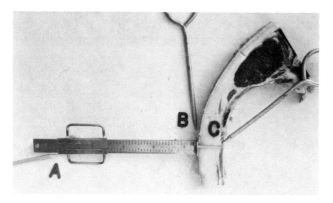

Figure 15-4 Measuring body wall thickness with a fat probe developed at the University of Wyoming. The tip of the probe can rest on the twelfth rib or it can measure body wall depth between the twelfth and thirteenth ribs.

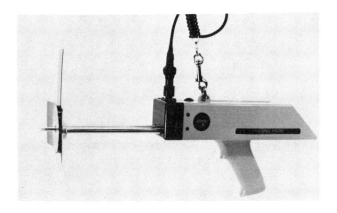

Figure 15-5 The Hennessy and Chong grading probe distributed by Precision Instrument Manufacturers, Auckland, New Zealand. The probe can be attached to a recorder to give an automated printout of body wall thickness for every carcass as it comes off the kill floor.

age of retail cuts decreases; a change of 1% of the carcass weight in kidney and pelvic fat changes the yield grade by one-fourth of a grade.

The conformation grade of the legs is scored by evaluating the thickness of muscling. The evaluation is made in terms of thirds of a grade and coded using 15 for high Prime and 4 for low utility. An increase in the conformation grade of the legs increases the yield of retail cuts; a change of one-third of a grade changes the yield grade by one-twentieth of a grade. Leg conformation has much less influence on yield grade than does fat thickness or kidney and pelvic fat percentage.

The equation for determining yield grade is

Yield grade = 1.66 − (0.05 × leg conformation grade score) + (0.25 × percentage kidney and pelvic fat) + (6.66 × adjusted fat thickness over the rib-eye in inches)

Therefore, the yield grade of a lamb with a low prime leg score (code 13), with 3% kidney and pelvic fat and 0.18 in. (4.6 mm) of adjusted fat thickness over the rib-eye is 2.9:

$$\text{Yield grade} = 1.66 - 0.65 + 0.75 + 1.20 = 2.96$$

The application of this equation usually results in a fractional grade like the 2.9 given in this example. However, in normal grading operations, any fractional part of a yield grade is dropped. In this example the computation results in a yield grade of 2.9 but the final yield grade is 2; it is not rounded to 3.

Yield grade descriptions found in the "Official United States Standards for Grades of Lamb, Yearling Mutton, and Mutton Carcasses," effective October 17, 1982, give descriptions for each of the five yield grades. These descriptions facilitate the subjective determination of the yield grade without making detailed measurements and computation. Therefore, it is possible for a grader to assign yield grades rapidly. Lamb grading standards should be amended to make it possible to assign yield grades as the carcasses are weighed hot on the kill line. Information on yield grade along with the weight of each carcass would give an excellent description of the value of lambs in a particular lot and make payment based on actual value of the lambs easier. When lambs are graded in the cooler, it is harder to record yield grade information by lot of lambs.

To date the yield grades for lamb (adopted in 1969) have not been widely used. Emphasis on low-calorie meat, coupled with the need for the producer to be paid for producing consumer-preferred lamb, may eventually result in the commercial use of yield grades.

Yield grades could, if used, benefit all segments of the sheep industry:*

1. For the producer: They provide a means to identify breeding and slaughter animals for differences in yields of salable meat. When the value differences associated with yield grades are reflected in prices to producers, this will encourage producers to produce high-yielding, high-quality lambs.
2. For the packer: They can help evaluate more precisely differences in value among the animals he buys and the meat he sells.
3. For the retailer: They provide a means for purchasing lamb with assurance as to its yield of retail cuts and, therefore, its value.
4. For the consumer: They provide an indirect means of reflecting consumer preferences for lamb—thick muscling, a minimum of excess fat, and fewer calories.

Yield Grading Live Lambs

The challenge of lamb producers is to identify and produce meaty, high-value lambs. The lamb producer and feeder can select these consumer-preferred lambs while they are still alive by evaluating the amount of fat cover on the animal at the following body parts:

1. *Ribs:* Put ends of fingers into the wool in the middle of the lamb's side. Move the fingers forward and back across the ribs. The less fat the lamb is carrying, the more prominently the ribs can be felt. On a grade 1 lamb, the ribs can be plainly felt and will feel sharp. The ribs on a desirable grade 2 lamb can be easily felt, but will have enough fat on them to keep them from feeling sharp. If the lamb has more than 0.5 in. of fat over the loin eye, the ribs cannot be felt at all.
2. *Edge of loin:* There is little or no muscle covering the ends of the lateral processes of the lumbar vertebrae. On a lamb carrying a thin cover of fat, the ends of these bones and the spaces between them can be easily felt. If the ends of the bones cannot be felt, the lamb is definitely too fat.
3. *Hip bones:* The top of the hip bones can be plainly felt and all feel rather sharp on a lamb carrying a thin cover of fat. As the lamb becomes fatter, more

---

*Official United States Standards for Grades of Lamb, Yearling Mutton, and Mutton Carcasses. 1982. Agricultural Marketing Service, USDA, Washington, D.C.

fat will cover these bones and they will become blunt and less easy to feel. They cannot be felt at all on a very heavily finished lamb.
4. *Dock:* The vertebrae in the dock of a lamb with a thin cover of fat can be easily felt. As the lamb becomes fatter, the dock gets progressively wider and the vertebrae become less easy to distinguish.
5. *Rear flank:* The thickness of the rear flank can be felt between the thumb and fingers. The flank will feel rather thin on a correctly finished lamb.

A grade 1 lamb with less than 0.1 in. (2.5 mm) of fat cover and less than 8% trim fat will feel very thinly covered with fat when handled over the parts outlined above. A grade 5 lamb with more than 0.5 in. (13.7 mm) of fat cover and more than 18% fat trim will feel very thickly covered with fat over these parts.

The best body parts to use in finding differences in muscling are those parts where there is the smallest amount of fat. There is less fat on the leg than any other body part. Therefore, the leg should be used to evaluate differences in muscling.

Desirable muscling is best indicated by the following:

1. A wide natural distance between the hind legs and between the forelegs.
2. A long rump as measured from hip bones to pin bones.
3. A long leg as measured from the hip bone to the hock.
4. A long leg as measured from the stifle joint to the back of the leg.
5. A thick, plump leg is due to the thickness of the muscles on the inside and outside of the leg.

Yield grades, which place greater emphasis on high-cutability lambs, may result in important changes in the type of lamb that is produced. As producers and feeders become more familiar with selecting live animals for breeding and for market on a yield-grade basis, carcass improvements should result, which will mean greater returns for the producer in the long run.

Hothouse Lambs

Hothouse lambs are usually born during the winter months and marketed at 6 to 10 weeks of age. They are sold for a much higher price per pound than spring or old crop lambs. Live weights range from 25 to 60 lb (11.2 to 27 kg) and they are hog-dressed with the head and pelt on but with the feet and viscera removed.[2] The pluck, consisting of the liver, heart, lungs, gullet, and windpipe, is left in. The object of this method of dressing is to prevent moisture loss from the carcass and to maintain the pink color of the lean. The New York City market consumes many of the hothouse lambs. Other areas where the Greek population is high also consume hothouse lamb especially at the time of their Easter. Only a fraction of 1% of the total lambs produced are sold as hothouse lambs.

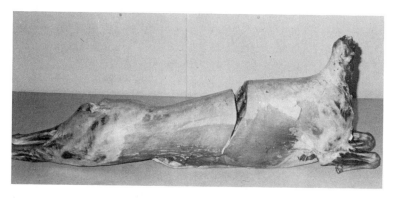

Figure 15-6 Lay the carcass on the table. Remove the kidney and pelvic fat. From the inside of the carcass insert a knife blade between the twelfth and thirteenth ribs near the backbone. Use a handsaw to separate the foresaddle from the hindsaddle where the knife mark is located.

CUTS OF LAMB

There are a number of ways to divide the lamb carcass into wholesale, retail, and institutional cuts. One of these ways is outlined in Figures 15-6 through 15-19. This cutting procedure combines the new and old cutting procedures outlined by the American Lamb Council (Figure 15-20). Loin and rib chops have shorter tails and less fat than those cut by the old method. Sirloin chops are made as described under the new method, but the sirloin is left on the leg prior to making the chops. Lamb, like other meats, is being made into more boneless, closely trimmed cuts, as shown in Figures 15-21 and 15-22.

LAMB AND MUTTON PROCESSING

Current changes in meat processing and marketing have the potential of increasing the acceptability of lamb and mutton to merchandisers and consumers. These

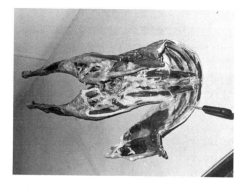

Figure 15-7 Remove the flank by following the natural contour of the leg. Continue the cut to a point 1 in. (2.5 cm) lateral to the loin eye muscle. Trim the fat from the flank and use the lean for ground lamb.

Lamb and Mutton Processing 277

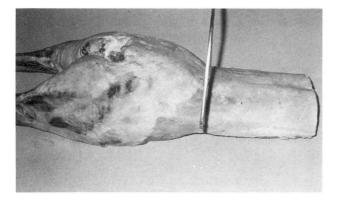

Figure 15-8 Cut the leg from the loin, leaving one lumbar vertebra on the leg.

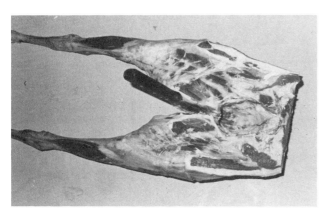

Figure 15-9 Split the aitch bone with a knife and then saw through the center of the vertebral column.

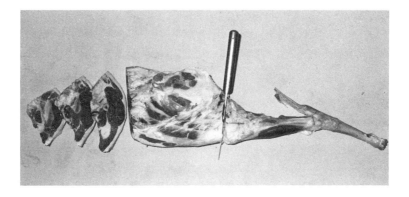

Figure 15-10 Trim the leg to 0.25 in. (6.3 mm) outside fat cover and remove all but three sacral vertebrae. Make an American leg by cutting through the stifle joint. Put lean from the shank into ground lamb. Cut three sirloin chops (1 in. or 2.5 cm thick) from the leg.

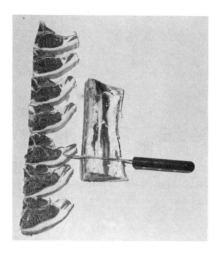

Figure 15-11 Split the double loin with a saw and cut chops 1 in. (2.5 cm) thick starting at the thickest end. Trim the chops to 0.25 in. (6.3 mm) fat cover.

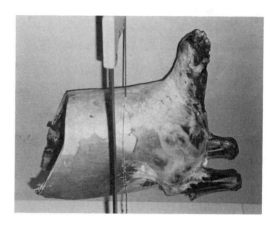

Figure 15-12 Cut a seven-rib rack from the shoulder by cutting between the fifth and sixth ribs.

Figure 15-13 Split the rack with a saw. Remove the ribs by cutting 1 in. laterally to the rib-eye on the small end and 1 in. (2.5 cm) laterally to the scapula on the blade end. Make rib chops 1 in. (2.5 cm) thick starting at the blade end.

changes provide an increasing variety of cuts to retail outlets and the food-service industry and eliminate the need to merchandise cuts that are in light demand.

Centralized cutting of carcasses into closely trimmed cuts is being adopted by the lamb industry. Breaking carcasses in a central location and shipping the cuts in boxes (Figure 15-23), rather than shipping whole carcasses, increases cutting and handling efficiency, decreases weight loss, and improves meat quality. Freshness of meat has been assured in primal and subprimal cuts by vacuum packaging and by use of carbon dioxide pellets as a refrigerant.

The sheep industry has been slow to adopt centralized cutting because cuts like

Lamb and Mutton Processing

Figure 15-14 Make the breast into trimmed lamb spareribs by removing the outside fat.

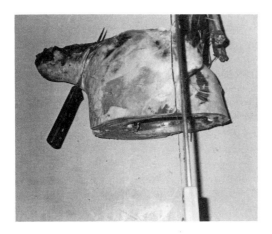

Figure 15-15 Remove the shank and breast from the shoulder by sawing across the distal end of the humerus and making the cut parallel to the top of the shoulder. The neck is removed next to the shoulder. Bone the neck and the small piece of breast (brisket) for ground lamb.

necks and flanks are often difficult to sell at a profit when they are removed from the carcass. The analogous parts of beef carcasses find a ready market in ground beef, but ground lamb and mutton are not popular items. One suggested solution for this problem is to include lean from the less demanded cuts in processed meat.

In comparison with other meats, lamb and mutton are not used to any great extent for processing beyond wholesale or retail cuts. Processed meat in other species includes sausages, restructured steaks, and cured, canned, dried, and roasted products. Very few further processed lamb products are on the market. This is due to excessive fat in most heavier lamb carcasses at weights where such processing is profitable and to the undesirable processing characteristics of lamb fat. Another reason why very little processed sheep meat is on the market is that the processing cost per pound of boneless meat is high for lamb and mutton compared to that for pork or beef because sheep carcasses are small, giving a low yield of meat for the

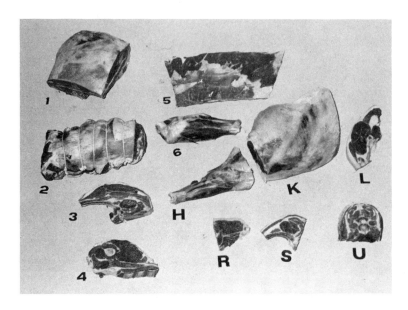

Figure 15-16 Prepare the foreshanks as shown (item number 6.) Cuts shown are as follows: 1, square cut shoulder; 2, rolled boneless shoulder; 3, blade chop; 4, arm chop; 5, spareribs; 6, foreshank; H, hindshank; R, loin chop; K, American leg; S, rib chop; L, sirloin chop; U, neck slice.

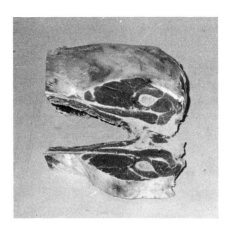

Figure 15-17 Separate the shoulders down the midline. Make arm chops 1 in. (2.5 cm) thick until the humerus enlarges.

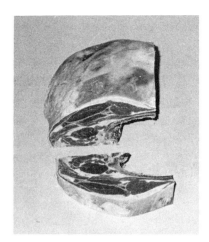

Figure 15-18 Turn the shoulder and make blade chops.

Figure 15-19 Put the neck portion that is not suitable for chops into lean trim or make neck slices as shown in Fig 16-16, item U. An alternative to blade and arm chops is a boneless shoulder roast made by removing the rib cage and then the blade bone and humerus.

labor expended. Processed sheep meat, particularly mutton, is also higher in connective tissue than other meats. Excess connective tissue can cause processed sheep meats to shrink and lose moisture during cooking. This not only makes the product less palatable but also represents a considerable economic loss to the processor.

Research is underway to solve problems of undesirable fat, high boning costs, and connective tissue in lamb and mutton. For fat, mechanical defatting procedures are being developed. In addition, producers can help by reducing the amount of fat on the animals they send to market. Mechanical deboning is an obvious solution to the high cost of hand boning, and machines similar to deboners can also be used to remove sinews from the meat. Another solution to the connective tissue problem is to prepare products in such a way that the objectionable characteristics are minimized. A chunked and formed mutton roast (Figure 15-24) has objectionable connective tissue in it, but when it is sliced extremely thin, much of the toughness is eliminated. Pressure cooking of thicker slices can also eliminate the toughness of connective tissue, but this often results in high cooking losses and in shrunken, distorted products. Lamb and mutton will encounter increasing competition for the consumer's dollar because meat from other species is being improved, and there has been a virtual explosion of convenience foods and products suitable for the food service industry. To survive, the sheep industry must compete successfully with other foods. The fast-food industry and the consumer could benefit from the variety processed sheep meat adds to the menu. Processed sheep meats will be more available to consumers in future years as problems with fat, labor costs, and connective tissue are solved.

BY-PRODUCTS

Pelts are by far the most valuable of the by-products from sheep. They often make up around 5% to 6% of the value of the live lamb but have made up a higher or lower proportion. They are used for skins in the garment industry in addition to

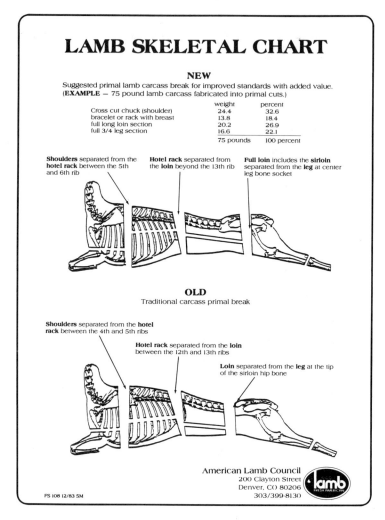

Figure 15-20 New and old cutting procedures outlined by the American Lamb Council. (Courtesy of American Sheep Producers Council)

their use for wool. Other inedible by-products include tallow for soap, cosmetics, and glycerine; blood for pharmaceuticals and animal feed; and intestines for ligatures and strings. Glands for pharmaceuticals are seldom saved from sheep.

Edible by-products include the heart, liver, tongue, and small intestines for sausage casings. There is very little market for sweetbreads, kidneys, fries, brains, cheek and head meats, and edible tallow from sheep, and as a result most packers do not save these items on a regular basis. Most gelatin is produced from pig skins. Gelatin is used in desserts, marshmallows, jellied meat, baking foods, ice cream,

The new look of American Lamb

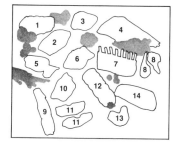

1. Boneless Rolled Shoulder
2. Shoulder Blade Portion
3. Boneless Shoulder Round Bone Section
4. Steamship Leg
5. Boneless Saratoga Roll
6. Dinner Ribs
7. French Rack
8. Double Rib French Chops
9. Tenderloin
10. Boneless Sirloin
11. Loin Chops Tenderloin Removed
12. Boneless Saddle
13. Double Boneless Loin Chop
14. Denver Rib

FS1/7-83

AMERICAN LAMB COUNCIL
200 CLAYTON ST. DENVER, CO 80206
303-399-8130

Figure 15-21 Modern retail cuts of lamb. (Courtesy of American Sheep Producers Council)

and other products. In addition, several new industrial uses for gelatin from skin collagen have been found by the photographic, metallurgical, cosmetic, and pharmaceutical industries. With increased cutting and boning of lamb and mutton in central locations, more bones will be available for use in gelatin and tallow, and in production of mechanically separated lamb and mutton, the residue of which can be used for gelatin. Mechanically separated lamb and mutton have been shown to increase total lean yield from the carcass, and they are useful in the production of processed meats.

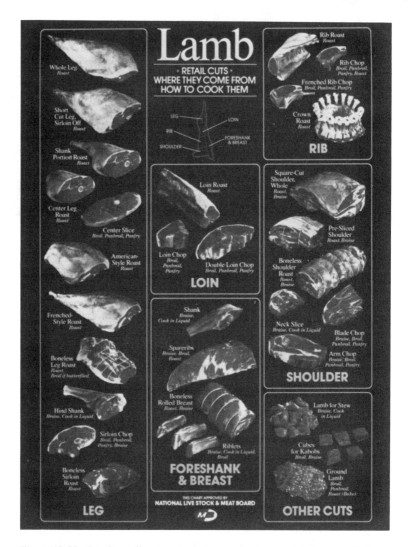

Figure 15-22 Lamb retail cuts, where they come from and how to cook them. (Copyright © 1984 by National Live Stock and Meat Board. Used with Permission.) From Smith et al. 1969.

LAMB—A HEALTHFUL FOOD

Variety is the key to a balanced, health-promoting diet, for no one food contains all the nutrients needed. Only by eating a wide variety of foods from different categories is it possible for the body to receive the more than 50 nutrients it needs daily. To help the consumer make the right choices, foods are divided into four groups:

Lamb—A Healthful Food

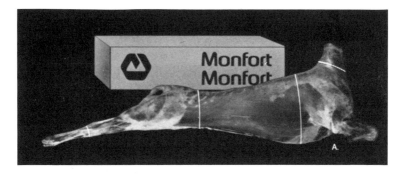

Figure 15-23 (A) Three-piece whole lamb carcass: one double shoulder, one untrimmed back, and one double leg is packed per box. (B) Six-piece whole carcass: two single legs, one trimmed double loin, one trimmed rack, and two single square-cut shoulders are packed per box. All pieces are vacuum packed. (Courtesy Monfort Packing Company.)

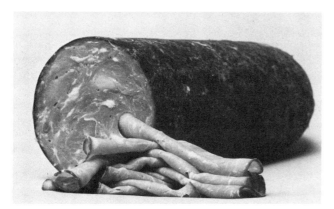

Figure 15-24 This restructured mutton roast was made with chunks of muscle from mutton leg, back, and shoulder, which has been ground through a 1.5-in. kidney plate. Finely ground breast and flank or mechanically separated mutton was added at the 10% level. Problems with tough connective tissue were overcome by cutting the cooked roast into thin slices.

meat, milk, vegetable-fruit, and bread-cereal. A variety of foods from each group should be eaten every day.

The meat group is especially important for it is the major dietary source of high-quality protein and many essential vitamins and minerals. It is recommended that two 2- to 3-oz (56- to 85-g) servings from the meat group should be included in the diet every day. Lamb is a valuable member of the meat group for it is not only a source of good nutrition, but it also adds interest and appetite appeal to mealtime.

The protein in lamb and all red meats is vital to good health for it is complete; that is, it contains all the essential amino acids in the proper proportions to build, maintain, and repair body tissue. Lamb's protein also strengthens the body's defense mechanism against infection and disease.

The B vitamins found in lamb (thiamine, riboflavin, niacin, B-6, and B-12) help the body make better use of other nutrients in food. They also help the body use energy and promote good vision, normal appetite, and healthy skin and nervous system.

Iron and zinc are two important minerals in lamb. Zinc is needed to form enzymes and insulin, while iron helps red blood cells carry oxygen to and carbon dioxide away from other body cells. The heme iron in lamb is in a readily usable form and enhances the absorption of nonheme iron in meat and other foods.

While lamb is high in these many necessary nutrients, it is relatively low in calories. A 3-oz (85-g) serving of cooked, lean lamb contains just 221 calories, yet provides 40% of the protein requirement, 17% of the iron requirement, and 31% of the required zinc for an adult male. This high nutrient density (ratio of nutrients to calories) makes lamb an excellent choice for everyone, even those counting calories.

In spite of all the nutritional benefits from lamb and other meats, some people avoid meat because of cholesterol and saturated fat. The extent to which their presence in the diet is associated with such health problems as heart disease and cancer is a matter of considerable disagreement among scientists.

Nevertheless, it is agreed that cholesterol is an essential part of our body chemistry and that the body manufactures from 800 to 1500 mg of cholesterol a day, even if one doesn't eat any cholesterol at all. One 3-oz (85-g) serving of cooked lean lamb contains 0.78 mg of cholesterol. The USDA estimates the amount of cholesterol in the U.S. food supply to be equal to about 500 mg per person per day. One guideline accepted by some doctors is the American Heart Association's recommendation of 300 mg of dietary cholesterol per day. The cholesterol contained in a 3-oz serving of cooked lamb amounts to only 26% of this recommendation.

The amount of saturated fat in the diet is an issue completely independent from that of cholesterol. A food that is low in saturated fat can be high in cholesterol, and vice versa. Cooked lean from retail cuts of lamb ranges from 6% to 15% fat and the fat is only 56% saturated. The rest is mono- or polyunsaturated fat. Therefore, cooked lean from lamb is low in fat, and the fat that is present is not all saturated.

TABLE 15-5 ANNUAL PER CAPITA CONSUMPTION OF MEAT IN THE UNITED STATES IN POUNDS (CARCASS BASIS)

Year	Beef	Veal	Lamb, Mutton, and Goat Meat	Pork[a]	All Red Meats	Poultry
1899–1910	69.6	6.4	6.6	70.1	152.6	
1911–1920	61.9	6.8	6.1	64.2	138.9	
1921–1930	54.8	7.4	5.5	67.6	135.2	
1931–1940	53.6	7.7	6.7	62.1	130.1	15.9
1941–1950	61.2	9.6	6.0	70.7	147.4	21.8
1951–1960	77.5	7.9	4.4	65.6	135.4	24.7
1961	87.8	5.6	5.1	62.0	160.5	37.8
1962	88.9	5.5	5.2	63.5	163.1	37.0
1963	94.5	4.9	4.9	65.4	169.7	37.5
1964	99.9	5.2	4.2	65.4	174.7	38.2
1965	99.5	5.2	3.7	58.7	167.1	40.8
1966	104.2	4.6	4.0	58.1	170.9	43.8
1967	106.5	3.8	3.9	64.1	178.3	46.3
1968	109.7	3.6	3.7	66.2	183.2	45.0
1969	110.8	3.3	3.4	65.0	182.5	47.6
1970	113.7	2.9	3.3	66.4	186.3	48.4
1971	113.0	2.7	3.1	79.0	197.8	48.6
1972	116.1	2.2	3.3	71.3	192.9	50.7
1973	109.6	1.8	2.7	68.9	178.0	49.0
1974	116.8	2.3	2.3	69.1	190.5	49.5
1975	120.1	4.2	2.0	66.1	182.4	48.6
1976	129.3	4.0	1.9	59.6	194.7	51.9
1977	123.8	3.8	1.7	60.5	189.8	53.3
1978	117.8	2.9	1.6	60.3	182.6	55.9
1979	105.5	2.0	1.6	68.7	177.8	60.5
1980	103.3	1.8	1.6	73.5	180.2	60.5
1981	104.3	1.9	1.6	69.9	177.7	62.4
1982	104.3	2.0	1.6	62.7	170.6	63.9
1983	106.4	2.0	1.7	66.1	176.2	65.0
1984	106.2	2.1	1.7	65.6	175.7	67.1
1985[b]	106.0	2.1	1.6	65.7	175.4	69.4
1986[b]	97.9	1.8	1.4	64.5	165.6	71.9

From *Agricultural Statistics* (various years), U.S. Department of Agriculture; *Livestock and Poultry Outlook and Situation,* U.S. Department of Agriculture, Economic Research Service, LPS-18, October 1985, and previous issues.

[a]Pork includes lard from 1971 to present. Prior to 1971, lard was excluded.
[b]Forecast.

That's the lamb nutrition story. It's a story that should be considered carefully by anyone concerned about good health. In comparison with other logical menu choices, lamb shows itself to be an excellent source of good nutrition.

LAMB CONSUMPTION

Lamb's image with the consumer is less favorable than that of any of its competitors. It is thought to be the most expensive, the least available, and the least likely to be served regularly. Lamb is visualized as being less versatile than other meat and poultry, with consumers feeling that it makes fewer good dishes, is least suitable for any meal, does not make good gravy, is not good for leftovers, and is not good cold. Consumers in general feel that most people do not enjoy eating lamb and that those who do tend to get tired of it. Finally, lamb is felt to be too fat, and many consumers feel that it does not smell good while cooking or when it is served. The undesirable image of lamb in the minds of consumers helps explain the declining consumption shown in Table 15-5.

The American Sheep Producers Council, Inc., the National Live Stock and Meat Board, and others involved in lamb promotion and consumer education continue to point out the fallacy of many consumer attitudes by bringing less expensive cuts to the attention of consumers and by making available recipes and cooking instructions that point out the variety of ways lamb can be used. Publicity, promotions, and advertising to combat the negative image of lamb are continuing. Some are advertising specific brands of lamb as natural because their lamb is produced without hormones, antibiotics, or other additives. Greater amounts of boneless, lean, branded lamb with specific health claims to meet the demands of a particular market segment are likely to be marketed in the future.

REFERENCES CITED

1. Council for Agricultural Science and Technology. 1982. *The U.S. Sheep and Goat Industry: Products, Opportunities and Limitations.* Report No. 94. The Council, Ames, Iowa.
2. Romans, J. R., and others. 1985. *The Meat We Eat,* 12th ed. Interstate Printers and Publishers, Danville, Ill. pp. 161-176, 497-528.
3. U.S. Department of Agriculture. 1982. Official United States Standards for Grades of Lamb, Yearling Mutton and Mutton Carcasses. Code of Fed. Reg. Title 7, Ch. I, Pt. 54, Sec. 54.121-54.127.

16 Lamb Growth and Carcass Composition

INTRODUCTION

The number of animals or live weight of animals prior to slaughter is often used as a measure of the amount of meat produced. However, sheep and lambs vary greatly in weight and animals of the same weight vary in composition. Measures of composition are needed because a knowledge of weight composition is vital to an assessment of the nutrient requirements of growing animals and because a knowledge of composition is essential before the true market value can be determined.

Chemically, the carcass contains protein, fat, water, and ash, and these components, when expressed as percentages of the entire carcass, total 100%. When fat increases, percentages of protein, ash, and water decrease.

Fat Distribution

Fat can be found on the outside of the carcass (subcutaneous fat), between the muscles (intermuscular fat), within the peritoneal cavity (kidney and pelvic fat), within the bone (marrow fat), and within the muscle (intramuscular fat), as shown in Figure 16-1. To determine where fat is located within a carcass, researchers have physically separated carcasses into lean and bone and subdivided fat into subcutaneous, intermuscular, and kidney and pelvic fat.

Changes in distribution of fat with changes in carcass weight are shown in Table 16-1. The values are expressed as allometric growth coefficients. These coefficients are the ratio of specific growth rates of the carcass component listed to the carcass as a whole. A coefficient greater than 1 indicates that the carcass component grows faster than the carcass as a whole, and a coefficient less than 1 indicates that

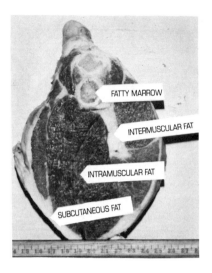

Figure 16-1 A cross section of a lamb leg showing where different fat deposits are located.

the carcass component grows slower than the carcass as a whole. As weight increases, the percentage of fat increases at a faster rate than carcass weight. Within fat depots, subcutaneous fat increases at a faster rate than intermuscular fat or kidney and pelvic fat.

Fat distribution in a lamb carcass is important because it can influence the value of retail cuts. Subcutaneous fat is easily trimmed during fabrication of the carcass, but intermuscular fat is more difficult to remove. Retail cuts left in the counter are often those with excess intermuscular fat. Boneless lamb shoulders and legs can often be trimmed of intermuscular fat, but it is more difficult to trim intermuscular fat in cuts like necks, breasts, and rib chops. Trimming of fat from any area of the carcass is expensive in terms of time and reduced retail yield. Some subcutaneous fat on the carcass is desirable because it increases attractiveness of retail cuts, that is, sales appeal. Marbling or intramuscular fat is almost always an asset because it is a mark of quality.

TABLE 16-1 ALLOMETRIC GROWTH COEFFICIENTS FOR THE MAJOR LAMB CARCASS COMPONENTS

Carcass Component	Ewes	Rams
Intermuscular fat	1.30	1.26
Kidney and pelvic fat	1.44	1.17
Subcutaneous fat	1.76	1.69
Total fat	1.46	1.37
Bone	0.69	0.77
Muscle	0.94	0.97

Adapted from Fourie, Kirton, and Jury.[9]

Ideal Amount of Fat

An ideal lamb should have 3% to 5% intramuscular fat, 0.12 in. (3 mm) of subcutaneous fat evenly distributed over the entire carcass, very little intermuscular fat, and no kidney fat, but completely eliminating undesirable fat deposits is probably impossible. The 0.12 in. of subcutaneous fat on an ideal lamb carcass is optimum. It is exceeded in areas like the dock, brisket, and midribs where excessive amounts of subcutaneous fat often accumulate. Carcasses with the optimum amount of fat will have a high proportion of lean and meet the increasing demand for good eating quality demanded by the modern lamb consumer. These carcasses will usually be choice, yield grade 2 based on USDA standards.

INFLUENCE OF AGE AND WEIGHT

Age and weight are interrelated. As lambs become older, they grow and increase in weight. Obviously, full-fed lambs will be heavier than lambs on restricted diets at the same age. If lambs on a restricted diet are kept longer until they reach the same weight as lambs on full feed, they will have a similar carcass composition. Data in Table 16-2, which is based on a survey of lambs of known age, show that USDA yield grade, an indication of fatness, increases only slightly as lambs of approximately the same weight increase in age from 120 to 269 days.

In contrast to age, weight is closely associated with carcass composition. Percentage of fat increases and percentages of lean and bone decrease with increases in weight. In sheep of a given breed and sex, body weight accounts for 88% to 95% of the variance in total body water, fat, and protein.[13] Percentages of the total ether extract, moisture, protein, and ash in the carcass that are found in each dissected tissue also change as carcasses become fatter and heavier (Table 16-3). For example, the proportion of the total carcass ether extract found in subcutaneous fat increases as carcasses become heavier, but the proportion found in muscle and bone de-

TABLE 16-2 USDA YIELD GRADES FOR LAMBS OF DIFFERENT AGES WITH SIMILAR SLAUGHTER WEIGHTS

Age in Days	No. of Lambs	Live weight (lb)	Live weight (kg)	Yield Grade
120–149	3,490	98.8	44.8	3.3
150–179	12,717	100.5	45.6	3.3
180–209	5,761	98.6	44.7	3.4
210–239	1,881	100.8	45.8	3.4
240–269	3,194	103.8	47.1	3.5

Adapted from Field.[7]

TABLE 16-3 DISTRIBUTION OF ETHER EXTRACT, MOISTURE, PROTEIN, AND ASH IN LEAN, INTERMEDIATE, AND FAT OVINE CARCASSES

Item	Light (N = 5)	Intermediate (N = 5)	Heavy (N = 5)
Fat depth, mm	1.0	4.6	12.6
Body wall, mm	8.2	16.8	34.8
Ether extract weight	1.98 lb (0.90 kg)	4.67 lb (2.12 kg)	11.0 lb (4.99 kg)
Side weight	11.60 lb (5.26 kg)	19.2 lb (8.71 kg)	30.19 lb (13.69 kg)
Ether extract, %			
Subcutaneous fat	28.3	36.7	48.0
Intermuscular fat	35.6	34.6	31.7
Muscles	22.0	18.9	14.9
Flat bones	6.3	4.2	2.7
Round bones	7.8	5.6	2.7
Moisture, %			
Subcutaneous fat	2.5	3.8	4.2
Intermuscular fat	6.5	6.7	7.1
Flat bones	78.3	80.6	82.6
Round bones	3.6	2.3	1.1
Protein, %			
Subcutaneous fat	3.3	4.2	5.0
Intermuscular fat	5.4	5.8	6.2
Muscles	69.5	71.4	71.4
Flat bones	14.7	12.4	12.6
Round bones	7.1	6.2	4.8
Ash, %			
Subcutaneous fat	.6	0.7	0.9
Intermuscular fat	1.1	1.1	1.2
Flat bones	51.9	50.4	56.5
Round bones	33.8	34.9	30.0

Adapted from Field and others.[8]

creases. In general, the desire for increased leanness has been accompanied by a desire to increase slaughter weight. It is possible to increase both weight and leanness if enough selection pressure against fatness is applied.

Advantages of Increasing Weight

An increase in weight could increase product availability and allow lamb to compete more successfully with other red meats. Net returns to producers could improve if heavier lambs were sold without price discounts. It costs the same to feed and care for the ewe flock regardless of the weight of the lamb sold. Only the cost of the feed is higher for heavier lambs.

The cost per unit of weight to slaughter heavy lambs is lower than for light lambs. Some packers kill up to 400 lambs per hour regardless of the lamb weight. In addition, costs to buy, truck, and sell a 110-lb (50-kg) lamb are nearly the same as the costs for a 126-lb (57-kg) lamb. The authors are aware of one lamb slaughter plant that dresses four lambs per worker hour and another that can dress up to seven lambs per worker hour. Theoretically, 16-lb (7-kg) heavier lambs would result in 64 to 112 lb (28 to 49 kg) more of lamb being dressed per worker hour at these plants.

Dressing percentage is variable for lambs of all weights, but it is common for 110-lb (50-kg) lambs to yield 50% and for 126-lb (57-kg) lambs to yield 52%. For every 110 lb (50 kg) of live lamb purchased, the packer often has one additional kilogram of carcass to sell when slaughter weights are increased from 110 to 126 lb (50 to 57 kg).

Carcass shrinkage during chilling and shipping is reduced when heavier lambs are processed. Evaporation of moisture from the carcass causes weight loss between the time the carcass is dressed and the time it is cut, and weight loss means dollar loss.

Loin chops, leg roasts, and other cuts from heavy lambs are trimmed and packaged in the same time it takes to package cuts from light lambs. Therefore, per unit weight, labor and packaging costs for heavier lambs are lower than they are for lighter lambs. Savings in labor could reduce the retail price for lamb, thereby increasing consumer demand.

Southam and Field[15] compared consumer selection of cuts from light and heavy lambs in retail counters. When unidentified retail cuts similar to those in Figure 16-2 were placed side by side in self-service counters, consumers selected rib and loin chops from 66-lb (30-kg) carcasses by a ratio of 6 : 5 over similarly finished rib and loin chops from 50-lb (23-kg) carcasses. Leg roasts from the lighter carcasses were selected by a ratio of 3 : 1 over leg roasts from heavier carcasses; but when extra sirloin steaks were cut from the heavier legs, they weighed approximately the same as the lighter legs, and under these conditions no difference in consumer selection of leg roasts was noted.

The preceding advantages for producing heavier lambs can be summarized by saying that the cost for producing and processing lamb would be reduced; more

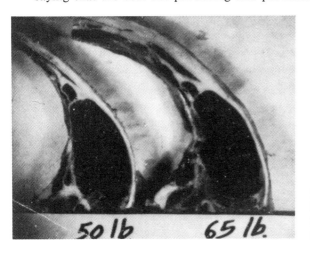

Figure 16-2 Rib chops from heavier lambs are preferred over rib chops from lighter lambs when the chops have a similar amount of subcutaneous fat.

lamb would be available for purchase and prices would be more competitive with other meats.

Disadvantages of Increasing Weight

The disadvantages of producing heavier lambs must be evaluated before an individual producer can decide to increase the weight of lambs going to market. Heavy lambs are discounted when there are large supplies of lambs. Discounts are common when old crop lambs are still available in feedlots and the movement of spring, milk-fat lighter-weight lambs starts. In the summer months when very few heavy-weight lambs are on the market, no discount for heavy lambs exists.

Feed efficiency decreases in lambs as weight increases. In one study, feed per kilogram of gain increased from 6.0 to 7.1 kg of feed needed to produce 1 kg of gain as live weight increased from 43 to 65 kg. The amount of feed needed to produce 1 kg of gain will vary depending on the kind of feed and the feeding conditions, but lighter lambs will always be more efficient than heavier lambs. In addition to lower feed efficiency, average daily gain is also lower in heavier lambs.

A major problem associated with marketing heavier lambs is that on the average they are fatter than lighter lambs. Fatter lamb carcasses require more trimming of retail cuts, and the greater amount of discarded fat lowers the carcass value per hundred weight when heavier lambs are compared to lighter lambs. When the cutting and trimming method used at the University of Wyoming is followed, 51-lb (23-kg) carcasses average 70% of the carcass as retail cuts, while 66-lb (30-kg) carcassses average 66% of the carcass as retail cuts. It is obvious that lamb weights cannot be increased further until heavy lambs that are lean can be produced. Lower cutability in heavier lambs is the reason why many retailers limit buying specifications to lighter carcasses weighing 55 lb (25 kg) or less. In addition, some change in cutting method is needed to make leg cuts an acceptable weight and price for consumers. If yield grades for lamb were used to identify heavy lambs that are not fat, heavier lambs would gradually be accepted and the price of these lambs would increase. On the average, yield grade 1 lambs have 4.6% of the carcass that must be trimmed and yield grade 5 carcasses have 19.0%.

Heavier lambs are often older, and this becomes a problem when both break joints fuse at around 10 to 15 months and the lambs are classified as yearlings. There is little demand for yearlings and the price is much lower. The yearling problem is already a common one when the tail end of the old crop lambs is leaving the feedlot.

Another disadvantage for the production of heavier lambs is that some smaller breeds will not be able to produce 57-kg lambs efficiently, and if 126 lb (57 kg) is reached, the smaller breeds of sheep will be too fat. Some small breeds of sheep have advantages such as a high lambing percentage, which cannot be overlooked. Before everyone in North America jumps on the "heavy lamb bandwagon," we need to remember that leading lamb producing countries such as New Zealand are still producing 29- to 35-lb (13- to 16-kg) lamb carcasses. Nevertheless, many in New Zealand are also considering production of heavier lambs to improve efficiency.

In summary, lamb producers, packers, retailers, and food service establishments will need to evaluate the "pros" and "cons" of producing, processing, and merchandising heavy lambs before deciding what weight of lambs to handle. It is the author's opinion that, overall, the pros outweigh the cons, and that live weights for market lambs will continue to increase gradually as more heavier but meatier lambs are marketed.

INFLUENCE OF SEX

Ram lambs produce red meat more efficiently than wether and ewe lambs, through more rapid gains, improved feed efficiency, and a higher percentage of carcass lean. This has resulted in an expression of concern by animal scientists over the policy of discounting ram lambs. The price discounts can occur at the feedlot or at slaughter. In contrast to castration of the male, spaying of the female is not a practical procedure, is seldom practiced, and produces no advantages in lambs.

The purpose of this section is to review briefly the advantages and disadvantages of lean meat production from ram lambs.

Advantages of Producing Ram Lambs

Ram lambs when fed a high level of nutrition outgain wether lambs by 15%. A low plane of nutrition can suppress the differences in growth rate, but even under these conditions rams often show some advantages in growth when compared to wethers. A difference in carcass yield in favor of wethers when compared to rams offsets part of the growth advantage for ram lambs, but even on a carcass basis rams are usually superior to wethers. The difference in favor of rams becomes larger when expressed as gain of lean meat per day of age.

Because the correlation between rate of gain and feed efficiency is high, rams are 12% to 15% more efficient in converting feed into live-weight gain than wethers.

The current problem of excess fat associated with lamb carcasses is one major reason why some are encouraging production of ram lambs. Measures of leanness such as separable muscle, percentage protein, fat depth, and weight of trimmed retail cuts all confirm that rams are leaner than wethers and that wethers are leaner than ewes when lambs of the same weight are compared (Table 16-4). Heavier ram lambs often have less fat in their carcasses than lighter wethers. Therefore, ram lamb production can improve leanness and weight at the same time.

Disadvantages of Producing Ram Lambs

Disadvantages of producing ram lambs are not as clear-cut as the advantages, and scientists do not always agree with persons in the meat trade on the importance of the disadvantages.

TABLE 16-4 MEANS FOR WETHER, RAM, AND EWE LAMBS AT A CONSTANT CARCASS WEIGHT OF 46 LB (21 KG)

	Wethers	Rams	Ewes
No. of lambs	276	207	202
Carcass trait			
Fat depth, mm	4.8	4.4	5.9
Kidney fat, %	1.9	1.5	2.4
Retail leg, loin, rib and shoulder, %	64.4	66.0	64.2

Adapted from Carpenter and others.[2]

Lamb feeders are opposed to feeding ram lambs because slaughterers pay feeders less for the lambs when they leave the feedlot. The aggressive sexual behavior of ram lambs in the feedlot plus unwanted matings when ram and ewe lambs are in the same pen are other reasons for price discrimination. Lamb feeders often buy older lambs that have come off drought areas, and these lambs can approach a year of age at slaughter. Because of the older ages, some ewe lambs are pregnant and some ram lambs exhibit secondary sex characteristics at slaughter.

Lamb slaughterers discriminate against ram lambs because grading systems permit grade discounts for lambs having secondary sex characteristics such as enlarged necks and shoulders, and there is little market demand for these lower-grade carcasses. The slightly higher proportion of the ram carcass that sells at a lower price as neck and shoulder and the slightly lower proportion of the carcass that is back and leg when compared to wethers do not justify the difference in price between rams and wethers, but it is another reason often given for price discrimination against rams. The problem of grade differences is restricted almost entirely to older, more mature ram lambs since numerous studies have shown that in younger lambs only minor grade differences between rams and wethers exist.

Most slaughter lambs are bought on a live-weight basis. Therefore, carcass yield based on live weight is an important consideration. Ram lambs always have a lower dressing percentage than wether lambs because they are leaner, and the ram's testicles, and horns in some breeds, are part of the live weight but not part of the carcass weight. At the same weight and age, the pelt from ram lambs often weighs more than the pelt from wether lambs, and heavier pelts further increase the differences between carcass yields of ram and wether lambs. Average dressing percentages of 49% for ram lambs and 51% for wethers are common.

Reasons for lower prices paid for live ram lambs when compared to wethers include difficult pelt removal and more damaged pelts when they are removed. Mechanical pelt removal (Figure 16-3) will reduce the problem with difficult dressing because less hand labor is required. In addition, less fells are broken when pelt pullers are used on ram lambs than when pullers are used on wether lambs, even though more force is required to remove the pelt.

Influence of Sex 297

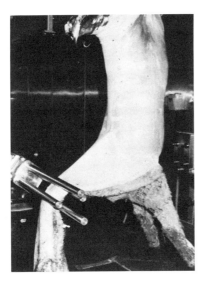
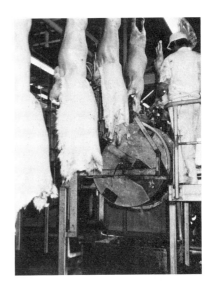

Figure 16-3 Mechanical pelt removal. The drum puller on the right removes pelts from the shoulder; the pelt puller on the left pulls pelts from the hind legs. Advantages of pelt pullers are cleaner carcasses and labor savings.

Status of Ram Lamb Production

Ram lamb production is increasing in farm flocks where lambs are marketed at less than 6 months of age. The disadvantages of lean meat from older ram lambs have been of sufficient magnitude to suppress production on Western ranges and in feedlots. The only ram lambs produced on Western ranges are those that escaped undetected at the time of castration. Castration is probably an asset to the sheep industry in the Western states because it decreases management problems and eliminates soft fat and lower grades associated with heavy ram lambs, 10 to 12 months of age.

Ram lambs are being produced in many Midwest and Eastern farm flocks and to a lesser extent elsewhere. Farmers who fatten ram lambs before they are 6 months old help the sheep business. These lambs are finished and marketed during the spring and summer months and do not compete with the fall run and feedlot lambs of the September to December period. Because lambs are in short supply in the spring and summer and because lambs are young, price discounts for these ram lambs are less severe. Minimal problems with grade discounts are experienced, but some slaughterers do complain about difficult pelt removal and lower dressing percentages. Leaner, younger ram lambs will always have a lower carcass yield than fatter, more mature wether lambs, but the meat they produce is more desirable because it contains less fat. Dressing percentage problems in leaner ram lambs can only be overcome by selecting for heavier muscling and by selling on a carcass-weight basis rather than on live weight, as discussed in Chapter 14.

INFLUENCE OF DIET

Of the facts discussed in this chapter that influence lamb carcass composition, diet is certainly the most controversial. Some feel that increasing the level of feed intake increases the amount of fat; some feel that level of feed has no effect on amount of fat; and some indicate that there is a decrease in amount of body fat with increased feed level.[1] Similarly, changing the protein content of diets has been shown to either increase, not affect, or decrease the weight of fat in lambs of the same weight.

Extremes in Level of Feeding

Variation in fat content of lambs of similar sex, weight, and genotype can often be attributed to extremes in level of feeding. When a lamb is fed at maintenance, there is a net gain in body protein and a net loss in body fat. Therefore, when a 55-lb (25-kg) lamb with normal growth is compared to a 55-lb (25-kg) lamb fed near maintenance, the lamb fed near maintenance will be leaner. Lambs coming from areas of severe drought often fit into the leaner category even though they are much older than lambs with normal growth. Black[1] suggests that a maximum difference in fat weight of about 4.4 lb (2 kg) can be obtained by altering the intake of a well-balanced diet and that the differences are generally less than 1.1 lb (0.5 kg) in sheep with an empty body weight over 88 lb (40 kg).

The growth rate of lambs following refeeding after periods of prolonged undernutrition is greater than it is in lambs of the same weight that have had unlimited access to feed. The most likely explanation is that there is a change in the partition of retained energy, with an increase in deposition of protein and water and a decrease in the retention of fat.[6] An example of lambs that could have slightly less fat as a result of rapid growth from refeeding would be lambs on low planes of nutrition that are then placed in feedlots where the gain is faster than it would have been if the lambs had always been on full feed.

Means typical of those obtained for carcass fat content of lambs on different feeding levels are found in Table 16-5. Fat thickness and percentage of ether extract in the boneless carcass did not vary when different energy levels were fed. Percentage of kidney and pelvic fat is often the most variable of fat deposits, but kidney and pelvic fat should not be considered part of the carcass since it is easily removed on the kill floor and does not contribute to the weight of retail cuts.

Extremes in Level of Protein Fed

In addition to level of feeding, the composition of the diet can influence fat content of carcasses of similar weight. As protein absorption increases, the proportion of energy deposited in body proteins increases and the amount of energy deposited in fat decreases.[1] The ratio of fat to protein in the body gain decreases markedly as protein absorption increases from deficient levels. Thus lambs fed diets in which

Influence of Diet

TABLE 16-5 MEANS TYPICAL OF THOSE OBTAINED FOR CARCASS FAT CONTENT OF LAMBS ON DIFFERENT ENERGY LEVELS

Item	Energy Level		
	Low	Medium	High
No. of lambs	46	45	47
Diet composition, %			
Alfalfa	89.3	51.6	14.3
Corn	8.4	33.4	59.4
Soybean meal	0	8.8	18.2
Protein	19.4	18.3	21.0
Digestible energy in Mcal/kg	2.7	2.9	3.3
Fat in carcass[a]			
Fat depth, mm	3.4	3.4	3.8
Kidney fat, %[b]	1.6	1.8	2.2
Ether extract, %	22.3	22.1	22.3

Adapted from Crouse and others.[4]

[a]Carcass weight adjusted to 43.3 lb (19.6 kg) for all energy levels.

[b]Kidney fat was not included with ether extract.

protein absorption is below requirements grow more slowly and contain more fat than lambs of the same weight receiving adequate protein.[11]

In contrast to diets that are deficient in protein, there is some evidence that diets well in excess of the protein requirements produce leaner lambs that grow more slowly than those receiving the correct amount of protein.[14] Because slower growth and more expensive protein supplements are required to produce leaner lambs, feeding excess protein to produce leaner lambs is not a common practice. Even if it were economically feasible, the possibility of excess protein feeding to produce leaner lambs is tenuous because the activity of rumen microbes in feeder lambs often makes the relationship between the amount of protein eaten and the amount of protein absorbed low.

The lack of differences in carcass composition between diets of lambs containing 10.5% and 13.5% protein in Table 16-6 emphasizes the difficulty of altering carcass composition with diet. The live traits of average daily gain, feed efficiency, and length of time on feed were influenced by protein level.

Addition of Fat or Acetate to the Diet

Variation in the level of fat in the diet can also affect fat deposition in the carcass because the energy in long-chain fatty acids is converted to body fat with a greater biochemical efficiency than the energy in other absorbed nutrients. For example, lambs fed a polyunsaturated fat supplement have more carcass fat than pasture-fed lambs.[12] The amount of acetate absorbed in relation to other nutrients can also increase the amount of fat deposition in lambs fed hay or concentrate diets.[1]

TABLE 16-6 MEANS FOR CARCASS FAT CONTENT OF LAMBS ON DIFFERENT PROTEIN LEVELS

	Protein, %	
Variable	10.5 (N = 32)	13.5 (N = 31)
Days on feed	152.12	131.31
Average daily gain	0.40 lb (0.18 kg)	0.46 lb (0.21 kg)
Feed per kg gain	9.21	8.41
Hot carcass weight	66.0 lb (29.9 kg)	66.0 lb (29.9 kg)
Fat depth, mm	8.10	9.10
Carcass analysis		
Moisture, %	49.70	49.11
Fat, %	35.29	35.91
Protein, %	13.52	13.54
Bone, %	13.85	13.85

Adapted from Craddock, Field, and Riley.[3]

Practical Application of Changes in the Diet

Although the nutritional treatments discussed can change carcass composition in lambs of the same weight, the change is small when compared with the effects of body weight, sex, and genotype, and it is questionable if changing fat deposition by utilizing these techniques will become popular in future years. Changes in the amounts of subcutaneous and intermuscular fat relative to one another are also largely independent of nutritional effects but are related to specific growth rates of these tissues.[16] Unlike sheep or cattle, body composition of the pig, rat, chicken, and duck can be changed by nutritional manipulation.[13]

INFLUENCE OF BREED

The economic pressure imposed on lamb producers in the United States by the decline in lamb consumption has stimulated interest in carcass improvement through selection. Reducing carcass weight to improve leanness often results in a lower rate of return per ewe, and elimination of castration where feasible is of no value in reducing fat in ewe lambs. Next to weight reduction, the best method of reducing fat is through genetic improvement. Our discussion will center on selection of maximum lean, minimum bone, and optimum levels of fat as methods of carcass improvement. It is recognized that other aspects of sheep improvement including fertility, prolificacy, survival, live weight, and palatability are also of economic significance to the sheep producer, but these aspects are discussed elsewhere in this text.

Bases for Breed Comparisons

Evaluation of differences in fat, lean, and bone have been made at a constant weight, constant amount of fat, and constant maturity. The basis chosen for breed comparison will influence interpretation of the results. Where breeds of similar mature size are compared, differences in fat content of the carcass at a constant weight are valid. When breeds of different mature weights are compared at constant weights, the differences in fat content are often a result of stage of development (Figure 16-4). Therefore, a smaller breed like Southdowns will always appear fatter at heavier slaughter weights than will a larger breed like Suffolks. Nevertheless, comparisons are often made at a constant weight because variations in levels of fat, lean, and bone are readily interpreted in commercial terms. In addition, breed comparisons at similar weights minimize environmental differences.

Kidney Fat Comparisons

Even when the total amount of fat in carcasses of different breeds is constant, distribution of kidney and pelvic fat within the carcasses may vary considerably. At a total body fat of 28%, kidney and pelvic fat weights of 3.3 lb (1.5 kg) in Rambouillets, 3.7 lb (1.7 kg) in Dorsets, and 7.3 lb (3.3 kg) in Finnish Landrace ram lambs have been reported.[10] At a constant live weight of 93 lb (42 kg), the differences between breeds were even larger, but smaller differences between breeds were observed when the data were adjusted to 85% of the mature sheep weight. Other data on variation in percentage of kidney fat by age and breed are found in Table 16-7. Corriedales and Targhees possessed more kidney fat than Suffolks, Hampshires, Dorsets, and Rambouillets at both 22 and 26 weeks of age.

Large differences in kidney and pelvic fat exist when all breeds are included.

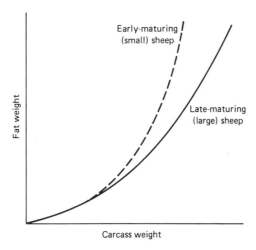

Figure 16-4 Larger, later-maturing breeds reach heavier weights before depositing excess fat.

TABLE 16-7 VARIATION IN FAT BY BREED OF SHEEP WHEN SLAUGHTERED AT A CONSTANT AGE OF 22 OR 26 WEEKS

	22 Weeks				26 Weeks			
	Carcass Weight		Kidney Fat	Fat depth	Carcass Weight		Kidney Fat	Fat Depth
Breed	(lb)	(kg)	(%)	(mm)	(lb)	(kg)	(%)	(mm)
Suffolk	57.1	25.9	2.6	3.3	69.4	31.5	2.8	3.5
Hampshire	48.5	22.0	2.4	4.2	61.5	27.9	3.0	4.5
Dorset	41.0	18.6	2.4	2.6	52.0	23.6	2.7	3.1
Rambouillet	45.4	20.6	2.4	2.4	56.0	25.4	2.9	4.1
Targhee	47.0	21.3	2.8	3.6	58.2	26.4	3.6	4.4
Corriedale	42.3	19.2	3.0	4.2	53.4	24.2	3.6	5.5

Adapted from Dickerson and others.[5]

If only the more traditional breeds and crossbreds are compared, the large differences in amount of kidney fat among breeds diminish. In countries like New Zealand, kidney fat is removed on the kill floor. Therefore, it is not part of the carcass and it does not add to the amount of carcass fat. Where kidney and pelvic fat is removed, there are savings in transportation of fat and carcasses are more attractive to retail and food service operators. In addition, the variations in carcass composition among breeds or variations in fat distribution among breeds that results from kidney fat being left in the carcass are eliminated. Sheep producers might do well to ignore variation in kidney and pelvic fat when selecting for leaner carcasses because it is simply a matter of time until the outdated practice of including kidney fat as part of the carcass ceases. Kidney and pelvic fat will then become viscera, which could influence dressing percentage, but not carcass fat percentage.

Lean, Fat, and Bone Comparisons

Examples of the variation in percentages of lean, fat, and bone that can be found between breeds are shown in Table 16-8. Lean in physically separated carcasses

TABLE 16-8 VARIATION IN CARCASS COMPOSITION BY BREED OF SHEEP[a]

	Physical Separation		
	Lean (%)	Fat (%)	Bone (%)
Rambouillet	50.5	30.0	19.4
Romnelet	47.9	32.6	19.2
Columbia	46.5	34.4	18.9
Targhee	49.6	31.5	18.8
Suffolk	49.2	32.4	18.3

Adapted from Vesely and Peters.[17]

[a]Carcasses approximately 22 kg with kidney and pelvic fat removed.

varies from 46.5% in Columbias to 50.5% in Rambouillets. Perhaps the most important point to emphasize from these data is that at market weight the lamb carcass is only approximately 50% lean, with fat and bone making up the balance. Therefore, there is a need to increase lean percentage in the carcasses of all breeds. Percentage of lean in the carcass is often the most difficult to increase of the three major tissues because there is a tendency for bone percentage to decrease as fat percentage increases. For example, Rambouillets (Table 16-8) had 30% fat and 19.4% bone, while Suffolks had 32.4% fat and 18.3% bone. While fat varied by 2.4 percentage points, lean only changed by 1.3 percentage points, because the remainder of the increase in fat between Rambouillets and Suffolks was offset by a decrease in bone.

Retail Cut Comparisons

Variation in weight and percentage of retail cuts by breed of sheep slaughtered at a constant age is another method of comparing differences between breeds (Table 16-9). At both 22 and 26 weeks, Suffolks produced a greater weight of retail cuts than the other breeds, while Dorsets had the highest percentage of retail cuts at both ages.

Effect of Breed on Fat Distribution

Differences between breeds in both subcutaneous and intermuscular fat at constant weights of total depot fat have been reported, but the differences are of little practical or economic significance. In one study the ratio of subcutaneous to intermuscular fat only ranged from 1.03 to 1.06 when eight of the most popular British breeds were considered.[18] Only in the case of a ninth breed, the Wensleydale, was there a significant difference when the subcutaneous to intermuscular fat ratio of 1.11 was compared to those of the other eight breeds. Small but economically unimportant

TABLE 16-9 VARIATION IN WEIGHT AND PERCENTAGE OF RETAIL CUTS BY BREED OF SHEEP WHEN SLAUGHTERED AT A CONSTANT AGE OF 22 OR 26 WEEKS

	22 Weeks			26 Weeks		
	Retail Cuts		Retail Cuts (%)	Retail Cuts		Retail Cuts (%)
Breed	(lb)	(kg)		(lb)	(kg)	
Suffolk	44.8	20.3	78.5	53.4	24.2	77.2
Hampshire	38.8	17.6	79.3	47.0	21.3	77.0
Dorset	32.8	14.9	79.9	40.8	18.5	78.6
Rambouillet	36.2	16.4	79.2	43.8	19.4	76.7
Targhee	36.4	16.5	77.4	43.2	19.6	74.7
Corriedale	32.8	14.9	77.2	39.5	17.9	74.8

Adapted from Dickerson and others.[5]

effects of breed on bone distribution have also been shown through physical separation of bone, fat, and muscle.

Effects of Breed on Lean Distribution

At constant weights of total muscle within the range at which lambs are normally marketed, breed differences in muscle distribution are small.

Factors Influencing Selection for Leanness

Rate of gain and weight at slaughter are often more important factors influencing the choice of breed than carcass leanness because of a lack of premium payment for leaner carcasses and because large differences in rate of gain and optimum weight at slaughter between breeds exist. Selection of larger, faster-growing, later-maturing breeds of sheep that can be slaughtered at heavier weights without being overfat is thought by many to be the most economical way to produce leaner lambs under today's marketing system. Nevertheless, selection for growth rate alone cannot be used to improve leanness because a low correlation between growth rate and carcass leanness exists.

Heritability of Lean and Fat

It is clear from estimates of heritability of lamb carcass traits that selection within breeds can be used as a method of improving leanness. Heritability estimates of 0.40 for percent of lean in the carcass have been reported, while heritability of percent of fat ranges from 0.20 to 0.50. In addition, fat depth over the longissimus muscle has a heritability of 0.20 to 0.50. This easily obtained measurement, when combined with carcass weight, is a good indication of the percentage of lean or fat in a lamb carcass. Fat depth over the longissimus muscle can also be obtained by ultrasonics on live animals and used in conjunction with live weight to make improvement in carcass leanness. In practice, fat depths in faster-growing lambs will need to be adjusted to a constant weight to avoid selection of lighter, slower-growing lambs.

Use of Conformation to Improve Leanness

Conformation of the live lamb or of the carcass varies between and within breeds of lambs (Figure 16-5). Nevertheless, selection for superior conformation is relatively ineffective as a method of improving leanness in lambs because the relationship between subjective measures of conformation and carcass lean is low. Conformation based on visual assessment in selection schemes could have a deleterious effect on progress in more important traits because carcasses with superior conformation are often fatter. On the other hand, extremely heavily muscled animals often have a higher dressing percentage than angular animals, and the muscle-to-bone ratio in

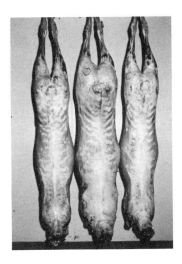

Figure 16-5 These lamb carcasses differ in conformation, but their carcasses contain a similar lean percentage.

heavily muscled animals is often improved. Therefore, superior conformation may have some economic benefits, in addition to the higher prices traditionally paid for more heavily muscled animals.

REFERENCES CITED

1. Black, J. L. 1983. Growth and development of lambs. In *Sheep Production,* W. Haresign, ed. Butterworths, London.
2. Carpenter, Z. L., and others. 1969. Indices for estimating cutability of wether, ram and ewe lamb carcasses. *J. Animal Sci.* 28:180.
3. Craddock, B. F., R. A. Field, and M. L. Riley. 1974. Effect of protein and energy levels on lamb carcass composition. *J. Animal Sci.* 39:325.
4. Crouse, J. D., and others. 1978. Effect of dietary intake on carcass composition and palatability of different weight carcasses from ewe and ram lambs. *J. Animal Sci.* 47:1207.
5. Dickerson, G. E., and others. 1972. Genetic resources for efficient meat production in sheep. Growth and carcass characteristics of ram lambs of seven breeds. *J. Animal Sci.* 34:940.
6. Drew, K. R., and J. T. Reid. 1975. Compensatory growth in immature sheep. III. Feed utilization by sheep subjected to feed deprivation followed by realimentation. *J. Agr. Sci. Camb.* 82:215.
7. Field, R. A. 1971. Survey and identification of management systems which are producing high cutability choice and prime lambs. Report to American Sheep Producers Council, Inc. Denver, Colo.
8. Field R. A., and others. 1985. Distribution of ether extract, moisture, protein and ash in dissected tissues from ovine carcasses. *J. Animal Sci.* 60:977.

9. Fourie, P. D., A. H. Kirton, and K. E. Jury. 1970. Growth and development of sheep. II. Effect of breed and sex on the growth and carcass composition of the Southdown and Romney and their cross. *New Zealand J. Agr. Research* 13:753.
10. Notter, D. R., C. L. Ferrell, and R. A. Field. 1983. Effects of breed and intake level on allometric growth patterns in ram lambs. *J. Animal Sci.* 56:380.
11. Orskov, E. R., and others. 1976. The nutrition of the early weaned lamb. IV. Effects on growth rate, food utilization and body composition of changing from a low to a high protein diet. *J. Agr. Sci. Camb.* 86:411.
12. Purchas, R. W. 1978. Some effects of nutrition and castration on meat production from male Suffolk cross (Border Leicester–Romney cross) lambs. I. Growth and carcass quality. *New Zealand J. Agr. Research* 21:367.
13. Reid, J. T., and others. 1968. Some peculiarities in the body composition of animals. In *Body Composition in Animals and Man.* Publ. 1598, National Academy of Sciences, Washington.
14. Searle, T. W., N. McC. Graham, and J. B. Donnelly. 1982. The effect of plane of nutrition on the body composition of two breeds of weaner sheep fed a high protein diet. *J. Agri. Sci. Camb.* 98:241.
15. Southam, E. R., and R. A. Field. 1969. Influence of carcass weight upon carcass composition and consumer preference for lamb. *J. Animal Sci.* 28:584.
16. Tulloh, N. M. 1964. The carcass composition of sheep, cattle, and pigs as functions of body weight. In *CSIRO Symposium: Carcass Composition and Appraisal of Meat Animals,* D. E. Tribe, ed. CSIRO, Melbourne.
17. Vesely, J. A., and H. F. Peters. 1972. Muscle, bone and fat and their interrelations in five breeds of lamb. *Canad. J. Animal Sci.* 52:629.
18. Wolf, B. T., and C. Smith. 1983. Selection for carcass quality. In *Sheep Production,* W. Haresign, ed. Butterworths, London.

17 Palatability of Lamb and Mutton

If it tastes good people will buy it. Lamb taste is composed of tenderness, juiciness, flavor, and aroma. Lamb can be made tough through rapid chilling or freezing immediately after slaughter, but most properly cooked lamb is tender. Lamb can be made extremely dry by overcooking, but consumers usually perceive lamb as being a juicy meat. With respect to factors influencing tenderness and juiciness, lamb is like any other meat. Lamb flavor and aroma are not like any other meat. For most people, they are distinctively different.

Whether the distinctive flavor of lamb is exquisite or peculiar depends on the preferences and psychology of the individual evaluating it. Food habits of individuals, appearance of the food, mental and emotional states, and so on, influence evaluation of flavor. These factors also influence likes or dislikes for lamb, and they may be more important than taste in determining whether or not people will buy it. Flavor is often cited as one reason for low levels of lamb consumption in the United States, but lamb flavor is not objectionable in New Zealand and Australia where lamb consumption is high. Indeed, the author has observed U.S. citizens eating lamb in New Zealand. Many of these consumers failed to purchase lamb at home, but they commented on the exquisite taste of the product in New Zealand. It is likely that the consumer, not the flavor, changed as the consumer traveled from the United States to New Zealand. While consumers in the United States often object to the intensity of lamb flavor, consumers in the Middle East countries often object to the lack of flavor in the same lamb. Overall, likes and dislikes related to flavor vary by individuals, by country, and by the environment or situation in which individuals find themselves.

In this chapter we will discuss some factors that affect the palatability of lamb and mutton. Much of the emphasis will center on flavor.

INFLUENCE OF AGE AND WEIGHT

Historically, three concepts have formed the basis for grade identification in the United States (1) increases in age are negatively associated with palatability; (2) physiological maturity indicators in the carcass are indicators of increases in age at slaughter; and (3) subjective estimates of physiological maturity can be used to detect differences in the palatability of lamb and mutton. The USDA meat-grading branch uses subjective estimates like muscle color, degree of bone ossification, and shape of bone to identify six maturities of A-minus to B-plus for lamb, to separate lamb from yearlings and to distinguish between yearling mutton and mutton carcasses.

The usefulness of subjective estimates of maturity to distinguish between lamb and mutton and to detect differences in connective tissue toughness has not been questioned. Nevertheless, several researchers have questioned the validity of using age or maturity to distinguish palatability differences not related to connective tissue. Wenham and others[11] reported that lean from 3- to 9-year-old ewes that was relatively low in connective tissue gave the same low shear values as that from lambs. Smith and others[10] reviewed studies in which more mature lamb carcasses were more tender than younger carcasses and other studies in which carcasses varying significantly in USDA maturity score did not differ appreciably in flavor or tenderness. When lambs of A and B maturity and yearling mutton were included in the same study (Table 17-1), juiciness, tenderness, and overall satisfaction scored slightly higher in younger A maturity lambs than in yearlings, but there were no differences in flavor scores. When these same palatability traits were ranked by chronological age instead of maturity, differences between age groups were slightly larger. Nevertheless, age must be estimated by changes in physiological maturity in intact carcasses under packing house conditions, and many factors affect muscle color and bone characteristics. Therefore, the usefulness of maturity in detecting differences in palatability is limited.

Because weight is highly correlated with age in growing lambs, the statements for the effects of age or maturity on palatability also apply to weight. Overall, very

TABLE 17-1 MEAN PALATABILITY SCORES FOR PRIMAL CUTS (UNKNOWN HISTORY) GROUPED ACCORDING TO USDA MATURITY CLASSES

	Maturity Class		
Trait	A	B	Yearling
Flavor	5.82[a]	5.88[a]	5.78[a]
Juiciness	5.95[a]	5.65[b]	5.76[b]
Tenderness	6.26[a]	5.86[b]	5.96[b]
Overall satisfaction	5.89[a]	5.73[a,b]	5.70[b]

From Smith and others.[10] Higher numbers indicate more desirable palatability traits.
[a,b] Values bearing the same superscript are not different ($P < 0.05$).

few consistent differences in tenderness, juiciness, flavor, aroma, or overall satisfaction that can be attributed to weight have been reported. Changes in weight have been shown to increase and to decrease each of these palatability characteristics. The major factor influencing tenderness of lamb is not weight but how rapidly the lamb carcass is chilled or frozen after slaughter. Normally, 8 to 12 hours are required for completion of rigor after slaughter. If the temperature of lamb muscle drops to 32°F (0°C) or below before rigor mortis is complete, the meat will always be much tougher than meat from carcasses chilled at a slower rate.

This discovery resulted in New Zealand requiring all lamb carcasses to be electrically stimulated immediately after slaughter and allowing at least 90 minutes to elapse from slaughter to freezing. Electrical stimulation plus the 90-minute holding period allows for acceleration of glycolysis, depletion of adinosine triphosphate, fall of muscle pH, and installation of rigor mortis, enabling rapid cooling without toughening due to cold shortening.

In the United States, where lamb carcasses are not frozen, there is much less need for electrical stimulation to prevent toughening of muscle. Lambs are usually slaughtered and then placed in a holding cooler where 8 to 12 hours are required for the internal muscle temperature to drop to 32°F (0°C). For this reason, cold shortening is not considered to be a serious problem. If lightweight lamb carcasses with little fat cover are placed in a 28°F (-2°C) cooler with high air velocity, cold shortening of lamb muscle could result in tough meat and in this case electrical stimulation would be of greater value (Figure 17–1). However, the current practice of slaughtering highly finished 110-to 125-lb (50- to 56-kg) lambs, coupled with temperatures normally found in most chill coolers, prevents cold shortening from occurring. If the lamb industry responds to consumer appeals for leaner meat, it may be necessary to use electrical stimulation or to carefully control cooler temper-

Figure 17–1 Electrical stimulation of lamb carcasses prevents cold shortening and improves tenderness in lighter-weight lambs or lambs with less external fat.

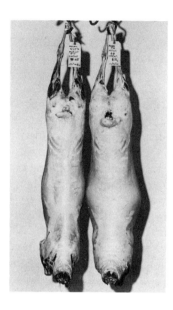

Figure 17-2 The carcass on the left would be expected to chill faster than the one on the right because it is lighter and has less finish. Therefore, it would be expected to be tougher if chilled rapidly.

atures. At present, fat cover and kidney fat act as insulators against cold protecting the muscle from rapid chilling (Figure 17-2).

In studies where light, young lambs have been reported to be tougher than heavier, older lambs or where ram lambs with little finish have been reported to be tougher than wether or ewe lambs with greater finish, the differences are probably due in part to the rate at which the muscle chilled and toughened. This toughening is referred to as myofibular toughness, and it differs from connective tissue toughness, which is not related to chilling rate. Both types of toughness are important to the consumer, but only connective tissue toughness is related to age and to be detected the age difference must be approximately one year or more.

INFLUENCE OF SEX

The scientific evidence for the existence of sex-related flavor and aroma in ram lambs is inconclusive. Nevertheless, lamb buyers in the United States continue to discriminate against ram lambs, particularly those over 5 or 6 months of age. Their reasons are often related to factors other than palatability. It is clear that lower dressing percentages, slightly heavier necks and shoulders, and softer unattractive fat are common on older, heavier, grain-fed ram lambs. These are serious problems because they result in lamb carcasses that are difficult for the packer to sell. Even when cuts from ram lambs are vacuum packaged, the softer fat wrinkles are unsightly. Objections are passed from the retailer or food service establishment to the packer and back to the feedlot operator, who uses this information, plus his own bad experience with ram lamb behavior, to encourage producers to castrate ram lambs.

While we are sympathetic with lamb feeder and lamb buyer concerns, we cannot use these objections to state that meat from all ram lambs tastes bad. Indeed, the preponderance of information states just the opposite. Because electrical stimulation and post-mortem handling can control any minor tenderness differences that exist between ram and wether lambs, we will focus on differences in flavor and aroma. A review of the literature[4] concluded that differences in flavor between ram and wether or between ram and ewe lambs were small and that no trends were apparent. The studies were based on young market-weight lambs weighing less than 110 lb (50 kg). In addition, meat from pasture-fed rams and ewes up to four years of age does not differ in flavor or aroma.[6]

In contrast, several studies have shown that heavy ram lambs fed high levels of concentrate have more intense flavor and aroma than meat from ewe or wether lambs.[3] Therefore, a diet/sex interaction exists for flavor in heavy concentrate fed lambs over approximately 110-lb (50-kg) live weight. We believe that the diet-produced and sex-related soft oily fat in heavy ram lambs is related to the more intense flavor that some consumers find objectionable.

On the other hand, some consumers, particularly some heavy lamb users, prefer a more intense flavor in lamb meat. When cuts from heavy ram lambs fed concentrate are sold at retail, it may be good merchandising practice to separate cuts from rams and advertise that the cuts possess a more intense natural flavor than other lamb. This might be feasible in markets, both domestic and foreign, where a more intense flavor is preferred, and it could increase the number of satisfied lamb consumers because those preferring intense flavor would purchase cuts from ram lambs, while those preferring mild aroma and flavor would purchase cuts from wether or ewe lambs.

A final point to be made regarding sex-related flavor in lamb and mutton is that most reported differences between rams and wethers or ewes have been based on trained panel evaluation. Trained panels such as the one shown in Figure 17-3

Figure 17-3 A trained panel such as this one can detect small differences in palatability of meat, but the differences may or may not relate to consumer demand for the product.

are an excellent method of detecting small differences in meat, but these small differences may not be detected by consumers. Even when they are detected, consumers may consider other factors, such as amount of fat associated with the cut, more important.

INFLUENCE OF DIET

Some of the earliest references to flavor indicate that the diet of the lamb influences the flavor of its meat. *Douglas' Encyclopedia,* published in 1907, states that the exquisite flavor of English and Scottish mutton comes from the aromatic wild herbs in their pastures, but this reference and others like it are not very well documented. In this section, we will discuss alterations in characteristic lamb flavor that have been attributed to diet. We will also discuss level of energy in the diet as it relates to lamb flavor.

Legumes

It is well established that white clover, alfalfa, and tropical legumes grazed or fed in drylot can result in rapid, efficient, and economical gains in lambs, but feeding these legumes immediately before slaughter may cause flavor problems.[5] Researchers have reported that rib chops of lambs fed white clover had greater flavor intensity than those from lambs fed rye grass and that the flavor develops after only 3 weeks on feed. Differences in intensity of flavor and aroma between cuts from the shoulder, loin, and leg were small, and data from hoggets (yearling sheep) supported those for lambs. Some evidence is available to show that if lambs are removed from pasture and fasted for 24 hours before slaughter, the aroma and flavor resulting from white clover feeding diminishes.

In addition to problems with white clover, rape has been described as producing lamb meat with a nauseating aroma and flavor, vetch as producing a sweetish odor and stronger flavor than meat from grass-fed sheep, and alfalfa as being responsible for more intense flavors. Fasting lambs for up to 3 days before slaughter did not reduce the alfalfa taint. In Australia, two weeds most often suspected of producing off flavor in lamb are parthenium and wild turnip. When lambs fed legumes have been compared to those fed concentrate mixtures in drylot, the same general conclusions have been reached as were reached for those fed legumes versus grass. More undesirable or more intense flavors result from legume feeding.

Corn silage, the grazing of oats, and some grains have also been reported to result in variable flavors in lambs, but these are isolated studies and the results are not very conclusive. Most of the off-flavors in lamb have been associated with the fat. Indeed, characteristics of the fat, including flavor, are often used by panelists to distinguish lamb or mutton from other species.

The same diets that have been implicated in off-flavors in lamb have also been shown to cause off-flavor in milk. The flavor and odor of milk are generally most affected when cows are fed 30 minutes to 2 hours before milking. If feeding follows milking, most of the substances responsible for off-flavor are destroyed prior to the next milking period.

It is evident that odors and flavor in milk are associated with the fat, since milk with high fat content has more pronounced off-flavor than milk with low fat content. Rapid metabolism of milk fat in the mammary gland of cows, compared to the relatively low metabolism of subcutaneous and intermuscular fat in lamb carcasses, is probably the reason why changing feed or fasting cows for a few hours before milking can eliminate off-flavor in milk, while 1 to 3 weeks on a different diet is necessary to eliminate objectionable flavor in lamb.

It seems possible that meat could have an off-flavor if lambs are slaughtered immediately after the feed responsible for the flavor is ingested and large amounts of blood, which transports the flavor, are retained in the muscle. This would not be likely to happen in commercial slaughter of lambs because they are fasted prior to the kill, and most of the blood is removed from muscle when routine slaughter procedures are followed.

There is no doubt that certain legumes affect the flavor of lamb. Nevertheless, the significance of these findings is open to question. The flavor and aroma differences have often been obtained from controlled studies where diets containing 100% legume or 100% grass were fed. However, most lambs marketed receive mixed diets. In addition, consumers eat meat from only one lamb at a time and are therefore unable to compare flavor differences, especially when the flavor is disguised by other components of the meal. The authors do not recommend altering lamb diet for the purpose of making lamb flavor more acceptable. Certainly, any differences in flavor that can be attributed to the lambs' diet are masked by cooking and serving lamb with other ingredients of the meal. Curing also appreciably alters the flavor of lamb (Figure 17-4). Cured meat from lambs fed corn silage has been reported to taste like cured pork.

Figure 17-4 Processing lamb by making bologna or other cured cooked sausages can eliminate flavor differences. If lamb or mutton fat is replaced with beef or pork fat, very little difference between mutton and beef bologna can be noted.

Degree of Finish

Lipids modify the taste of other compounds in foods. They also influence the physical state of the food, which affects movement of compounds to taste and odor receptors. The major differences in aroma between heated lamb and heated beef originate from the fatty tissues. Therefore, higher planes of nutrition that produce fatter lambs should result in different meat flavors than lower planes of nutrition that produce lambs with less fat. However, most of the available data do not support this conclusion. The few studies indicating that increased flavor desirability is associated with fatness of the lambs are mostly from early research conducted in the 1930s when consumers preferred fat meat.

Since the 1930s, there has been a convincing wealth of information indicating that lamb flavor is not associated with amount of fat.[5] Prime lamb contains more fat than choice lamb and choice more than good but differences in flavor are small. In addition, no differences in flavor or aroma of meat from lambs on high- and low-energy diets exist.

Overall, there is very little evidence that plane of nutrition or fatness has any influence on lamb flavor (Figure 17-5). Since lambs on low-energy diets are usually fed grains, there is a confounding of much of the data because the effects of energy level and type of feed cannot be separated. Nevertheless, it appears that any differences in flavor that exist are more likely to be related to type of feed than they are to energy level.

Soft, Oily Fat

High-energy grain diets fed over extended periods alter the fatty acid composition of subcutaneous fat and it becomes soft and oily. This is a result of high proportions of relatively low melting methyl-branched-chain and odd-numbered fatty acids and low proportions of relatively high melting saturated fatty acids, such as palmitic and stearic acid. It is known that methylmalonyl-Co A, a metabolite of propionate, not only is incorporated into methyl-branched-chain fatty acids but also inhibits the

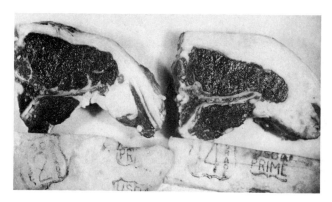

Figure 17-5 Loin chops from a yield grade 2 (left) and yield grade 4 carcass are similar in flavor.

incorporation of malonyl-Co A into saturated even-numbered fatty acids. It has been proposed that increased propionate from concentrate diets increased methyla-malonyl-Co A, which in turn caused an increased synthesis of low-melting methyl-branched-chain fatty acids and a decreased synthesis of high-melting evennumbered saturated fatty acids in the fat.[7] This process occurs in lamb but not in beef, and the soft lamb fat can be observed in late winter in almost any lamb slaughter plant located in the area of large feedlots.

Hormonal effects may also be involved because subcutaneous fat from old, heavy ram lambs is higher in methyl-branched-chain fatty acids and odd-numbered straight-chain fatty acids than is fat from wethers (Table 17-2). Higher proportions of some unsaturated fats, particularly linoleic, in soft, oily lamb fat are more liable to undergo oxidative rancidity, which results in objectionable flavor. Therefore, a high grain diet that results in soft, oily fat could decrease lamb flavor scores of meat kept in frozen storage by making it possible for the fat to turn rancid sooner.

It is clear that numerous chemical factors contribute to flavor, and much more is involved in "off-flavors" or "intense-flavors" of lamb than firmness of the fat.

TABLE 17-2 FATTY ACIDS IN SUBCUTANEOUS FAT OF HEAVY RAM AND WETHER LAMBS

Variable	Rams ($n = 32$)	Wethers ($n = 32$)	SE
Fatty acids, mg/g total fat			
10:0	1.28	1.31	0.07
4-Me 10:0	1.26[b]	0.93[a]	0.08
11:0	0.30[b]	0.18[a]	0.02
2-Me 11:0	1.38[b]	0.48[a]	0.08
12:0	1.50[b]	1.00[a]	0.08
2-Me 12:0	2.54[b]	1.41[a]	0.13
13:0	1.97[b]	1.36[a]	0.10
4-Me 13:0	2.16[b]	0.77[a]	0.12
14:0	16.85	16.79	0.78
4-Me 14:0	10.94	9.47	0.53
15:0	14.33[b]	8.78[a]	0.47
4-Me 15:0	8.29[b]	4.96[a]	0.37
16:0	129.75[a]	152.15[b]	4.60
16:1	15.40[b]	10.09[a]	0.69
17:0	24.96[a]	27.34[b]	0.76
12-Me 17:0/17:1	35.62[b]	18.68[a]	1.32
18:0	52.28[a]	90.22[b]	3.38
18:1	327.69[a]	348.72[b]	6.77
18:2	26.84	25.41	1.74
18:3	10.35	10.46	0.46
Branched 10-17	62.20[b]	36.73[a]	2.20

Adapted from Busboom and others.[1]

[a,b]Means on the same line with different superscripts are different ($P < 0.05$).

Indeed, some lambs with firm fat have been reported to possess objectionable flavors. Nevertheless, it is our observation that off-flavors in cooked lamb are common when the lambs possess soft, oily subcutaneous fat. Lambs fed protected lipid supplements possess soft, oily fat and off-flavor. Our work with heavy ram lambs fed high levels of concentrates in their diets shows that these lambs also possess softer fat and often possess more intense flavors than wethers. As early as 1936, Richardson and Dickson[8] reported that wheat screenings containing many hard weed seeds produced a greasy lamb fat having a peculiar flavor. Overall, diet-produced soft, oily fat may contain substances that contribute to objectionable or more intense lamb flavors directly or diet could contribute to objectionable flavor by producing more unsaturated fatty acids or other compounds that undergo chemical changes and produce off-flavors during storage or cooking.

According to Cramer[2], undesirable lamb flavors or flavor precursors are probably water-soluble components of adipose tissue. The total flavor is undoubtedly a blend of many components by thermal degradation when meat is cooked. Ammonia, hydrogen sulfide, carbonyl compounds, sugars, nucleotides, and amino acids can react to form a variety of compounds. Lactones, intermediate-branch-chained fatty acids, pyrazines, pyridines, pyrroles, furan and pyran derivatives, thiophenes, thianes, thiepanes, mercaptans, and organic sulfides have all been implicated in lamb flavors. However, the components causing the undesirable aspects of mutton odor and flavor have remained elusive and are yet to be identified.

INFLUENCE OF BREED

Although the relationship of sulfur compounds to mutton flavor is based largely on conjecture, this conjecture offers an interesting framework in which to consider flavor differences by breed. Sheep differ from other animals because they have wool instead of hair. Because wool is about 4% cystine, sulfur requirements for sheep are higher than those for any other species. Hydrogen sulfide and glutathione produced by lamb during cooking are also higher than those produced by beef.

Because meat from fine-wooled sheep contains a more intense mutton flavor than meat from coarse-wooled sheep, and because fine wool is higher in cystine than coarse wool, Cramer[2] proposed the following hypothesis: Sheep having high sulfur requirement for wool growth may have evolved a mechanism unique to the species for storage of sulfur to maintain wool growth during low-sulfur stress periods. Selection for quantity of wool in the fine-wooled breeds may have caused this mechanism to develop to a higher degree than normal by placing an artificial demand on the mechanism. Storage of sulfur would probably be affiliated with adipose tissue, since lamb adipose tissue releases about four times as much hydrogen sulfide as does lean during cooking. Specialized sulfur stores could supply compounds that would make the cooking odor of lamb different from other meat and explain why fine-wooled sheep have a more intense mutton flavor than coarse-wooled sheep. It should be made clear that not all research supports the contention that fine-wooled breeds have a more intense flavor than coarse-wooled breeds. However, if there is a breed difference in palatability of lamb, it probably is related to flavor.

There is no evidence that connective tissue or myofibular proteins vary by breed. Indeed, they are very similar among species. Therefore, if one breed of lamb is less tender than another, the difference is probably related to weight or amount of fat, not breed as such. Muscles from heavy or fat carcasses undergo less cold shortening than muscles from lighter or thinner carcasses. Therefore, they would be expected to be more tender when chilled rapidly. Under normal or slower than normal chilling conditions, no differences due to breed would be expected. Although there are several articles in the literature on the effect of breed of lamb on tenderness of the meat, composite scores of all articles support our contention that no breed differences exist.

INFLUENCE OF FROZEN STORAGE

Prior to writing this section, the author went shopping for a turkey. Fresh turkeys were selling for twice as much as frozen turkeys. It was clear from the price differential that most consumers placed greater value on the fresh turkeys, yet the author bought a frozen turkey. While differences in frozen and fresh turkey do not relate well to fresh and frozen lamb, it should be clear to the reader that either the author knew something that the average consumer did not know about frozen meat or that he is a "cheapskate."

There is considerable variation in opinion regarding the tenderizing effects of freezing on the palatability of lamb. Within one study,[9] freezing resulted in tougher loin chops and more tender rib chops. The influence of freezing on flavor also varied by cut. At least three variables were considered in attempting to resolve the differences. First, time in frozen storage differed; second, the temperatures used for frozen storage differed; and, finally, the two sets of chops were from different anatomical locations. Differences in anatomical location probably cannot explain tougher frozen loin chops and more tender frozen rib chops. However, storage time could be important, because lamb that is stored for longer periods of time has greater moisture loss, resulting in tougher muscle fibers.

A major factor influencing storage time is the temperature at which meat is kept in frozen storage. Freezer temperatures of 17°F (-8°C) will cause lamb to have a much shorter frozen storage life than lamb stored at 0°F (-18°C). Colder temperatures extend frozen storage life for even longer periods of time. In one trial, lamb frozen and stored at -3°F (-20°C) immediately after slaughter was considered fresh after 2 years of frozen storage. This result would not be expected with any other species of meat normally found in retail meat counters. The reason why lamb can be frozen for longer periods than other meat is because the fat has less unsaturated fatty acids; hence, less rancidity develops during frozen storage.

Differences in packaging materials are also important. Packaging that allows moisture to escape results in tough, desiccated lamb. Differences in flavor can also be related to packaging. In aerobic packaging, oxygen combines with unsaturated fat, causing rancidity. Anaerobic packaging of frozen meat when compared to aerobic packaging results in an extension of storage life of about 20%. In oxygen-impermeable vacuum packaging, residual oxygen is scavenged from the air by fresh meat and microbiological flora, creating an anaerobic environment. This is not the case for frozen meat where temperatures below 17°F (−8°C) will not support microbial growth, and the slightest trace of oxygen in the package negates the effects of good anaerobic packaging. Therefore, anaerobic packaging is more important for fresh than for frozen meat.

Another significant effect on the storage time of frozen lamb is the aging period prior to freezing. New Zealand researchers have shown that a 24-hour chill at 32°F (0°C) reduces the frozen storage of lamb by about 25% when compared to carcasses that are electrically stimulated and chilled 2 hours before freezing. Carcasses or cuts stored for 7 to 13 days prior to freezing have a much shorter frozen storage life than those frozen after shorter periods.

Animal to animal variability as it relates to firmness of the fat can also influence the storage life of frozen lamb. Ram lambs and all lambs fed a high-concentrate diet have much higher levels of unsaturated fat than ewe or wether lambs or lambs fed forages, and the frozen storage life of these cuts is shorter.

In summary, time in frozen storage, temperature of frozen storage, length of aging prior to frozen storage, type of packaging materials, and amount of unsaturated fat are all important in determining the frozen storage life of lamb and mutton. Because all these factors must be considered, it becomes difficult, if not unrealistic, to give one maximum time for frozen storage of all lamb and mutton. Frozen storage times to be realistic must be established for different products under specified conditions, and this is difficult in commercial practice.

The discussion thus far has centered on intact muscle meats. Before ending the discussion for the influence of frozen storage on lamb palatability, it should be mentioned that ground lamb has a shorter frozen storage life than intact meat because oxygen is incorporated into the meat during grinding. Other factors reducing frozen storage life of lamb and mutton include cooking and the addition of salt or other meat additives that speed up development of rancidity. A frozen lamb dinner that has been cooked and salted prior to freezing would not be expected to keep as long as fresh frozen lamb. Nevertheless, the frozen storage life of lamb and mutton in any form is longer than it is for other meats cooked and processed in a similar manner because lamb has less unsaturated fat.

VACUUM-PACKAGED FRESH LAMB

The popularity of chilled, vacuum-packaged lamb is increasing. Primal cuts are vacuum packaged to reduce contamination, increase storage life, and increase the ease and economics of handling (see Chapter 15). This system of packaging makes imports and exports of fresh lamb possible. Attempts to market case-ready vacuum-packaged cuts are also in progress. Storage life at 32°F (0°C) for primal cuts is approximately 50 days and that of case-ready retail cuts is 10 to 14 days. Treatment of carcasses with a spray of 1.5% to 3% acetic acid either immediately after dressing or after an overnight chill increases storage life when the cuts are vacuum packaged. Other processes yet to be approved can also successfully increase storage life.

Major factors limiting the storage life of fresh vacuum-packaged lamb are microbial growth, odors, and flavors. Green discolorations on the surface of primal cuts are also present at times on cuts stored for 10 weeks or longer.

The overall flavor acceptability of meat from cuts stored fresh in vacuum packages declines with storage time, but the decline is minimal with 50 days of storage or less at 32°F (0°C). The aroma of freshly opened vacuum packages can be intense and objectionable in cases where bacterial growth is high. Leaks in the vacuum packages increase bacterial growth and aroma problems. Much more subtle aromas not associated with bacterial growth are also present in vacuum-packaged meat, but they are not detected by panels after the meat is cooked. In contrast to a decline in aroma and flavor during storage of vacuum-packaged lamb, there is an increase in tenderness because proteolytic enzymes continue to tenderize the meat during storage. The increase in vacuum-packaged lamb, both primal and case-ready cuts, will accelerate as problems associated with microbial growth, flavor, and aroma are solved.

REFERENCES CITED

1. Busboom, J. R., and others. 1981. Characteristics of fat from heavy ram and wether lambs. *J. Anim. Sci.* 52:83.
2. Cramer, D. A. 1983. Chemical compounds implicated in lamb flavor. *Food Technol.* 37(5):249.
3. Crouse, J. D. 1983. The effects of breed, sex, slaughter weight and age on lamb flavor. *Food Technol.* 37(5):265.
4. Field, R. A. 1971. Effect of castration on meat quality and quantity. *J. Animal Sci.* 32:849.
5. Field, R.A., J. C. Williams, and G. J. Miller. 1983. The effect of diet on lamb flavor. *Food Technol.* 37(5):258.
6. Kirton, A. H., and others. 1983. Palatability of meat from electrically stimulated carcasses of yearling and older entire-male and female sheep. *J. Food Technol.* 18:639.
7. Miller, G. J., J. E. Kunsman, and R. A. Field. 1980. Characteristics of soft subcutaneous fat in ram lambs fed corn and corn-silage diets. *J. Food Sci.* 45:279.
8. Richardson, J. E., and W. F. Dickson. 1936. The effect of feeds on the quality and palatability of lamb. *Montana Agr. Exp. Sta., Bozeman, Bulletin 320.*

9. Smith, G. C., and others. 1968. The effects of freezing, frozen storage conditions and degree of doneness on lamb palatability characteristics. *J. Food Sci.* 33:19.
10. Smith, G. C., and others. 1969. Lamb palatability studies. *Proc. Recip. Meat Conf.* 22:69.
11. Wenham, L. M., and others. 1973. Eating quality of mutton compared with lamb and its relationship to freezing practice. *J. Animal Sci.* 36:1081.

18 Wool Evaluation and Marketing

THE WOOL SITUATION

Many accounts of the history of wool (the outer covering of sheep) have been written. A popular historical approach has been to demonstrate a close association between wool and the progress of mankind. Skins of sheep slaughtered for food and wool removed from live sheep have contributed to the well-being of people since the Stone Age. Also, wool has been shown as a source of economic and political power for the development of nations in a manner similar to that of oil during the twentieth century.

Some writers have used a biological approach to the history of wool. Changes in wool fibers and in follicular structures in the skins of sheep have been identified, beginning with wild sheep and continuing through domestication and the formation of breeds (Chapter 19). Documentation of these changes has been essential for continuing constructive wool research.

On the American scene, sheep and wool production systems have changed dramatically. Early colonists kept small flocks to provide both food and fiber for their families. In contrast, during the settling of the West, large flocks of wethers were kept for wool production only. This system has been abandoned in favor of the two-crop approach, where sheep are raised for the production of both meat and wool. Well-written histories of sheep and wool production are quite readily available in most libraries and are recommended as stimulating reading for anyone interested in sheep.

Because of the importance of supply and demand as they affect price, manufacturing companies, marketing agencies, and producers are particularly interested in where, what kind, and how much wool is grown annually on a local, national, and world basis. *Agricultural Statistics,* published annually by the U.S. Department of Agriculture, U.S. Printing Office, Washington, D.C. 20402, is suggested as a good reference for these kinds of data and is available in many libraries. Other less readily available yet reliable sources of statistical information concerning both the production and consumption of wool, as well as other fibers, include the Wool Bureau, Inc., 360 Lexington Avenue, New York, N.Y. 10017, and Cotton and Wool, Outlook and Situation, U.S. Department of Agriculture, Washington, D.C. 20250.

A review of the data presented in current issues of the preceding publications leads to the following general observations:

1. The bulk of the world's wool supply is produced in relatively sparsely populated areas. This is also the case within the United States, where more than 70% of the domestic wool has been grown in the 11 western states and Texas (Table 1-4). The cost of transporting wool from producer to processor to consumer is a major economic factor.
2. In terms of the world wool market, the United States has little bargaining power. It produces less than 2% of the world's wool.
3. Annual wool production in the United States has declined steadily at approximately 10% per year since the mid-1940s. An apparent leveling off or upward turn was noted beginning in 1981 but did not persist.
4. At least since 1930, the United States has been a net importer of wool. Even at relatively low annual consumption rates of 1 to 3 pounds per capita, the United States has not been able to produce enough wool to meet its needs. The world wool market is directly reflected in prices received by producers of domestic wool.

The relative importance of returns from the sale of wool and lambs in sheep production has been studied carefully for various management systems. Frequently published estimates indicate that wool income normally represents from 10% to 40% of gross returns. Table 18-1 demonstrates that, for individual operations, income from wool is inversely proportional to the *efficiency* of lamb production. This table can be applied to individual production situations by substituting current prices and actual reproduction and replacement rates.

Although income from wool is less important than from lamb to most producers, it represents a more predictable return. On most farms and ranches, the lamb crop weaned can vary a great deal from year to year due to reproduction problems, disease, weather, feed supply, predators, and the like. Many managers of range sheep operations report that, in years when spring storms have caused large lamb losses or during years when losses to predators have been especially severe, their wool income has "kept them in business."

Whether wool production in the United States is compared to that of other countries, to other agricultural commodities, or to income from lamb within individual sheep production units, it is normally portrayed as a minor product. Because

TABLE 18-1 RELATIVE INCOME FROM LAMB AND WOOL[a]

Percent Lambs Weaned	Lamb				Wool		
	Per Ewe[b]	Percent of Total[b]	Per Ewe[c]	Percent of Total[c]	Per Ewe[d]	Percent of Total[b]	Percent of Total[c]
160	$56.00	83.6	$70.00	86.5	$10.96	16.4	13.5
140	48.00	81.4	60.00	84.6	10.96	18.6	15.4
120	40.00	78.5	50.00	82.0	10.96	21.5	18.0
100	32.00	74.5	40.00	78.5	10.96	25.5	21.5
80	24.00	68.6	30.00	73.2	10.96	31.4	26.8

[a]Theoretical situation.
[b]Income from lambs marketed with enough lambs saved to replace 20% of ewe flock and lambs sold at 80 lb (36 kg) at $50/cwt (45 kg). No return from salvage ewes included.
[c]As in note b, but lambs sold at 100 lb (45 kg) at $50/cwt (45 kg).
[d]Estimated at national average of 8 lb (3.63 kg)/head and the 1982 national average grease price, plus incentive payment.

wool represents a minor source of income to sheep producers, preparation for marketing and selection for quality and quantity are frequently neglected. Domestic wool is often correctly described as poorly grown and poorly marketed. Sheep producers are reminded that, because wool does represent a source of income, neglecting it is considered poor business management.

THE LANGUAGE OF WOOL

The language of the wool industry is unique in many respects. A glossary of some of the words used by wool producers, marketing people, manufacturers, researchers, and consumers is presented following Chapter 22. Readers are encouraged to study these terms and their definitions carefully at this point as a basis for better communication in this and the following chapters.

Examples of the more readily misunderstood terms are "clothing," which means wool fibers too short to comb, as opposed to "apparel," a term used to describe wool used for apparel fabrics; "fine" wool refers to fiber diameter rather than a relative description of superior quality; "shrink" refers to the proportion of nonwool component in grease wool, not fabric shrinkage, which is correctly called felting.

DESCRIPTION OF RAW WOOL

The relative value of grease wool is a function of both qualitative and quantitative evaluation. *Clean price* is based primarily on qualitative factors that determine the end use of the raw fiber. The most important qualitative factor is grade.[6] Others normally considered are length (which is also included as a quantitative factor), uniformity, strength, crimp, handle, color, character, purity, wastiness, and contaminants. Because wool leaves farms and ranches with varying amounts of nonwool components, the *quantity* of usable fiber must also be described before realistic grease prices can be determined.

Grade

In the American marketing system, "grade" refers to average fiber diameter or fineness only. "Quality," sometimes used to describe fineness alone, is more correctly used as a comprehensive term to include all the many qualitative measures of wool. Grade is considered wool's most important value-determining characteristic because it governs the minimum thickness (spinning limit) of yarns that can be spun. Regardless of grade, a minimum number of fibers is required in a cross section of yarn for it to be strong enough for knitting or weaving. The finer grades of wool can be spun into finer yarns than coarser grades.

The relative desirability reflected as differences in price of grades is directly related to the type of yarn required for manufacturing a given product. For bulky sportswear, the coarser grades are preferable, but lightweight gabardine suiting materials require finer grades. Long-term comparisons of prices paid for the various grades of raw wool indicate that particular grades are not consistently more valuable than others. Price changes related to grade are considered a reflection of changes in consumer attitudes toward the style and utility of the many types of fabrics that are produced from each grade, along with supply-and-demand aspects of the domestic as well as world wool trade.

Several systems have been developed for describing the diameter of wool fibers for marketing purposes. Grades of wool grown and marketed in the United States are normally described by the American blood system, the spinning count, or by microns (micrometers).

Blood system. The blood grades were originally used to describe the proportion of fine wool (Merino or Rambouillet) breeding represented in a given fleece, clip, or lot. Wool from purebred Merinos or Rambouillets was called *fine.* Wool from sheep that were one-half Merino or Rambouillet and one-half some other breed or combination of breeds was called *one-half blood.* As the proportion of finewool breeding represented in fleeces became smaller, it was described accordingly as *three-eighths, one-quarter,* or *low-one-quarter blood.* Wools most suitable for carpets, lumbermen's socks, or similar very bulky products were called *common* and *braid.*

As greater accuracy in the description of average fiber diameter became necessary, the blood system changed to classifications based on "typical" fiber diameter rather than breeding background. These grades are normally determined by visual evaluation based largely on crimp frequency (Chapter 20). One limitation of the blood grade system is that it permits only six categories for relating price to fiber diameter. It has essentially been dropped as a system of describing grade when marketing domestic wool.

Spinning count system. The spinning count or Bradford system, which originated in Great Britain, relates fiber diameter to length of yarn and weight of wool top. It describes wool fiber diameter as an estimate of the number of *hanks,* each 560 yards (512 m) long, of yarn that can be spun to a minimum thickness from 1 lb (0.45 kg) of top.

There are more fibers in a pound of top made from fine fibers than one made from coarse fibers. Therefore, because a minimum number of fibers are required for yarn to be strong enough for weaving or knitting, more yarn can be spun from finer than coarser wools. In common usage, the word hanks has been omitted from the description. Consequently, 64 hanks is written as 64's. An average 64's-type wool is one that can be expected to produce 64 hanks (more than 20 miles) of yarn when spun to a minimum thickness. Some writers and market reporters use the term *quality number* when referring to spinning counts.

The spinning count system is similar to the American blood grade system in the sense that the different grades are normally determined visually, based largely on crimp frequency. However, the spinning count system is more specific because it provides 16 categories for describing fiber diameter as compared to only 6 blood grades.

Micron system. The micron grade is considered the best method available for describing average fiber diameter of wool. It is a familiar term in all phases of the wool industry, which has been derived by common usage from micrometer (10^{-3} mm or 1/25,400 of an inch). The primary advantage of the micron system is that it is determined by objective measure, and it describes wool fiber diameter on a continuous scale rather than by categories.

The USDA recognized that the many systems of describing wool grades were causing a great deal of confusion among wool industry people from producers to fabric makers. In 1966, a set of official standards was adopted.[11] Table 18-2 shows the relationship of three different systems used for describing wool fiber diameter in the United States.

As might be expected, the micron (μ) ranges for each spinning count grade have been challenged as *too specific*. Objective measures of wool fiber diameter are normally considered accurate to no more than ± 0.4 μ and spinning count grade breaks are described at the 0.01 μ level. This problem has not been, and probably will not need to be, resolved because the international wool trade has moved almost entirely to using the continuous *micron* scale for describing wool fiber diameter. The American blood grade system is essentially obsolete, and it is anticipated that the British spinning count system will be abandoned very quickly in favor of microns.

TABLE 18-2 SPECIFICATIONS FOR GRADES OF WOOL

Blood Grade	Spinning Count Grade[a]	Range for Average Fiber Diameter (μ)	Standard Deviation Maximum (μ)
Fine	Finer than 80's	Under 17.70	3.59
	80's	17.70–19.14	4.09
	70's	19.15–20.59	4.59
	64's	20.60–22.04	5.19
1/2 Blood	62's	22.05–23.49	5.89
	60's	23.50–24.94	6.49
3/8 Blood	58's	24.95–26.39	7.09
	56's	26.40–27.84	7.59
1/4 Blood	54's	27.85–29.29	8.19
	50's	29.30–30.99	8.69
Low 1/4 Blood	48's	31.00–32.69	9.09
	46's	32.70–34.39	9.59
Common and Braid	44's	34.40–36.19	10.09
	40's	36.20–38.09	10.69
	36's	38.10–40.20	11.19
	Coarser than 36's	Over 40.20	—

Adapted from USDA *Official Standards*.[11]

[a] Spinning count is assigned on the basis of average fiber diameter and standard deviation. If the measured standard deviation exceeds the maximum specified for the grade to which the measured average fiber diameter corresponds, the wool shall be placed in the next coarser grade.

Type system. A system based on differences in wool grade along with other qualities such as length, softness, and luster is frequently used to classify breeds of sheep in terms of the "type" of wool they grow (Figure 19-2). It is used to group wool types rather than breeds per se in market reports, which sometimes leads to misunderstanding among wool industry people. It is nonspecific and seldom used as a basis for pricing wool.

When used to describe shorn wool, *fine wool* implies wools that grade 64's or finer. In market reports of international trading, these wools are frequently described as Merino and may be classified further as superfine or fine. However, some strains of Merinos have been selected for increased fiber diameter, and their wools are classified as medium, strong, or extra strong, indicating progressively coarser grades.

Wool that grades 54's to 60's is usually described as *medium wool,* regardless of breed. In this sense, it includes the medium, strong, and extra strong Merino as well as crossbred and comeback types and wool from the English Down (meat) breeds. It has also been used to describe only wool grown by the meat breeds, which typically ranges in grade from 58's to as low as 44's or coarser (Figure 19-2). Their fleeces are relatively harsh to the touch and contain colored fibers if grown by those with black or brown faces and legs. Because of the several ways it is used, the term medium wool should either be avoided or qualified when used.

In the wool industry, the term *crossbred* is normally used to describe wool from sheep produced by mating long-wool with fine-wool breeds. The Columbia and Corriedale are typical and grades are in the 60's to 50's range, free of colored fibers, and relatively soft to the touch. Some writers and reporters of international trading include any wools other than Merino in this category. Therefore, as in the case of medium wool, the term should be qualified when used.

Although sometimes included in the crossbred or medium types, the term *comeback* is most frequently used to describe grades on the fine side of medium in the 62's to 60's range. It usually refers to wools from specific breed combinations where crossbred types are bred back to the finewool parent. The American Targhee and Australian Polworth are examples.

Wool described as *longwool* might be more correctly called coarse wool because it tends to grade 48's and coarser. However, annual staple lengths normally measure 8 in. (20.32 mm) or longer, so it has been described as longwool. Typical breeds are the Lincoln, Leicester, Cotswold, and Romney.

Coarse, harsh-handling types that contain varying amounts of medulated fiber are called *carpet* wool. Wools from the unimproved breeds such as the Navajo, Spanish Churra, some of the mountain breeds of Great Britain, and the fat-tailed and fat-rumped breeds of Asia, North Africa, and the Mediterranean area are included. On the other hand, some breeds such as the Romney are being selected for quality carpet wool.

Staple Length

The second most important value-determining characteristic of wool is average length of fibers. As a general rule, "the longer the staple, the more it is worth." However, extra-long-stapled wool may actually be detrimental in normal combing operations.

For marketing purposes, the classification of staple length within each grade is usually directly associated with the processing system for which it is destined. The wool manufacturing industry consists of two distinctly separate divisions: apparel and carpet. Most of the wool produced in the United States is for apparel and generally finer in grade than carpet wool. It is processed in one of two apparel systems, *worsted* and *woolen*.

The worsted system requires combing where fibers are laid parallel in a thick strand called *top* and the short fibers, second cuts, leg and face hair, and the like, are removed as a waste called *noil*. The yarn produced is relatively smooth. Yarns manufactured on the woolen system are usually more bulky than worsted yarns. The fibers are not laid parallel, resulting in proportionately more ends projecting outward (Figure 18-1). Minimum lengths for each grade are required for combing, but the woolen system can make use of virtually all lengths of wool fibers. Noils and recycled wool, as well as either long- or short-stapled virgin wool can be included in the manufacture of woolen yarn of different qualities. However, it is understood

Worsted yarn Woolen yarn

Figure 18-1 Fiber arrangement in wool yarn.

that, although not combed, woolen yarns made from long (staple or French combing) fibers are stronger than those made from the shorter classes.

Although objective measures are superior (Chapter 20), staple length can be estimated to fit the relatively broad classes used in the U.S. marketing system without laboratory facilities or sophisticated instruments. Breakdowns in communication frequently occur because of lack of understanding or agreement related to the descriptive words used. Length within grade is an important factor when combing wool in the worsted system. Traditionally, *staple, French combing,* and *clothing* have been used in turn to describe the length of wools suitable for processing with a minimum of waste on British Noble combs, French combs, or too-short-to-comb so must be used in the manufacture of woolen yarn. Virtually all wool combers in the United States use French combs, but staple wools still command the highest prices because longer fibers produce stronger yarns. Suggested length classes for each grade are shown in Table 18-3. The good French and average French classes shown in this table are often combined and described as French combing.

Standards that specify lengths for the various terms used have not been established as official for either domestic or international wools marketed in the United States. For example, the USDA *Wool Market News* (Table 18-4) describes 64's staple as 2 3/4 in. and up, French Combing as 2 1/4 to 2 3/4 in., and does not include categories for average French or short French and clothing because sales of

TABLE 18-3 LENGTH CLASSES OF WOOL BY GRADE (INCHES)[a]

Length Class	64's, 70's/ 80's	60's/62's	56's/58's	50's/54's	46's/48's	36's/40's/ 44's
Staple	≥2.75	≥3.0	≥3.25	≥3.5	≥4.0	≥5.0
Good French	2.25–2.75	2.5–3.0	2.25–3.25	2.5–3.5		
Average French	1.25–2.25	1.5–2.5				
Short French and clothing	<1.25	<1.5	<2.25	<2.5	<4	<5

Adapted from Pohle.[(9)]

[a]Based on measuring unstretched locks of grease wool.

TABLE 18-4 DOMESTIC WOOL QUOTATIONS*,a

USDA grade	Graded territory wool (1)				Graded fleece wool (2)			
	Clean delivered*	%	Grease Equivalents** (5)	%	Clean delivered*	%	Grease Equivalents (5)	%
64s (20.60-22.04 microns)								
Staple 2-3/4" and up	$1.80-2.05	52	$ .94-1.07	48	.86- .98			
	1.60-1.75	44	.79- .90	38	.68- .78			
Fr. Combing 2-1/4"-2-3/4"		50	.80- .88	46	.74- .81			
62s (22.05-23.49 microns)								
Staple 3" and up	1.60-1.75	47	.75- .82	43	.69- .75		50	45
60s (23.50-24.94 microns)								
Staple 3" and up	1.40-1.50	47	.66- .70	43	.60- .64		50	45
58s (24.95-26.39 microns)								
Staple 3-1/4" and up	1.30-1.40	52	.67- .73	46	.60- .64		53	46
56s (26.40-27.84 microns)								
Staple 3-1/4" and up	1.25-1.35	52	.65- .70	46	.57- .62		53	46
54s (27.85-29.29 microns)								
Staple 3-1/2" and up	1.25-1.35	53	.66- .72	47	.58- .63	$1.20-1.25	54	.65-.68	47	.56-.59

	Texas wool (3)					Territory wool (1)			
					Original bag wool				
64s (20.60-22.04 microns)									
Staple 2-3/4" and up	$1.85-2.10	54	1.00-1.13	50	.92-1.05	$1.80-1.90	53	.95-1.01	48	.86-.91
		46	.83- .96	38	.70- .80		43	.77- .82	39	.70-.74
Fr. Combing 2-1/4"-2-3/4"	1.65-1.80	52	.88- .94	48	.79- .86					

(1) Wool grown in the range areas of Washington, Oregon, the intermountain states, including Arizona and New Mexico, and parts of the Dakotas, Nebraska, Kansas, and Oklahoma. This wool covers a wide range in shrinkage and color.

(2) Most wool grown in the farm areas of the Dakotas, Nebraska, Kansas, Oklahoma, and parts of Minnesota, Wisconsin, Iowa, and Missouri is somewhat dark in color and of moderately heavy shrinkage. Most of that grown in the states east of the Mississippi River and in parts of Minnesota, Iowa, and Missouri is bright in color and of relatively light shrinkage.

(3) Wool grown in the range areas of Texas, mostly bright in color and moderate in shrinkage, except in the Panhandle, where it is considerably darker in color and heavier in shrinkage.

(4) Grease wool finer than 46s grade is dutiable at 10¢ clean content basis, scoured wool at 11¢ clean content basis, and carbonized wool at 13¢ net weight basis. All wool not finer than 46s grade is presently duty-free.

(5) In order to assist in estimating grease wool prices, delivered to mill clean basis market prices have been converted to grease basis equivalents using the various yields quoted. Prices determined in this manner are largely nominal and not necessarily those received by individual growers, as allowances for freight, handling, etc., must be considered.

*Clean price based on actual sales when available, otherwise nominal.
**Grease basis, FOB Warehouse.
aAdapted from U.S.D.A. Wool Market News.[12]

these lengths are sporadic and volumes low. On the other hand, some buyers consider 3 in. as a minimum for fine wool to be described as staple.

Marketing agencies reporting world wool market trends currently describe staple length as either combing or carding to differentiate combable from noncombable wools of each grade. A continuous scale description of average staple length along with other objective measures such as grade and yield is becoming more prevalent in international trading and is expected to replace the traditional length categories currently used to describe U.S. wools.

Uniformity

To the manufacturer, uniformity of both fiber diameter and staple length is very important. No fleece or clip is absolutely uniform in either measure, and differences, both within and between fleeces, are great. Uniformity of length in a clip can be increased by breeding and careful shearing, whereby second cuts are minimized.

In a lot or clip of wool, the sources of variation in *fiber diameter* are the following:

- Between fleeces
- Between body regions within fleeces
- Between fibers within staples (locks)
- Between points along fibers

Fiber-to-fiber variation within staples has been shown as most important, accounting for 58% to 92% of the total variation in fiber diameter within fleeces and 34% to 82% in lots.[10] The proportion of variation due to body region was generally low for all lots studied, ranging from 1% to 14%. Variation along fibers was responsible for 3% to 6% in lots from well-fed flocks and up to 29% in a lot from a flock where seasonal variation in nutrition was great. Between fleece variation in fiber diameter accounted for only 9% to 13% of total lot variation in fleeces from three purebred flocks and increased to 29% to 57% in lots from flocks that were less systematically bred.

These and similar data suggest that when fleeces in a lot of grease wool are shown to vary excessively, market desirability can be enhanced by dividing them into groups of similar grades. Also, in fleeces where average diameter varies excessively (3 to 4 spinning counts) between body regions, extensive skirting may be appropriate but is usually not cost-effective in the United States unless conducted at shearing. The negative effects of end-to-end variation along fibers can be minimized at the producer level by controlling nutrition and/or management practices such as time of shearing. The degree of fiber-to-fiber variation is a function of inheritance and can only be corrected by selective breeding (Chapter 21).

The greater the uniformity of fleeces in a clip the more valuable it is to the manufacturer. Uniform clips require less sorting and blending to meet specifications for diameter and fiber length at various stages of processing.

Description of Raw Wool

Strength

Fiber strength is particularly important to processors that comb wool in preparation for the manufacture of worsted yarns. The term *sound* wool is used by the trade to describe wool that is considered strong enough to withstand the stresses of carding and combing.

Fleeces vary in relative strength from *sound* to *tender* to *breaks present*. Sound fleeces are normally those that have been well grown throughout the year so that the end-to-end variation in fiber diameter is small.

Tender fleeces are the most difficult to describe correctly for marketing purposes. When small, pencil-sized locks of tender fibers are tugged sharply, a subjective measure, they show some weakness, but the weak areas do not occur at the same point along each fiber (Figure 18-2). It is encouraging to note that an objective test for measuring strength (Chapter 20), which indicates the *degree* of soundness as well as the position of the weakest point, is available and is being added to basic presale test data by some of the major wool-producing countries.

Several causes of tenderness in wool fibers have been suggested. It may be due to one or a combination of the following: A temporary reduction in nutrients, primarily energy, which reduces fiber diameter enough to cause a "weaker than the rest of the fiber" section but is not severe enough to cause a definite break. Genetic and/or environmental influences, such as limiting specific dietary amino acids or minerals, may result in temporary changes in the chemical structure rather than diameter of fiber. In either situation, fibers grown from some follicles, probably primary versus secondary, may be affected more or less severely, resulting in the typical uneven separation. Microbial degradation of fibers during storage has also been shown to cause tenderness.

Whatever the cause, tender wools are worth less to top makers than sound wools. Fleeces are frequently described as "tender but strong enough to comb," which means they can be combed with somewhat more noilage than sound wools but are much less wastey than wools with breaks.

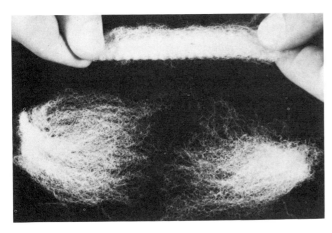

Figure 18-2 Tender wool. Note uneven ends. (Reproduced with permission from Pexton, Johnson, and Craddock[8])

The incidence and position of *breaks* in a lot of wool are the most important concerns related to fiber strength in the manufacture of top. Breaks are due to distinct reductions in fiber diameter at a given point along each fiber and are identified readily by tugging small locks (Figure 18-3). These constrictions are often called "fever" breaks because they occur when sheep have been ill and running high temperatures. They may also form when sheep are subjected to severe stress (Chapter 19). Breaks have been observed at the same point along fibers in nearly all fleeces shorn from large range bands of ewes as a result of storms, where they have been subjected to severe cold, high moisture, and wind without access to feed for two or more days. Entire flocks may develop fleece breaks due to severe infestations of internal parasites combined with poor nutrition.

In many management situations, occasional severe stresses, particularly those due to extremes in weather, cannot be avoided and breaks occur near the midsection of the staple. However, the economic importance of breaks can be minimized by timely shearing. Fevers in sheep are frequently associated with lambing problems, such as uterine infections or mastitis. Breaks from these causes occur near the base or tip of the staple in sheep shorn either shortly before or shortly after lambing. When ewes are shorn close to lambing, noilage (waste) from tip or base breaks in their fleeces tends to be lower than if they are shorn at other times during their production year.

Color

A creamy white (bright) color in raw wool is always considered superior to wool with any discoloration. Under certain environmental conditions, primarily high humidity and/or high temperature, the yolk in wool tends to turn yellow. This can occur while the wool is on the sheep or during storage. Variations in the degree of yellowing in fleeces have been reported among sheep within as well as between breeds and flocks in similar environments, and may occur in well-defined bands

Figure 18-3 Wool with a break. Note distinct separation. Reproduced with permission from Pexton, Johnson, and Craddock[7])

along the fiber or over the entire staple. Either way, yellowness in raw wool acts as a dye and is not removed in scouring. Scoured wool with a yellow hue must be dyed a color at least as dark as the hue, and variation in degree of yellowness can result in lack of color uniformity in yarns and fabric.

Stains from drugs such as phenothiazine, branding fluids that *bleed,* manure and urine, or certain soils are not removed by commercial scouring solutions and become a limiting factor for processors in the selection of color. Wools containing stained fibers cannot be used in the manufacture of solid, pastel-colored fabrics. White fleeces, free of stained fibers, are therefore worth more than those with stains and are priced accordingly.

Crimp

Crimp is the term used to describe the wavy appearance of locks of raw wool. Crimp frequency (number of crimps per unit of length) is used as an indicator in the visual assessment of grades, because coarser wools tend to have fewer crimps per unit of length than finer wools. However, prediction of objective measures of fiber diameter based on crimp frequency has been shown to be inconsistent (Chapter 20).

It has been suggested that the spinability and several characteristics of fabric made from well-crimped wool are superior to that of poorly crimped wool.[15] Regardless of whether or not crimp is a major concern for processors, grease wools that show a uniform, clear crimp formation are definitely more attractive to buyers than those without it. Fleeces from well-managed flocks of white-faced breeds that have been carefully selected normally exhibit a distinct, uniform crimp and lock formation.

Handle

The relative softness or harshness of wool is described as *handle*. Fine wools are normally softer to the touch than coarse wools, and fleeces grown by the meat breeds tend to be harsher than those of the same grade from the wool breeds.

Whether a soft or harsh wool is considered superior depends on the type of fabric desired. If a stiff, firm fabric with a lot of body is needed, processors start with harsh-handling wools. On the other hand, soft wools are definitely superior to their harsher counterparts in the manufacture of flannels and felts.

Character

When evaluating grease wool, color, crimp, and handle are often combined in a single term—*character.* In this sense, character refers to the overall attractiveness of grease wool. It is also used to indicate a distinct uniform crimp or to imply grade suggested by crimp frequency, so a fleece or clip may be described as having fine, medium, or coarse character. Separate descriptions of grade, color, crimp, and handle are preferable to the general term.

Purity

The degree of naturally brown or black fibers in wool is described as *relative purity*. To the manufacturer, fleeces with poor purity are identical to those with stained fibers. These dark-colored fibers cannot be separated from white fibers either chemically or mechanically, so they must be used in dark, solid-colored fabrics or in relatively dark tweeds and plaids.

Fleeces from the meat breeds and their crosses and some individuals in the wool breeds that show spots of colored fiber either in the fleece or on the legs are considered poor in purity. In range bands where naturally "black" sheep are used as "markers" for rapid counting, or black-faced rams are used as terminal sires for producing feeder lambs, there is a cross-contamination with white-fleeced sheep due to mingling. In these situations, black fibers are found in the fleeces of sheep that produce only pure white wool. Occasionally, the black marker sheep are not sorted out for shearing *after* the rest of the band has been shorn, so a great deal of black fiber contamination occurs on the shearing floor. A few producers have tried to avoid this problem by using special brands, bright dyes, or some other means that make markers readily identified for counting. Whatever the method, fleeces from marker sheep should be shorn, bagged, and sold separately. Contamination of white fleeces with dark fibers and consequent price reductions due to poor purity can be avoided, and producers are encouraged to do so.

Other, but less important because they are less frequent, causes of poor purity in raw wool are hair and kemps. The short kemp or face and leg hairs are removed readily during the combing process. Medulated, hairlike long fibers are not removed and do present processing problems, particularly in relation to uniformity of dyeing (Chapter 19).

Wastiness

When shearers allow the combs on their power-driven handpieces to cut wool anywhere other than next to the skin, short tufts of wool are left on the sheep. For the shorn sheep to have a smoother appearance, the tuft is frequently removed in a "second cut." The effect of short fibers resulting from second cutting during shearing is the same as tip or base breaks (see Strength, this chapter). To processors, they represent increased noilage and therefore waste.

Contaminants

In some wool marketing situations, contaminants are included with colored fibers and stained pieces as part of color or purity. Regardless of when they are considered, contaminants are nonwool components of raw wool that are not removed or that require extra steps for removal during processing.

Short pieces of hay, straw, or weed seeds, described collectively as *vegetable matter,* are not removed readily during scouring. Burrs are especially troublesome.

Vegetable matter can be removed chemically by treatment with weak acid solutions, called *carbonizing* (Chapter 19), but it represents an added processing cost when required.

Many branding fluids are difficult and often impossible to scour out of wool. There is an apparent inverse relationship between their scourability and durability, and those that *bleed* are especially troublesome because they tend to dye entire fleeces during scouring and limit color choices. As a general rule, branding fluids that remain legible for a full season are less scourable than those that require mid-year repeat branding. Regardless of their relative scourability, the negative effects of branding materials can be reduced by minimizing the amount used.

Other methods of identification have been tried, such as face brands and ear tags, but have not been widely accepted by producers. In many areas of the United States, paint branding is considered the best method of identifying ownership of sheep. In these situations, flock managers are often willing to accept price reductions due to paint in their clips as fair trade for ease of identification of individual sheep and/or ownership of their flocks.

The single most important contaminant of domestic wool identified in recent years is *polyethylene twine,* described as "black poly" or "poly-contamination." It cannot be separated from wool fibers either chemically or mechanically. Polyethylene fibers do not accept dyes and when fabrics containing them are heated, these tiny fibers melt and the fabrics are blemished. Probably due in part to static electricity, poly fibers attach themselves to wool very readily. Modern hay balers use polyethylene twine because it is easy to handle and retains its strength even when exposed to weather for long periods. Because it does not affect digestion, poly-twines are often incorporated into feeds by way of choppers or grinders and are frequently found in pelleted feeds. These short pieces of polyethylene fibers pass through the digestive tracts of animals. Later, as the feces decompose, poly fibers emerge and attach themselves to the fleeces of any sheep that touch them. Knotter ends or twines removed from bales and later used for tying gates or even fleeces also lead to poly contamination. Careful attention to this problem cannot be overemphasized in the production of quality clips. Wool buyers normally do not reduce the price of clips contaminated with polyethylene. If they suspect it is present, they are instructed not to buy at any price.

Yield

Yield is a quantitative term used in the wool industry to describe the proportion, expressed as a percent, of clean wool (clean fibers present) in a given quantity of raw wool. *Clean wool* always refers to wool fibers at a *standard condition* (Chapter 20).

Traditionally, raw wool has been described in terms of its nonwool component, called *shrink* (100% minus shrinkage = yield). Although some wool buyers still refer to the shrinkage in wool, the more positive and direct term, yield, is becoming more popular.

Nonwool components are classified into three major types as follows:

1. Natural: the natural waxes in wool plus a mixture of sweat salts, commonly referred to as *grease* or *yolk*
2. Acquired:
    a. Animal: keds and other insects
    b. Vegetable: hay, grain, straw, burrs, weed seeds, corral dust, etc.
    c. Mineral: sand and soil
    d. Applied: paints and dyes
3. Moisture: called "hidden shrink"

From a marketing standpoint, the acquired materials have attracted the most attention because they are controllable to a degree in most management systems (see "Preparation," this chapter). For example, paint-free clips sell for higher prices within their respective grades than those containing branding materials. Also, small premiums are being paid for clips where paints that have been shown definitely scourable have been used.

Because it moves in and out of wool fibers so readily (Chapter 19), moisture is the most troublesome nonwool component in determining yield (Chapter 20). It is often called "hidden shrink" because it can vary widely from day to day and even from hour to hour, and relative amounts are not readily detectable tactily. Moisture in bagged wool usually ranges from 6% to 18%. Table 20-1 demonstrates the effect of fluctuating moisture in a single bag of wool where clean content remains constant and bag weight changes. If all bags described in this example were priced at $1/lb grease basis (with an estimated moisture content of 12%), the high-moisture bags would be overpriced and the low moisture bags underpriced.

PRICING RAW WOOL

Wool grown and shorn on farms and ranches is normally sold and moves to processing centers "in the grease." However, its value is always determined or estimated from the qualitative and quantitative aspects of the clean fibers present.

The first step in evaluating raw wool is to determine the price of its clean component and several factors are considered. A typical USDA Market Report (Table 18-4) indicates that a first step in determining clean price is to know where the wool was grown. Wools grown in different areas of the United States are often priced differently.

The three major geographic areas of wool production in the United States are Territory, Fleece, and Texas (see footnotes, Table 18-4). The majority of Territory wool is shorn from white-faced breeds managed under range conditions. Although many grades are included, it is typically bright in color. Texas wools are also usually bright and finer and more uniform than those grown in other areas of the United States. Occasionally, they may contain discolorations from certain soils or varying degrees of yellowing due to high humidity and/or temperature. Fleece wools are sometimes incorrectly described as domestic wool. They are grown primarily under farm flock conditions and tend to be variable in grade, length, color, and yield. They include a predominance of fleeces from the meat breeds or their crosses and often contain more vegetable matter and manure or urine stain than Territory or Texas wools.

The second step in determining the clean price of wool is to describe it as *graded* or *original bag*. Graded wools are those where individual fleeces from one or more clips are evaluated visually and combined to form larger quantities of fleeces that are similar. Original bag wools are those that tend to be less uniform in grade and length than graded lots but are considered uniform enough to require minimal or no sorting before processing. For example, an original bag 62's clip would be expected to contain small quantities of 64's and 60's and even a few 70's or 58's fleeces.

The final requirement for determining *clean price* is a description of average grade and length. Periodically, within a marketing year and between years, prices for given grades and lengths may vary greatly depending on consumer demand.

After clean price has been determined, an estimate of the proportion of clean fibers present (yield) must be made. Clean price and yield can then be combined to determine grease price in the following formula:

$$G \text{ (grease price)} = C \text{ (clean price)} \times Y \text{ (percent yield)}$$

This grease price is usually a delivered price. To determine grease price at the weigh point closest to a given farm or ranch, handling costs (H), such as freight, grading charges, storage, commissions, and coring, are deducted. Because these functions involve wool in the greasy state, appropriate charges are subtracted from the delivered grease price. Handling charges vary greatly from clip to clip depending on method of marketing. The formula for determining grease price paid to the producer becomes:

$$G = [C \times Y] - H$$

For example, a clip worth $2/lb clean and yielding 50% with $0.07/lb handling costs would be worth $0.93/lb in the grease as weighed.

Careful evaluation of this formula suggests that grease wool *must be described correctly before it can be priced correctly*. Accurate description by objective measure of at least the two most important value-determining characteristics (grade and yield) is considered basic. The addition of objective measures of length, strength, color, and purity, which are presently estimated by visual and tactile evaluation, is expected to enhance the accuracy of describing raw wool.

MARKETING SYSTEMS

There are several ways growers may sell their clips. Producers in the major wool-growing areas of the United States (Texas and the western states) usually have more options than those in the traditional farm flock (fleece wool) states. Predominant methods are described below.

Private Treaty

This has been referred to as the "Will you take?" method where a wool buyer asks a producer "Will you take?" a given price. The sale may take the form of a spot sale or preshearing contract. Preshearing contracts occur most frequently where relatively large clips are involved. Spot sales often reflect "area prices," particularly in the fleece wool states, where all the wools in a given geographical area are priced the same regardless of grade, length, yield, or other value-determining factors. In either case, there is often a maximum of one or two bids considered. The private-treaty method of marketing wool is considered the least competitive method available to wool producers.

Wool Pools

Primarily due to relative freight costs, large quantities of wool tend to attract greater buyer interest and more competitive bidding than small ones. Consequently, groups of producers of relatively small clips frequently *pool* their wool to increase the amount offered for sale at a given time and place. They may or may not function as official cooperatives with established bylaws, officers, and so on.

Some pools offer their members' clips "as is" from each flock. Others grade each clip and form lots made up of parts of different clips that are similar in grade, length, and other value-determining factors. Pools offering graded lots normally request bids (either sealed or private treaty) by grade. Growers are then paid for the amounts of wool they have represented in each grade.

A distinct advantage of pooling wool is that it tends to reduce handling costs as compared to marketing by consignment. Members often provide their own clerical, accounting, weighing, and (if included) grading needs. They frequently own a building, which is rented to other area businesses when it is not used for wool storage. On the other hand, if the group does not own a building, individual clips may be stored on the farm until the day they are delivered. Members often initiate programs to increase the quality of all clips involved, such as encouraging greater care in preparing fleeces at shearing time or control of burr-bearing plants in sheep pastures.

One of the negative aspects of wool pools, particularly where grading is not practiced, is that the price spread between the best and poorest clips is usually small. Producers of the superior ones become disenchanted because they are apparently

Marketing Systems

not rewarded adequately for their efforts in breeding and preparation. A cooperative attitude among members is essential for long-term success of wool pools.

Sealed Bid

Relatively large individual clips or groups of clips may be offered for sale by sealed bid. It is considered the most competitive method of marketing wool in the United States. In the call for bids, active buyers in a given area are usually contacted by letter. Conditions of the sale such as quantity and type of wool offered, allowable deductions for below average yielding wools, excessive vegetable matter content, and wools of poor purity are specified. The dates of inspection, opening of bids and terms of payment and delivery are also specified.

The sealed-bid method is most frequently used by pools and wool warehouses, but it is an option that should be considered by producers of original bag quality clips large enough to attract the attention of area buyers.

Consignment

Growers may consign their wool at a warehouse, which may be relatively large or small and either private or cooperative. In doing so, they pay someone who has more experience, buyer contacts, and the like, to sell their wool for them. A cash advance may or may not be provided. The clip is normally graded, and it may be several months before all fleeces involved are sold and final settlement is made.

Presale Testing

Presale testing is also called "specification marketing," "sale by sample," or "description marketing." It may be used in any of the four methods listed previously. A key feature has been that the primary value-determining characteristics of wool (grade and yield) are determined by objective measure of samples rather than visual estimates. Other characteristics are being added as suitable testing instruments and procedures are developed (Chapter 20).

Evaluation of individual clips by objective measure has not been used as frequently in the United States as in some of the major wool-producing countries. For example, it has been reported that 0.6% of Australian wool marketed in 1972 was presale tested. By 1981, 96% of their wool sales were based on objective measure.[14]

The relative size of U.S. clips represents a limit for this procedure because of costs involved and also because a minimum size sample (core) must be taken. For small clips, the cost for an official core as well as damage from excessive sampling prohibits this method of evaluation. As a rule of thumb, growers producing original bag quality wool of 5000 lb or more are encouraged to have their wool cored and sell it based on objective measure of its value-determining characteristics. It takes the guesswork out!

Market Information

Regardless of the method producers use to sell their clips, they are urged to be aware of price trends. The USDA *Wool Market News*[12] shown in Figure 18-4 also includes a sheep and lamb market report. It is published weekly February through June and biweekly July through January.

The annual charge for this publication is considered a good investment by most sheep producers. However, many livestock-oriented papers and magazines also carry regular reports of wool sales and the American Sheep Producers Council[1] provides 24-hour wool market news by calling 303-322-4499.

GOVERNMENT INFLUENCES

When evaluating world, national, or local wool prices, it is essential to consider the influences of governments as well as the economic situations in the countries involved. Because wool is economically important in international trade, governments often become involved in a regulatory manner. The price of wool at a given time and place is a function of supply and demand *plus* the attitudes of governments.

The prices growers receive for their wool may also be affected by industry influence. The Australian Wool Corporation is an example of an industry-supported program that essentially establishes a floor (minimum reserve) price for different qualities of wool.

Tariffs

The method most frequently used by governments to influence wool prices is the *duty*. When imposed on imports of wool, duties are an attempt to protect domestic wool from being undersold by wool from competing countries. For example, there is a duty on apparel wools entering the United States. However, wool used in carpets, lumbermen's socks, felt boots, and the like (not finer than 46's and typically 44's and coarser) is considered noncompetitive with domestic production and enters the United States duty free.

Three words pertinent to a meaningful discussion of government influence are as follows:

1. *Duty:* a specific levy imposed by law on imports or exports of goods.
2. *Tariff:* a schedule (list) of duties imposed by a government on exports or imports or goods.
3. *Ad valorem:* a compensatory duty, meaning in proportion to value, added on.

Government Influences

Prior to 1816, all wool, regardless of form, entered the United States duty free. The first official tariff act in 1816 imposed 15% ad valorem on raw wool and 25% on wool goods. Minor adjustments were made in specific duties between 1816 and 1930. The Hawley-Smoot Tariff Act of 1930, with periodic adjustments, continues in effect. A detailed current list of duties (Schedule 3, Textile Fibers and Textile Products) for each class of wool or wool product may be obtained by writing the Department of the Treasury, U.S. Customs Service, Washington, D.C. 20229.

It is important for producers to remember that governments can and do affect the world wool market. Manufacturers of wool goods must consider duties as well as monetary exchange rates, freight, and relative quality when choosing between domestic or foreign wool for their plants.

Incentive Program

Besides the imposition of duties on imported wool, the U.S. government has supported wool prices for U.S. producers since 1954. The National Wool Act of 1954 was initiated to encourage domestic production by offering financial assistance to growers in the form of direct payments. The source of funding for this act is revenue from duties on imports of wool. No more than 70% of gross receipts from duties can be used for this purpose.

The Food and Agriculture Act of 1965 extended the Wool Act through 1969. Since then, the program has been reviewed and adjusted periodically. Actual values for each year are based on a formula developed by the USDA that reflects production costs and therefore the price considered necessary to encourage wool production. Table 18-5 shows changes that have been made in the support prices and their relationship to average prices received by year.

The National Wool Act, referred to as the Wool Incentive Program, has not accomplished its purpose of increasing sheep and wool production in the United States. It has, therefore, been challenged frequently by congressmen, who describe it as ineffective. Leaders of wool growers' organizations concede that sheep numbers and wool grown in the United States have decreased steadily, but suggest that if the incentive program had not been functioning, the decrease may well have been even greater.

As long as the Wool Act is in effect, it is important that producers understand procedures for estimating their return from the incentive program when preparing budgets and/or cash flow statements. All values used in the following examples are based on grease weight rather than clean content. Producers who market their clips on a clean basis report their sales and receive incentive payments based on the equivalent grease prices received.

Shorn wool. To estimate income from the incentive program for shorn wool, producers need to know the *support price* for the calendar year in which the

TABLE 18-5 SUMMARY OF THE AMERICAN WOOL SUPPORT PROGRAM[a]

Year	Support Price[a] (cents/lb)	Average Market Price Received by Producers[b] (cents/lb)	Shorn Wool Payment Rate[c] (percent of market return)
1955	62	42.8	44.9
1956	62	44.3	40.0
1957	62	53.7	15.5
1958	62	36.4	70.3
1959	62	43.3	43.2
1960	62	42.0	47.6
1961	62	42.9	44.5
1962	62	47.7	30.0
1963	62	48.5	27.8
1964	62	53.2	16.5
1965	62	47.1	31.6
1966	65	52.1	24.8
1967	66	39.8	65.8
1968	67	40.5	65.4
1969	69	41.8	65.1
1970	72	35.5	102.8
1971	72	19.4	271.1
1972	72	35.0	105.7
1973	72	82.7	d
1974	72	59.1	21.8
1975	72	44.7	61.1
1976	72	65.7	9.6
1977	99	72.0	37.5
1978	108	74.5	45.0
1979	115	86.3	33.3
1980	123	88.1	39.6
1981	135	94.5	42.9
1982	137	68.4	100.3
1983	153	61.3	149.6
1984	165	79.5	107.5
1985	165	63.3	160.7
1986	178	66.8	166.5

Data adapted from *ASCS Commodity Fact Sheet,* USDA Agricultural Stabilization and Conservation Service, May 1981.

[a]Support prices and payments are for the marketing years beginning April 1 for the 1955-62 period, the nine months April 1 through December 31 for 1963, and calendar years beginning 1964.

[b]Prices for 1955-56 and 1964 through 1976 are calendar year prices; prices for 1957 through 1962 are for April-March marketing years; price for 1963 is for the period April-December.

[c]Payment rate for shorn wool is expressed as a percentage and is applied to each producer's dollar return from the sale of wool. For example, under the 1978 program, producers received $45 for each $100 received from marketing wool.

[d]No payments were required on 1973 marketings as the average price received exceeded the support level.

wool is sold or planned to be sold and estimate the *average price* of all wool sold in the United States the same year (national average). For example, using 1984 values:

Support price for shorn wool	$1.65
Average price received by U.S. growers for year	0.795
Difference	$0.855
Percentage necessary to bring the $0.684 national average up to $1.65	107.5

A simple formula for calculation is as follows:

$$\frac{\text{Support price} - \text{national average}}{\text{National average}} \times 100 = \textit{percent} \text{ incentive payment}$$

This proportion (percent) that the national average must be raised to reach the established price assumed to provide a fair return for wool is then multiplied by the *actual price each grower received* for his or her clip. This value is the incentive payment paid by the USDA to each grower submitting a claim each marketing year. Claims are filed at local Agriculture Stabilization Committee offices prior to January 31 for wool marketed the previous year.

A key feature of this program is that there is truly a built-in incentive for growers to increase the quality of their wool through breeding, feeding, and preparation and to market their wool as competitively as possible. It rewards superior wool production and marketing management.

For example, the payment in 1984 (Table 18-5) for an incentive base of $1.65 and a national average of $0.795 per pound was 107.5%. Actual prices received by growers may well have varied as much as $0.50/lb. If grower A was careless in his wool production management or marketing, he may have sold his wool for only $0.55 ($0.25 below average). His incentive payment would have been 107.5% × $0.55 or $0.591 for each pound of wool he sold. If grower B was an average producer, he received 107.5% × $0.795 or $0.855/lb. Another producer, grower C, may have been a "wool man" with a clip that represented careful selection, nutrition, preparation, and marketing. His sale price of $1.05 ($0.25 above average) was also multiplied by 107.5% for a payment of $1.129/lb.

Unshorn Lambs. A second feature of the Wool Act is a payment for wool grown on "unshorn" lambs while a producer owns them. Computation is based on an assumed 5 lb of wool that could be shorn from lambs weighing 100 lb. However, this wool is normally shorter and therefore less valuable than wool grown on sheep shorn annually, so a reduced payment is appropriate. The 5-lb production estimate is reduced by 20% to 4 lb. Or, the incentive payment per pound is reduced by 20%. For example, again using 1984 values:

Support price for shorn wool	$1.65
Average price received by U.S. growers for year	0.795
Difference	$0.855
80% of difference to adjust for reduced value	0.684
Wool (greasy, shorn basis) per cwt of lambs	5 lb
Payment per cwt of unshorn lambs	$3.42

Simple formulas for estimating income from unshorn lambs are as follows:

$$\text{Support price} - \text{national average} \times 80\% \times 5 \text{ lb} = \text{payment per cwt of unshorn lambs}$$

or

$$\text{Support price} - \text{national average} \times 4 = \text{payment per cwt}$$

A third and very important aspect of the Wool Act is a self-help system where producers contribute a portion of their incentive checks for promotion of lamb and wool. The American Sheep Producers Council[1] is the organization that administers these funds. Sheep producers are requested to vote by referendum each year the Wool Act is reviewed by Congress as to whether or not to support this program.

Originally, each grower contributed 1¢ per pound of shorn wool and 5¢ per hundredweight of unshorn lamb marketed. These deductions from each grower's annual incentive check have been increased to 1.5¢ and 7.5¢ in 1966; to 2.5¢ and 12.5¢ in 1978; to 4¢ and 20¢ in 1982; and to 6¢ and 30¢ in 1986. Approval of these referendums indicates strong support by sheep producers for promotion of their products.

HARVESTING WOOL

Shearing time is harvest time for wool growers. Proper shearing and fleece preparation are essential for successful wool production. All too frequently, fleeces from flocks that have been well bred and properly cared for throughout the year are damaged or even ruined by careless shearing and preparation.

Shearing

Wool must be removed from sheep properly to ensure minimum damage to both the fleece and the sheep. Cuts made in the skin of sheep by shearing equipment may be severe enough to require surgical stitches. If they are in areas close to leg tendons, the sheep may become permanently unable to walk properly. Deep cuts in the neck area result in excessive bleeding or death. One of the most serious problems is cutting of teats and vulva of ewes or the sheath of rams. These sheep must be removed from the flock as nonproductive. Small cuts or nicks in the skin leave the sheep vulnerable to fly-strike and maggot infestation or local and secondary infections. Skin pieces left in fleeces represent a serious problem in processing.

Second cuts in the fleece, those places where the shearer permits the comb of his handpiece to leave the skin and then goes back to remove the short fibers remaining in order to make the sheep appear smoother after shearing, are a primary problem associated with shearing. When a second cut is required because the sheep moved or the shearer was out of position, the short fibers are better left on the sheep. The appearance of the sheep is less attractive, but these fibers will be more useful to wool processors as part of next year's fleece than removed from the sheep as a second cut. These short fibers are a serious concern in the worsted system as they are removed in combing and result in high noilage. Also, average staple length is reduced in fleeces where second cutting has occurred.

For sheep flock managers, the problem of finding qualified shearers is a serious one, regardless of size of flock. Skillful sheep shearing requires a great deal of practice and good physical condition. Unless the shearer is located in an area where he can shear feeder lambs during the late summer, fall, and early winter, his opportunity for employment is limited to late winter and spring each year. Relatively few shearers can plan to be employed on a year-round basis. On the other hand, some farmers have found that shearing can be an excellent supplemental source of income. Many producers of small flocks have learned to shear their own and perhaps a few of their neighbors' flocks. Over a period of years, they can justify their investment in equipment and have the convenience of being able to have their own sheep shorn on a timely basis.

Because the availability of qualified shearers is a problem for sheep growers throughout the United States, manufacturers of shearing equipment, along with education systems such as high school vocational agriculture departments, junior colleges, universities, and extension services, have put forth a great deal of effort to train sheep shearers. An example of material available to people interested in learning how to shear sheep is a shearing chart prepared by Oster[5] (Figure 18-4). With careful study of this "Australian style" of shearing sheep, a few sheep for practice, and a great deal of determination, it is possible to become a sheep shearer. Anyone interested in learning how to shear sheep is encouraged to study this chart and, if possible, participate in a local shearing school or consult an area shearer for help.

Preparation

Innovative procedures for enhancing the quality of preparation of grease wool have been reported by individuals and groups of producers across the United States. However, it is generally agreed that most domestic wool is less carefully prepared than that imported from the major wool-producing countries.

A great deal of printed information, video tapes, and photos have been prepared emphasizing the importance of proper care of fleeces at shearing. A typical list of rules is as follows:

How to Shear Sheep

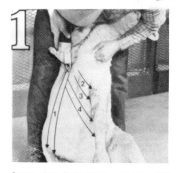

Wool is an important farm product, and the manner in which it is shorn has a direct bearing on the price the grower receives.

Machine shearing is preferred by both the grower and the woolen manufacturer because the machine gets all of the wool that can be safely removed from the animal, and gives full length fibres for conversion into yarn.

While shearing is done primarily to harvest a valuable crop, the work must be done without injury to the sheep.

One who is learning to shear sheep should always keep in mind those things which are fundamental to good workmanship such as holding the sheep in positions that insure the skin of the animal being stretched tight on the part of the body being shorn, making each stroke with the shearing handpiece as long as possible, and keeping the ends of the teeth of the shearing comb on the skin of the sheep throughout every stroke.

There are other factors that influence good shearing such as, keeping the shearing equipment in good condition, providing a clean, smooth surface on which to work, and penning the unshorn sheep close to the shearing space. These, however, are incidental to good workmanship, so the material on this chart deals with the fundamentals of good shearing, the technique of holding the sheep, and making the strokes with the shearing handpiece.

The strokes with the shearing handpiece in each position are numbered consecutively to show the order in which they were made. The type of sheep, condition of the sheep, and condition of the wool will have a lot to do with the number of

Support sheep firmly between your knees. Rest sheep's foreleg on your left side to tighten skin on belly. Make first stroke of the handpiece straight down on right side of brisket to the open flank area. Make the second stroke on the left side of brisket down under left front leg. Follow strokes three and four to the left flank area.

First stroke out inside of right back leg. Second stroke starts on the toe and goes up and back around the udder.

Using left hand apply pressure to the left stifle joint to make sheep hold leg straight out. Note right front leg of sheep has dropped between shearer's legs. Also shearer should move his right leg back a few inches to help turn sheep.

Shearer's left hand applies pressure on the stifle joint to make sheep hold leg straight out. First stroke goes out on top of leg. Second stroke starts at toe and goes back up cleaning the flank area.

Straighten sheep up, left foot close to sheep's hip, right foot in crotch between legs. As first stroke is ending by sheep's chin, turn handpiece to right causing wool along top side of stroke to break off so you can see where to make second stroke.

Turn sheep's head so the nose is up and the right side against leg. Hold body firmly between knees. Left hand is shifted toward nose as strokes are completed along jaw.

Complete last stroke up neck, which should remove wool to base of left ear, hold head against leg with heel of hand and grasp ear, shear around ear.

Figure 18-4 Self-teaching shearing chart. (Provided by Oster[5])

strokes required. The beginner, however, should strive to shear each position with the smallest number of strokes possible.

Keep the sheep's body, as nearly as possible, in the positions shown in the pictures. It should be maintained in these positions very largely by the legs. **In order that the legs may be used to advantage in holding the sheep, it is important that the feet be placed correctly in relation to the sheep's body, as shown.**

Note, that the shearer's left hand is used very little in holding the sheep, but aids in tightening the skin on the area being shorn. The left elbow and forearm, at times supplement the legs, feet, and knees of the shearer in maintaining the sheep's body in the correct position.

The experienced shearer takes advantage of the characteristic of sheep to give up easily. A sheep placed in nearly any position will remain quiet for some time providing the position is not decidedly uncomfortable. Remember, however, that the sheep's legs cannot be held forcibly in a cramped position beyond a limited time without causing the sheep to struggle. The front feet must be kept from touching the floor, otherwise the sheep will attempt to get to its feet.

With the things already mentioned well in mind, the beginner should be ready to undertake the task before him. He will learn by doing.

Study the pictures and the copy underneath each before shearing is attempted. Hang the chart close to where the shearing will be done so the pictures can be referred to as shearing progresses.

Note that sheep's left front leg is held by shearer's left wrist, leaving his left hand free to tighten skin on the belly area.

Hold sheep fully relaxed against your legs and apply pressure in flank as illustrated. Caution—slide comb at angle around ham string to avoid accidental cutting of ham string.

Drop wrist to position the comb to enter wool on top of tail and shear forward along backbone cleaning all wool from tail.

Without changing position of sheep, shift left hand to sheep's head for shearing top knot. Shear top knot only, ending strokes on line between front side of ears.

Hold sheep's body between knees, hold head against left leg with elbow, use left hand to stretch skin on shoulder, start stroke at knee shearing upward to clean tags on front shoulder.

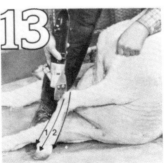

Take boots off with downward strokes.

Holding sheep's leg close to body shear down to shoulder as shown.

347

Lay sheep on its back, keep right foot between sheep's hind legs, left foot under right shoulder as shown, force left foreleg toward sheep's head to stretch skin on side.

On reaching backbone make sure sheep's tail is behind your right foot. Your left leg is under sheep's shoulder, with your left foot across sheep to give leverage. Apply downward pressure to sheep's head with your left hand to "roll" sheep up. Shear two strokes over backbone to save time on "last side." Avoid standing on or tearing fleece when swinging right leg out to shear "last side."

On final stroke of long blow, with left hand hold ear up and extend stroke under ear and out to jaw.

Using a full comb, make remaining strokes at same angle to toe.

With sheep in position, hold skin of right flank in left hand and apply pressure to stifle joint. Caution: slide comb at angle around ham sring to avoid accidental cutting of ham string. Hold sheep's head between left arm and right leg moving left foot back so sheep settles down on its side and moving right foot forward to allow more reach for last strokes. Left foot keeps sheep's shoulder and feet off floor.

Good Sheep Management Practices

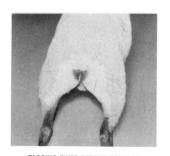

TAGGING EWES BEFORE BREEDING

This practice results in a larger percentage of lambs and a lamb flock of more uniform age.

The wool removed can be sold but if left on will become dung locks of no value.

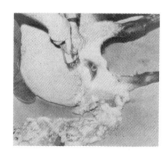

TAGGING EWES BEFORE LAMBING

Ewes tagged before lambing raise a larger percent of strong lambs. More lambs are saved because they get their first fill of warm milk when they need it most. Ewes with clean udders seldom fail to own their lambs.

Clamp sheep between your legs. Hold sheep's head with knees. After clearing wool from ear and face, shear straight down to point of shoulder, using left hand to straighten wrinkles. Allow sheep's right foreleg to come forward after the third blow.

Push down on point of shoulder with palm of hand. At the same time pull skin up with finger to clean tags under front leg. On final stroke turn handpiece under front leg and shear to toe.

Hold sheep firmly between legs, its head down on its left side which stretches skin on right side, as wool is taken off. As last stroke is made move left foot to other side of sheep's hind legs and raise head.

CORRECTING WOOL BLINDNESS (FACING)

The winter growth of wool on the faces of several popular breeds produces wool blindness.

Wool blind sheep are less alert, do not get their share of feed, and often ewes fail to own their lambs. It pays to correct wool blindness.

TAGGING FEEDER LAMBS

Experienced lamb feeder's have found it pays to tag all lambs when they are put on feed. Tagged lambs are cleaner, healthier and more attractive at the close of the feeding period.

SUMMER SHEARING

Ewe lambs will be retained as breeding stock should be sheared all over in early summer. Sheared lambs do better during hot weather and develop faster than unshorn ones.

1. Shear only when sheep are dry.
2. Shear on a clean floor.
3. Keep shorn fleeces clean.
4. Roll and tie fleeces with flesh side out.
5. Use only paper fleece ties.
6. Pack "off" fleeces or parts of fleece (those exceedingly different from the bulk of the clip) separately. These may include black fleeces, fleeces from black-faced sheep if only a few in a flock, fleeces shorn from dead sheep, as well as crutchings, dung locks, and sweepings.
7. Shear black and black-faced sheep last at a single shearing to avoid contamination of the shearing floor.
8. Pack fleeces in burlap bags.
9. Mark bags to identify type of wool in each.
10. Store in a dry place.

These lists may be longer or shorter, depending on where, when, or by whom they are prepared. They may be summarized in one sentence: *Keep wool as clean as possible all year long,* or, as the American Sheep Producers Council[1] suggests, IT IS AMERICAN WOOL—PACK IT WITH PRIDE.

It is not always possible to avoid some contamination of fleeces. For example, a producer's cheapest feed source may be a sandy, sparsely covered range or aftermath of corn, cereal grain, or sugar beets. Because these conditions are usually very dusty, fleeces will contain an excessive amount of mineral matter. Mineral matter is readily removed at scouring but does reduce yield significantly. In these situations, producers are urged to market their wool based on an objective measure of yield as buyers may tend to underestimate the yields of these "heavy" fleeces. On the other hand, there is little justification for handling hay, straw, or other feedstuffs carelessly over the backs of sheep resulting in excessive vegetable matter, permitting poly-contamination, mishandling fleeces at shearing, or using nonscourable branding paint.

Shearing facilities in the United States range from permanent shearing sheds to trucks or trailers with built-in handling systems to open pens with wood platforms or old carpets for shearing floors (Chapter 3). Regardless of the type of facility, U.S. wool growers, not shearing crews, are responsible for harvesting their wool. At shearing time, they should be "on the shearing floor" to make sure their sheep are handled correctly and their fleeces are cared for in a manner that will maximize value.

Excellent printed materials, slides and video tapes concerning wool preparation, marketing, and facilities for shearing and handling fleeces are available from the ASPC for use by producers as well as organizations as aids for detailed study and/or discussion.

Skirting and Grading

Parts that are different from the majority of the fleece, such as the belly, britch, neck and tags are referred to as *skirts* and their removal is called *skirting*. The purpose is to separate these less valuable parts, leaving a uniform *main sort* which can be combined with others to develop groups (lots) of similar fleeces. The smaller pieces are also combined with others to form larger groups of bellies, tags, britches, and so on.

Extensive skirting (called sorting when the fleece is broken into several parts) is a labor-intensive practice that is frequently challenged in countries where it has become routine. It has not been practiced widely in the United States, probably because of a lack of skilled people to do the job and because financial rewards have not been demonstrated.

Reports of research using fleeces from American white-faced breeds[3,7,13] have not suggested strongly that skirting should become a standard practice in U.S. sheep flocks. A study of the economic effect of skirting[4] indicates that table skirted fleeces (bellies, shin locks, low britch wool, and tags removed) must be sold at a premium of at least 7% and the skirts not discounted more than 11.3% from the clean price of the skirted fleeces. In this study, shearer skirting (only bellies, lower leg locks, and some tags removed) was shown impractical.

On the other hand, removal of bellies and tags on the shearing floor[2] is becoming more popular nationwide. These fleeces are higher yielding and freer from stain and vegetable matter, so they normally command higher prices than nonskirted fleeces of similar quality. Accumulations of bellies and tags have been reported to sell competitively.

Whether or not fleeces need to be graded at the farm or ranch is directly related to breeding management. Flocks that have been carefully selected for a given grade are probably uniform enough to be sold "original bag" (Table 18-4) without grading. For clips where fleeces vary widely, it may be more cost-effective to have them

Figure 18-5 Grading specialist sorts wool lots according to fiber length, fineness, and cleanliness of the fleeces. (Photo courtesy of R. C. Elliott and Co., Salt Lake City, Utah.)

graded at a warehouse by trained graders in facilities designed to do the job efficiently (Figure 18-5). Wool growers should consult their own marketing organizations (pool, warehouse, cooperative, etc.) or buyers before deciding if, when, and where their clips should be graded.

REFERENCES CITED

1. American Sheep Producers Council, Denver, Colo.
2. Bassett, J. W., and others. 1972. The effect of removing belly wool at shearing. *Proc. Western Section Amer. Soc. Animal Sci.* 23:76.
3. Hallford, Dennis, and others. 1984. Quality and Yield Characteristics of Skirted and Unskirted Fleeces Produced by Fine-Wool Ewes Maintained under Drylot Conditions. Research Report 547. New Mexico State University Agr. Exp., Sta. Las Cruces.
4. McFadden, William D. 1958. The economic effect of skirting fine-wool fleeces. Bulletin 442, Agr. Exp. Sta., New Mexico College of Agriculture and Mechanic Arts.
5. Oster Professional Products, Dept. 5055, N. Lydell Ave., Milwaukee, Wisc.
6. Pattinson, Russel. 1981. What characters determine the clean price of wool at auction. *Wool Technology and Sheep Breeding,* 29, 3:93. University of New South Wales, Australia.
7. Pexton, J. E., Alexander Johnston, and S. A. Larsen. 1969. Influence of Skirting upon marketing qualities. *Proc. Western Section Amer. Soc. Anim. Sci.* 20:409.
8. Pexton, Ed, LeRoy Johnson, and Frank Craddock. 1980. *4-H Wool Judging.* University of Wyoming Agricultural Extension Service, Laramie, No. 12500h-80.
9. Pohle, Elroy M. 1963. Grading and production of wool. In *Wool Handbook,* Vol. I, Werner von Bergen, ed. Wiley-Interscience, New York.
10. Stobart, Robert H., and others. 1986. Sources of variation in wool fiber diameter. *J. Animal Sci.* 62:1181.
11. U.S. Department of Agriculture. 1966. *Official Standards of the United States for Grades of Wool.* Service and Regulatory Announcement C & MS No. 135. USDA Consumer and Marketing Service, U.S. Government Printing Office, Washington, D.C.
12. U.S. Department of Agriculture. *Wool Market News.* Agricultural Marketing Service, LS Division, Market News Branch, 711 O Street, Greeley, Colo. 80631.
13. Willingham, T. D., M. Shelton, and J. W. Bassett, Fleece quality as affected by britch removal, *SID Research Digest,* Vol. I, No. 1:26.
14. Welsman, Sandra. 1981, Are universities, technical and agricultural colleges keeping abreast of developments and maintaining standards required to service new procedures in wool production and marketing? *Wool Technology and Sheep Breeding* 29, 11:80. University of New South Wales, Australia.
15. Yeates, N. T. M., T. N. Edey, and M. K. Hill. 1975. *Animal Science,* p. 323. Pergamon Press, Elmsford, N.Y.

SUGGESTED READING

American Sheep Producers Council. 1965, 1968. *The Story of Wool.* American Wool Council and the Wool Bureau, Inc., Wool Education Center, Denver, Colo.

Burns, R. H. 1962. Sensory judgments of greasy wool. *J. Text. Inst.* 53, 11:756.

Garside, Alton. 1955. *Wool and the Wool Trade,* 2nd ed. Wool Associates of the New York Cotton Exchange, New York.

Sachse, James M. 1985. *Shearing Facilities and Wool Preparation at Shearing.* New Mexico State University Extension Service, Las Cruces.

Von Bergen, Werner. 1963. *Wool Handbook,* Vol. I. Wiley-Interscience, New York.

19 Wool Growth and Structure

WOOL GROWTH

An understanding of the biological functions of wool growth is basic to management practices designed to increase both the quality and quantity of wool produced per sheep. Physical and chemical properties of wool fibers are nearly as important as dimensional measures in the manufacture of wool goods. A better understanding of these properties and the factors affecting them can lead to corrective measures that benefit not only consumers and processors but producers as well.

Domestication

A good study of wool growth begins with the evolution of sheep[10] (Chapter 1). Students of changes in wool growth in sheep recognize that much speculation is involved in attempts to "put together the pieces" of the development of sheep and their fleeces because archeological remains of sheep (fossils) that existed in the Middle Stone Age, when domestication began, are few.[14] However, remains of fibers, skin, leather, and parchment from the last few centuries B.C. have been useful in the study of changes in wool types and follicle structure.

The two most important changes in wool growth associated with domestication are the reduction of hairy (kempy) outer coat fibers along with an increase in production of true wool (Chapter 1), and the near elimination of the "shedding" characteristic that is demonstrated by comparing the nondomesticated *Ovis canadensis* (Rocky Mountain Big Horn) with the *Ovis aries,* which are mostly nonshedders. The whole coat of wild sheep is molted every spring in matted masses, but domestic sheep require periodic shearing.

Figure 19-1 suggests lines of evolution of main fleece types. Subsequently, as breeds were developed, dramatic differences in wool growth have become apparent. A diagram of the development of breeds based on wool type is shown in Figure 19-2. Differences among breeds suggest that suitable wool types are available to manufacturers for the production of a wide variety of wool goods, ranging from lightweight tropical worsted suiting material to carpets.

Follicle Development

The following discussion is presented as an aid to understanding the relationship of wool follicles to the genetic and environmental factors that influence wool production (Chapter 21) rather than a detailed study of the physiology or histology of sheep skin.

Basic skin structure. In the early embryonic development of mammals, three major types of cells have been described: (1) the *ectoderm* forms the outer epithelium and the nervous system; (2) the *mesoderm* becomes connective tissue, muscle, blood, and most of the urogenital system; and (3) the *endoderm* gives rise to lung and gastrointestinal tract epithelium and its derivatives.[7]

Epithelial tissue is derived from both ectoderm and endoderm. It can regenerate, its nutrition is avascular, and it covers the surface of the body and the lining of ducts and structures directly or indirectly connected with the outside. The primary function of epithelial tissue is protection.

Two distinct layers of skin (Figure 19-3) are the epidermis (outer layer) and dermis (inner layer). The *epidermis* is composed of a horny outer layer of dead protein cells, chemically similar to wool, hair, horns, and hooves. There are intermediate layers of dead and dying cells and an inner layer of actively dividing cells described as *basal* or *germinal*. The basal layer cells are of primary interest in the development of wool follicles. They not only provide replacement of outer epidermal cells that are removed by wear and tear of the skin surface but are also the cells that proliferate to form follicles and, subsequently, wool fibers. There is no vascular system in the epidermis.

The *dermis,* also known as the corium, is composed of loose connective tissue containing fibers of a collagen-type protein that is distinctly different from keratin. The dermal layer that lies immediately beneath the basal layer is called the papillary or thermostatic layer because it is well supplied with blood vessels and nerves and is important in the regulation of body temperature. The reticular layer is composed of a coarser, more open network of collagen fibers. A fatty layer, which is generally not well developed in sheep, lies below the reticular layer.[15]

Follicle formation. The development of wool follicles, the structures that produce wool in the skin of sheep, begins during the early fetal life of the lamb. The stages of follicle development have been described by many researchers. The diagrams in Figure 19-4 are classic and show stepwise changes beginning with the

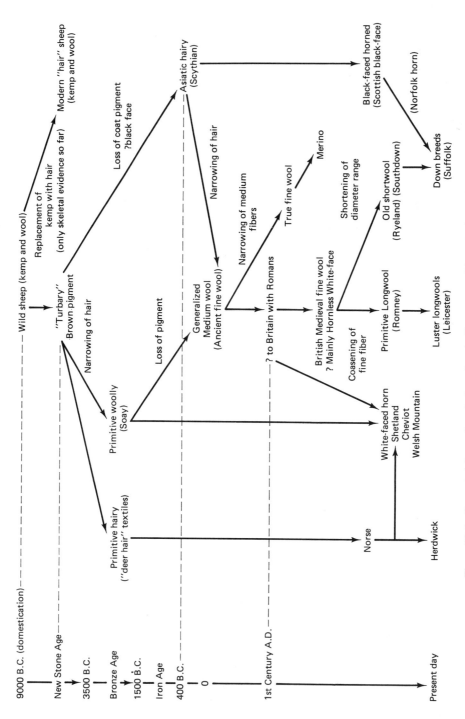

Figure 19-1 Suggested lines of evolution of main fleece types. (From Ryder and Stephenson.[15] Reprinted by permission from *Nature*, Vol. 204, pp. 555–559. Copyright 1964. MacMillan Journals Ltd.)

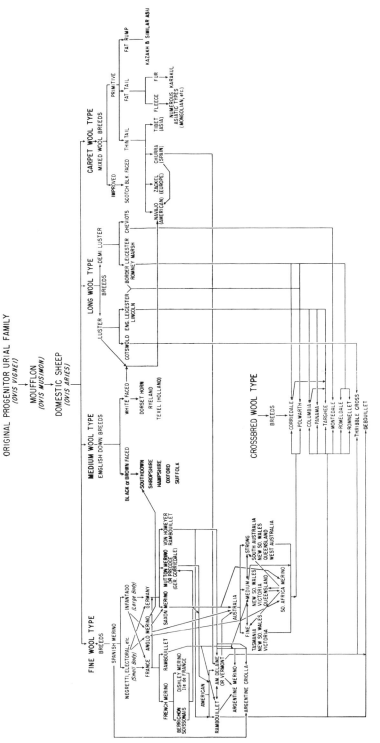

Figure 19-2 General pedigree of the domestic sheep. (From Burns,[4] with permission)

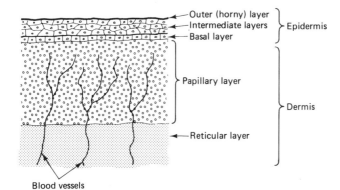

Figure 19-3 General structure of the epidermal and dermal layers of skin. (Adapted from Onions[12])

formation of a *follicle plug,* which results from rapid multiplication of basal layer cells at given points in the skin. Subsequently, these rapidly dividing cells force their way into the dermis, forming a follicle that reaches the mid-dermal vascular system. At this stage, the base of the follicle plug inverts to form the papilla (growth point for wool fibers) and the hair canal begins to form. Accessory structures—the arrec-

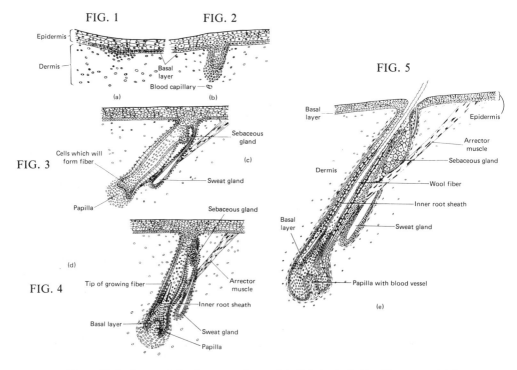

Figure 19-4 Stages of development of a wool follicle in fetal skin. The sequential figures represent vertical sections in skin magnified about 100x (Figures 1 to 4) and 150x (Figure 5) (After Wildman,[17] with permission from WIRA *Wool Science Review*[1])

tor muscle, sebaceous (wax) gland, and sudoriferous (sweat) gland—become identifiable. Basal layer cells are continuous around the follicle, but only those that form the papilla remain very active.

As the papilla develops further, capillary networks form around the follicle and the tip of a growing fiber appears. As the fiber continues growing, it becomes keratinized or hardened, beginning at about one-third of the distance from the papilla to the skin. It is in this zone of keratinization (mid one-third of the follicle) that previously soft fiber protein becomes hardened and also that scales (cuticle cells) become identifiable and probably most of the crimp formation takes place. The fiber continues growing until it forces its way through the epidermis.

Features of mature follicles and their associated structures of interest in relation to wool and its growth are as follows:

1. A rich supply of *blood* to the papilla (the growth point of wool fiber) provides nutrients for the formation of fibers. Vasodilation due to changing ambient temperature is thought to influence seasonal differences in wool growth (Chapter 21) by varying the flow of nutrients to the follicles.

2. The *arrector muscle* is a relatively inactive structure in sheep as compared to dogs and cats, but is thought to play a role in crimp formation.

3. The *sudoriferous (sweat) gland* is considered relatively inefficient as a cooling mechanism in sheep. However, its secretions of solutions (called suint) of mainly potassium salts of organic and inorganic acids tend to affect the apparent degree and consistency of yolk (wool wax plus suint). The relative activity of sweat glands can certainly affect the amount of moisture in wool at shearing.

4. The *sebaceous (wax) gland* provides a protective coating of waxy material for each fiber as it emerges from the skin. It is a complex mixture of esters that form fatty acids and alcohols when hydrolyzed.[2] Although the amount of grease in raw wool varies between sheep within breeds and/or grades, it is proportionately higher in the finer grades, which have more surface area per unit of weight than the coarser grades. It is a challenging nonwool component to remove in commercial scouring systems. However, when recovered from scouring solutions, it has a wide variety of uses as an additive in compounds ranging from anticorrosion preparations to ointments and cosmetics. When highly purified to meet rigid standards, it is called lanolin.

Primary follicles. There are some differences among reports by researchers as to the exact fetal age at which each stage of follicle development occurs. Breed as well as inherent differences among individuals are probably responsible for variations in fetal ages at which stages in the formation of follicles have been observed. Generalizations of the timing of these events are considered adequate for optimum sheep flock management related to wool production. It appears that central primaries are initiated beginning on the face and poll and spreading over the rest of the body at about 50 to 75 days after conception (early in the mid-third of gestation).

Some 15 days later, lateral primaries appear at either side of each central to form *trios*, which provide the framework for follicle groups. Follicle groups are normally composed of a trio of primaries and their associated secondaries separated from other groups by bands of connective tissue. All primary follicles are usually fully developed and producing fiber at birth.

Primary follicles tend to lie in distinct rows across the skin, with the arrector muscle and glands on one side and associated secondary follicles on the other. This relatively orderly arrangement of trios of primaries combined with their early formation in fetal skin suggests they represent the basic units in follicle bundles around which secondaries develop.

Secondary follicles. Secondary follicles develop in much the same manner as primaries, but there are some major differences. They are usually smaller in diameter than primaries, the sweat gland and arrector muscle are not present, and the sebaceous gland is smaller and single rather than bilobed as in primaries. Also, secondaries are often branched, in which case emerging fibers share the same orifice in the epidermis of the skin. The most important differences between primary and secondary follicles are shown schematically in Figure 19-5. Although most secondary follicles are initiated during *late gestation,* few of them are producing fibers visible on the skin at birth. Therefore, environmental effects may be important in

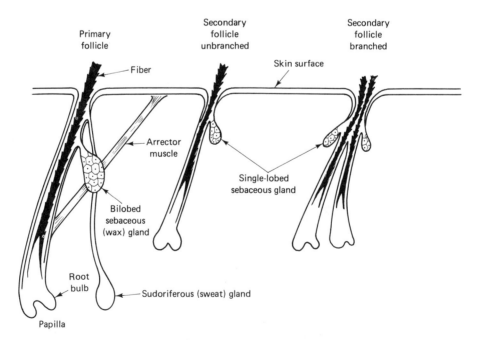

Figure 19-5 Major structural differences between primary and secondary follicles. (Adapted with permission from Hardy and Lyne[8])

Wool Growth

determining whether or not the genetic potential for initiation and branching of secondary follicles is reached (Chapter 21).

Secondary/primary (S/P) ratio. It is generally agreed among researchers that differences in density (number of structures per unit area of skin) of primary follicles among sheep and even among breeds is relatively small.[6] Real differences in the density of fibers growing from the skin of mature sheep are, therefore, due to the density of secondary follicles. Fleece density is often described in terms of the number of secondary follicles for each primary (S/P ratio). The higher the S/P ratio within grade and length, the denser and, therefore, more productive the fleece. Typical differences in S/P ratio among breeds are shown in Table 19-1. Familiarity with the grades of wool produced by these breeds indicates that fleeces with high S/P ratios tend to be not only denser but also finer than those with low S/P ratios.

Photomicrographs of cross sections of fibers from locks taken from the midside of sheep (Figure 19-6) show that finer grades tend to be more uniform than

TABLE 19-1 FOLLICLE DENSITY, S/P RATIO, AND FLEECE TYPE IN DIFFERENT BREEDS AND STRAINS OF SHEEP

Breeds and No. Sampled (n)	Age of Sheep Sampled (months)	Mean No. of Follicles per mm²		Mean No. of Secondaries per Primary (S/P)	Fleece Type (approx.)
		$\bar{n}(p+s)$	$\bar{n}p$		
Fine Merino (148)	10-15	71.7	3.56	19.1	Merino: 70-90's
Medium Merino (145)	12-20	64.4	2.93	21.0	Merino: 60-64/70's
Strong Merino (63)	12-15	57.1	3.27	16.5	Merino: 58-60's
Polwarth (63)	14-15	50.2	3.63	12.8	Crossbred: 58-64's
Corriedale (63)	7-12	28.7	2.43	10.8	Crossbred: 56-60's
Southdown (21)	11-12	27.8	3.9	6.3	Crossbred: 56-60's (Down)
Dorset Horn (21)	11-12	18.5	2.9	5.4	Crossbred: 56's (Shortwool)
Suffolk (21)	11-12	20.4	3.5	4.8	Crossbred: 56-58's (Shortwool)
Romney Marsh (21)	11-12	22.0	3.4	5.5	Crossbred: 48's (Longwool)
Border Leicester (21)	11-12	15.8	2.9	4.4	Crossbred: 46-48's (Longwool)
English Leicester (21)	11-12	14.4	2.5	4.9	Crossbred: 40-46's (Longwool)
Lincoln (21)	10-11	14.6	2.3	5.4	Crossbred: 36-44's (Longwool)

Adapted with permission from Ryder and Stephenson[15] (after Carter[5]).

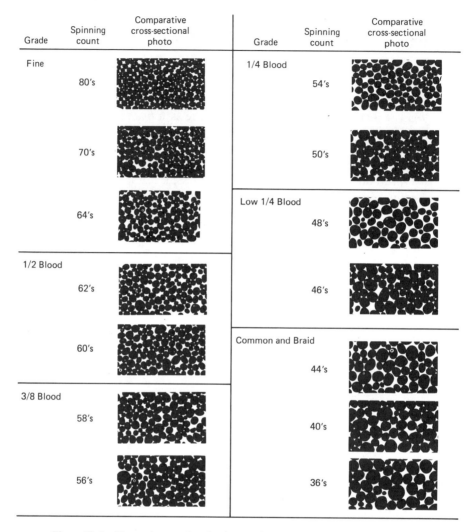

Figure 19-6 Photomicrographs of cross sections of wool fibers taken from the mid-side of fleeces of different average diameters. (From Pexton, Johnson, and Craddock,[13] with permission)

coarser grades. Because fibers growing from primary follicles are normally coarser than those from their associated secondaries, skin with a greater proportion of secondary follicles (higher S/P ratio) produces fibers with less variation in diameter.

The differences in fiber diameter shown in Figure 19-6 help visualize the concept that the majority of variation among fibers in a fleece occurs between fibers within locks (Chapter 21). This figure also shows that fibers are irregular in shape and often elliptical, with definite major and minor axes, an important consideration when measuring diameter (Chapter 20).

WOOL FIBER STRUCTURE AND PROPERTIES

The physical and chemical structure of wool fibers is relatively more important to manufacturers and consumers of wool goods than to producers because it is directly associated with the spinability, feltability, and dyeability of wool, as well as the handle or drape of fabrics made from it. However, more and more wool growers are becoming concerned with the quality of the wool they produce. They recognize that, although the fiber diameter, length, and yield of their fleeces are the primary criteria associated with price determination, other factors such as strength, softness, medullation, uniformity, and color are also important. Some of these can be altered by breeding and management changes in ways beneficial to the entire industry (Chapter 21).

Macrostructure

With the aid of a simple microscope, wool fibers can be described as made up of a *tip, shaft,* and *root*. The pointed tips of fibers from all sheep at first shearing are not present at subsequent shearings (Figure 19-7). A more important concern for manufacturers is the degree of *end-to-end variation* in diameter along fibers. Figure 19-8 shows variation in diameter due to relative availability of nutrients and stress. Other environmental influences (Chapter 21) may also contribute to this variation. End-to-end variation is important when measuring fibers for marketing or selection (Chapter 20). It is difficult to describe the average diameter of fleeces accurately when end-to-end variation along fibers is large.

The *crimp,* or wavy appearance, of wool fibers is readily visible. It may be uniplanar in the finer wools, but microscopic examination shows that wool fibers are more frequently coiled (Figure 19-9). The formation of crimps in wool takes place in the follicle. Several authors have attempted to describe in detail exactly how

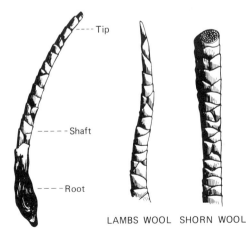

Figure 19-7 Illustrations of wool fibers showing their parts and tip types (From McFadden,[11] with permission)

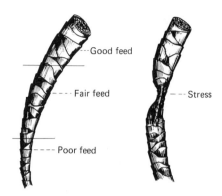

Figure 19-8 A diagramatic example of the effects of nutrition and stress on wool fiber diameter. (Adapted with permission from McFadden[11])

the crimpiness in wool fibers is formed.[15,16,18] It is probably due to a combination of cyclical movements of the follicles stimulated by the arrector muscle, differences in cortical cell types, and/or a difference in the rate at which keratinization (hardening) of cells occurs at opposite sides within the follicle. Regardless of the mechanisms involved, clarity of crimp and lock formation in fleeces grown by the meat breeds is much less distinct than in fleeces grown by the wool breeds.

Crimp frequency is probably associated with growth rate, as it is usually higher in the finer, slower-growing wool types and lower in the coarser, more rapidly growing types, so it is used as an indicator of fineness (Chapter 20). It may also be related to spinability and fabric quality (Chapter 18).

Microstructure

Wool fibers are composed of two basic cell types: cuticle and cortical. Occasionally, a medulla (hollow area) appears in the center of fibers (Figure 19-10).

Cuticle. Discussions of cuticle cells are frequently included as a part of the macrostructure of wool fibers because they are readily apparent at relatively low magnification (Figure 19-11). Also, when a lock of wool is passed through the fingers from tip to base, it feels more abrasive than when the fingers move along the

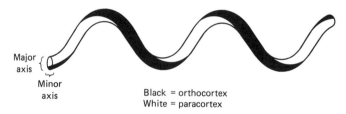

Figure 19-9 Schematic representation of a crimped fiber showing typical coil formation, elliptical shape, and ortho-para cell types.

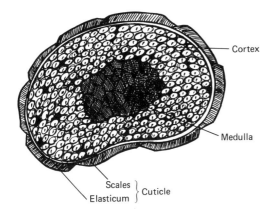

Figure 19-10 A cross-sectional diagram of the organization of cell types in a medulated wool fiber. (Adapted with permission from McFadden[11])

fiber from base to tip, a difference due to the scalelike formation of cuticle cells on the surface of wool fibers. On the other hand, cuticle is cellular in nature, originating in the follicle bulb, so it is more correctly included as part of the microstructure of wool fibers.

Cuticle cells are lapped similar to shingles on a roof in the finer grades of fibers and always in a manner whereby the raised edge is in the direction of the tip of the fiber. Three of several different types of scale formations that have been described are shown in Figure 19-12. The mosaic rather than overlapped pattern found in coarser wools probably contributes to their lustrous appearance. The function of cuticle cells is primarily a means of protection for the internal structure of the fiber. They tend to repel water on initial contact but are porous enough to permit the movement of vaporized moisture either into or out of the interior of the fiber readily.

In manufacturing processes, fibers are normally randomized so the raised

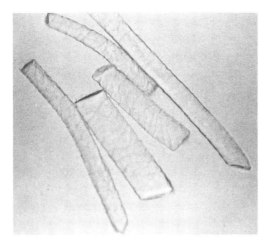

Figure 19-11 Images of wool fibers at 500x magnification. (University of Wyoming photo)

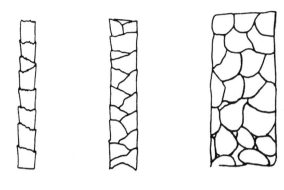

CORONAL   IMBRICATE   RETICULATE

Figure 19-12 Diagrams of different cuticle cell (scale) formations on the surface of wool fibers. (From McFadden,[11] with permission)

edges of cuticle cells are in opposition (Figure 19-13). This is a distinct advantage when spinning or felting is the desired process. However, this same relationship of fibers represents a disadvantage to the consumer who agitates wool fabric excessively when washing it. These fabrics are said to "shrink" due to felting of the fibers. Detailed study of the cuticle of wool fibers has led to treatments whereby wool fabrics can be machine washed and dried using ordinary household detergents.

Medulla. Microscopic examination of the medulla in wool fibers shows a network of apparent cell wall remnants filled with air (Figure 19-10). Medullation probably occurs at the papilla in certain follicles. It is most frequently noticed on the britch of the sheep, but medullated fibers may be observed scattered throughout the fleece. They are similar to guard hairs grown by primary follicles in wild sheep and apparently persist after centuries of selection against them. Medullation may appear as continuous or interrupted, occupying varying proportions of the center of fibers. Those with discontinuous medullas are described as *heterotypes* (Figure 19-14).

Kemp fibers are similar to medullated fibers except that the degree of medullation is greater and they are normally considered "shedding" or discontinuously growing fibers. They are readily visible, may vary in length from a few millimeters to nearly as long as the staple, and are often scattered throughout the fleece. Chalky in appearance, they can normally be distinguished visually from leg or face hair in shorn wool because they appear tapered at both ends (Figure 19-15).

Figure 19-13 Scales of randomized wool fibers interlocked to increase adherence to each other during processing.

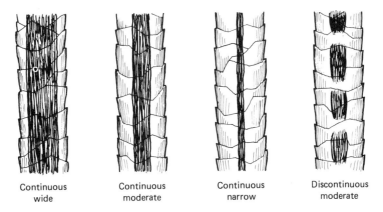

| Continuous | Continuous | Continuous | Discontinuous |
| wide | moderate | narrow | moderate |

Figure 19-14 Types and degree of medulation in wool fibers. (Adapted with permission from McFadden[11])

The incidence and degree of medullation in wool samples can be documented accurately when projection microscope procedures are used to measure fiber diameter (Chapter 20). Regardless of the relative thickness of medullae or whether they are continuous, interrupted, and so on, medullated fibers and kemps in wool are considered negative by processors because they are weaker and appear lighter after dying than nonmedullated fibers.

Cortex. Cortical cells are the most important in wool structure. Produced from basal layer cells at the papilla, they appear cigar-shaped and are variable in size, ranging from 80 to 100μ long and 3 to 5μ wide at the middle, keratinized, and cemented together to form the body of the fiber. Two types of cortical cells (ortho and para) have been described. Ortho cells tend to appear toward the outer portion of coiled fibers and para cells toward the inner side[12,15] (Figure 19-9). However, wide variations have been observed. Ortho cells are softer and contain a lower proportion of disulfide linkages (cystine) than para cells. Para cells are less alkali sensitive, less readily stained in basic dyes, and appear to swell more in length when wetted than ortho cells.

The electron microscope has contributed to a more detailed description of cortical cells. The components of a fiber are shown schematically in Figure 19-16.

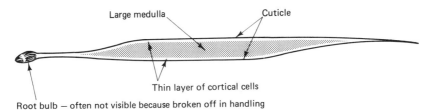

Figure 19-15 Diagram of a whole kemp fiber.

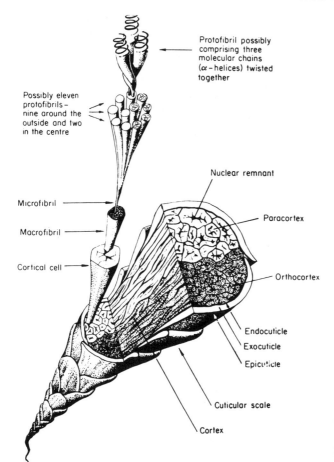

Figure 19-16 Diagram showing one possible way in which the α helix molecules and protofibrils are grouped within the microfibrils of wool. (From Ryder and Stephenson,[15] with permission)

Chemical. Chemically, wool is approximately 50% carbon, 22% to 25% oxygen, 16% to 17% nitrogen, 7% hydrogen, and 3% to 4% sulfur. Because it is composed almost totally of amino acids, wool is described correctly as 100% protein and is classified as an insoluble, sulfur-containing protein called *keratin*. Eighteen amino acids, characterized by an amino (NH_2) and a carboxyl (COOH) group plus one hydrogren atom attached to a carbon and differing in nature by an organic (R) group (Figure 19-17), have been identified in wool. They are sometimes classified in groups, such as monoamino, dicarboxyl (aspartic and glutamic), monocarboxyl, and diamino (lysine and arginine), to indicate their acidic or basic nature and the number of active groups present.

$$\begin{array}{c} \text{HOOC} \\ \phantom{\text{H}_2\text{N}}\diagdown \\ \text{H}-\text{C}-\text{R} \\ \phantom{\text{H}_2\text{N}}\diagup \\ \text{H}_2\text{N} \end{array}$$

Figure 19-17 Chemical structure common to all amino acids.

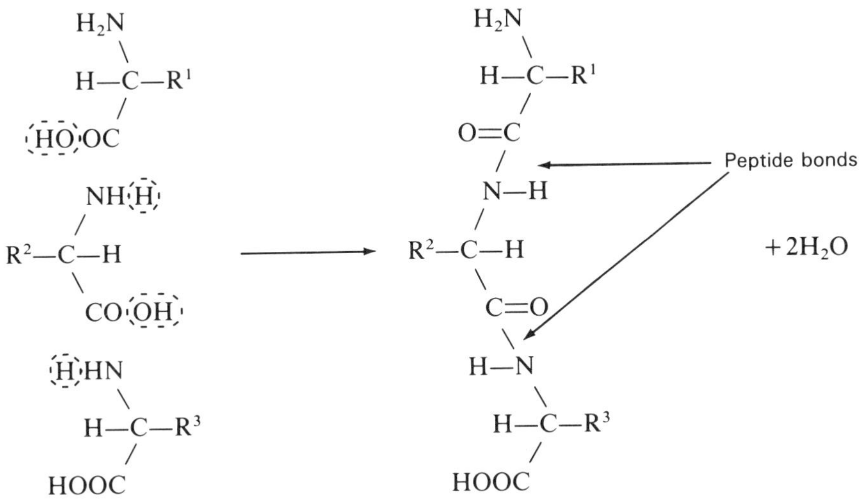

Wool Fiber Structure and Properties

$$
\begin{array}{c}
H_2N \\
\backslash \\
H-C-R^1 \\
/ \\
(HO)OC \\
\\
NH(H) \\
/ \\
R^2-C-H \\
\backslash \\
CO(OH) \\
\\
(H)HN \\
\backslash \\
H-C-R^3 \\
/ \\
HOOC
\end{array}
\longrightarrow
\begin{array}{c}
H_2N \\
\backslash \\
H-C-R^1 \\
/ \\
O=C \\
/ \\
N-H \\
/ \\
R^2-C-H \\
\backslash \\
C=O \\
\backslash \\
H-N \\
\backslash \\
H-C-R^3 \\
/ \\
HOOC
\end{array}
+ 2H_2O
$$

Peptide bonds

Figure 19-18 Formation of peptide bonds. Three or more amino acids $\xrightarrow[\text{(dehydration)}]{\text{polymerization}}$ polypeptide + water.

During keratinization (dehydration), the amino and carboxyl groups of different amino acids are linked together by peptide bonds to form polypeptide chains (Figure 19-18). In unstretched fibers, these polypeptide chains are usually in the alpha configuration of a helix. When stretched, the intrahelix (mostly hydrogen) bonds break (Figure 19-22b) and the turns of the helices open to the beta form. When pressure of extension is released, they return to the alpha configuration. This, along with the elasticlike nature of the cementing matrix between corticle cells, helps explain why wool is frequently described as having a "built-in memory." Polypeptide chains (helices) in cortical cells may be arranged in either an amorphous or crystalline manner as suggested in Figure 19-19. The disorganized amorphous zones are richer in sulfur (cystine) than the orderly crystalline zones.[3]

The aspect of primary importance in the unique chemistry of wool is the bonding that takes place between amino acids. Besides the peptide bonds already described, the sulfur-containing amino acid (cystine), an essentially double amino acid, both ends of which can form part of a polypeptide chain, is responsible for covalent crosslinking (Figure 19-20). This disulfide (dithio) bond is relatively strong and is considered extremely important in the strength of wool. It may occur between or within peptide chains (Figure 19-19).

Another bond of importance is formed when the ionized form of free NH_2 and COOH groups of diamino and dicarboxyl amino acids combine forming salt linkages (polar bonds) usually between amino acids of two different helices (Figure 19-21).

Hydrogen bonds are also formed due to electron sharing of hydrogen atoms with other atoms, such as the oxygen of $C=O$ groups within their orbits (Figure

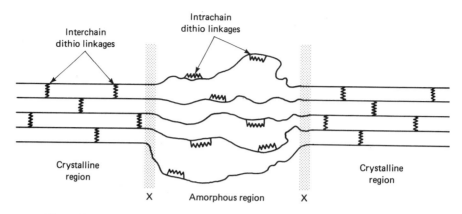

Figure 19-19 Theoretical structure of crystalline and amorphous regions in cortical protein. (From WIRA, *Wool Science Review*,[3] with permission)

19-22). Disruption of any of these bonds results in structural breakdown of fibers in varying degrees.

APPLICATIONS

The chemical nature of wool is important to researchers in understanding wool growth as well as its physical properties. The chemical behavior of wool in different environments is also important to producers and processors. A few of many possible examples are noted as follows:

The ultraviolet rays of sunlight have been shown to break disulfide bonds,

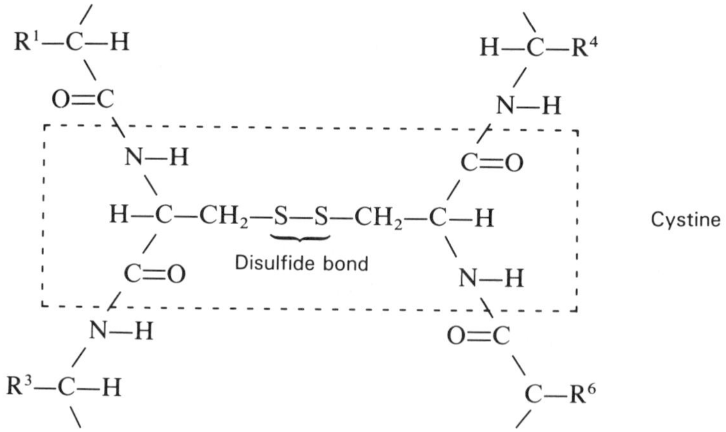

Figure 19-20 Cystine linkage between polypeptide chains in keratin.

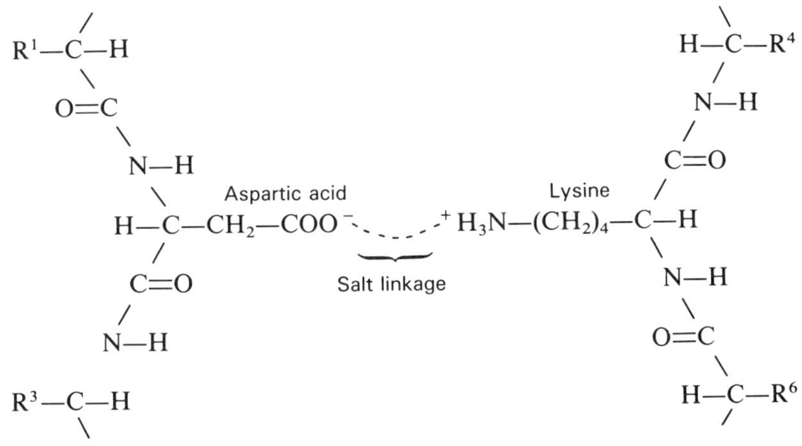

Applications

```
        /                                              \
 R¹—C—H                                         H—C—R⁴
        \                                              /
        O=C                                           N—H
          \                                           /
          N—H                                        C=O
            \      Aspartic acid       Lysine       /
            H—C—CH₂—COO⁻ ⋯⋯⋯ ⁺H₃N—(CH₂)₄—C—H
            /                                        \
           C=O              Salt linkage             N—H
          /                                            \
         N—H                                           O=C
           \                                             \
           R³—C—H                                       H—C—R⁶
              \                                         /
```

Figure 19-21 A between-chain salt linkage in keratin.

resulting in a condition called *weathering* of wool and subsequent slight tip breakage (waste) during processing. A small amount of weathering due to sunlight is unavoidable and considered normal in wools grown under range conditions. Dense fleeces tend to show less of this fault than those that are less dense. Also, excessive heat at drying following scouring or dyeing can result in some disruption of the natural chemical bonds and fibers become harsher and weaker than those dried at lower temperatures.

The relative acidity (pH) of solutions that reach wool fibers is critical to their processing performance and strength. When placed in mild acid solutions, wool fibers are damaged very little because the reactions at primarily the salt linkages are reversible. This permits *carbonizing* to remove vegetable matter (VM) as cellulosic materials become carbonized in acid solutions and are readily removed. Most dye solutions applied to wool fibers are slightly acid. On the other hand, basic solutions at pH's of 8.5 to 9.0 or higher damage wool fibers in essentially nonreversible reactions. For example, the standard test for VM content in wool requires boiling a bone-dry sample of known weight in 10% NaOH (Chapter 20). At this concentration and temperature, disulfide linkages and even peptide bonds are broken in complicated reactions. The wool is completely solubilized, but the VM is essentially unharmed and can be strained from the solution, dried, and weighed.

Because of its sensitivity to alkaline solutions, commercial scouring of wool must be monitored very carefully. Sodium carbonate (soda ash) is frequently added to scouring solutions as an aid in the removal of wool grease through saponification reactions. If these solutions are permitted to reach pHs of 8.5 to 9.0 or greater, permanent disruption of chemical bonds results and fibers become brittle and weak. A similar problem has been observed when sheep are grazed on highly alkaline soils. Fine particles of these soils attach themselves to the greasy protective coating of fibers on the sheep. When exposed to rain, the alkaline solution formed deteriorates

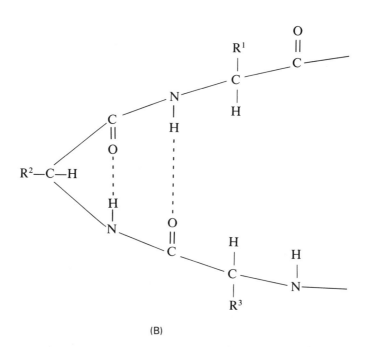

Figure 19-22 Hydrogen bonds between (A) and within (B) polypeptide chains represented by broken lines.

these fibers and the result is excessive weathering. On the other hand, not all wools are alike in terms of their vulnerability to alkaline solutions. It has been shown that wools with a greater proportion of orthocortex are more soluble in alkaline solutions than those with a predominance of paracortex. Those with a relatively low solubility in alkali or that are low in tyrosine are more resilient and harsh, so alkali solubility or tyrosine content can be used as tests for relative softness or harshness of wool[9] (Chapter 20).

Wool is described as *hygroscopic,* meaning that is is both hydrophilic (water loving) and hydrophobic (water hating). The hygroscopic property of wool fibers is the most important feature of wool chemistry. Water associates and disassociates itself at readily accessible reactive sites such as the free (unbonded) amino, carboxyl, and hydroxyl terminal groups of the side chains of individual amino acids. It also breaks into salt linkages and hydrogen bonds readily and at relatively high temperatures can even break disulfide linkages. These reactions are reversible. The effect of water entering or leaving wool results in weight change and also changes in the dimensional measures of wool fibers. When changing from an atmosphere of 0% to 100% relative humidity (r.h.), wool fibers increase in length by over 1.0% and over 16% in diameter.[12,18] In other words, a 21-μ fiber at "bone dry" would swell to over 24μ at 100% r.h.

The hygroscopic behavior of wool is important to producers from the standpoint of objective measures related to marketing and selection. To avoid variation due to moisture content, fibers measured in laboratory settings are conditioned at a constant relative humidity (Chapter 20). Because wool tends to pick up moisture slightly more readily than it releases it, samples to be conditioned are placed in the controlled environment so as to reach a constant moisture content from the dry rather than the wet side. Shorn wool picks up or releases moisture in a direct relationship to the relative humidity at which it is stored, so moisture content is considered one of the most difficult nonwool components to evaluate when estimating the yield of grease wool (Chapters 18 and 20).

For wool processors, moisture in wool is a constant concern. For example, the carrier of choice for dyes is water because it carries these dyes into the fiber for chemical binding. Drying wool following scouring or dyeing requires expensive energy. Static electricity developed during carding, combing, and spinning is lower when relative humidity is increased, so controlled environments are required for these processes.

REFERENCES CITED

1. Anonymous. 1950. The growth of the wool fiber. *Wool Science Review,* 6:13. International Wool Secretariate. Wool Industry Research Association, Technology Group, United Kingdom.

2. Anonymous. 1951. Wool grease, Part 2. *Wool Science Review,* 7:29. International Wool Secretariat, Wool Industry Research Association, Technology Group, United Kingdom.
3. Anonymous. 1963. Wool keratin: Its chemical structure and reactions, Part 2. *Wool Science Review,* 22:3 International Wool Secretariat, Wool Industry Research Association, Technology Group, United Kingdom.
4. Burns, R. H. 1955. The sheep family. *Wyoming Agr. Exp. Sta. Circular No. 61.*
5. Carter, H. B. 1955. The hair follicle groups in sheep. *Animal Breeding Abstr.* 23:101.
6. Chapman, R. E., and S. S. Y. Young. 1957. A study of wool production per unit area of skin in Australian Merino sheep. *Aust. J. Agr. Research* 8:723.
7. Frandsen, R. O. 1965. *Anatomy and Physiology of Farm Animals,* p. 40. Lea and Febiger, Philadelphia.
8. Hardy, Margaret H., and A. G. Lyne. 1956. The pre-natal development of wool follicles in Merino sheep. *Aust. J. Biol. Sci.* 9, 3:423.
9. LeRoux, P. L. 1959. Routine methods for determining quality in Merino wool. *Nature,* 184:917.
10. Lydekker, R. 1912. *The Sheep and Its Cousins,* pp. 8–45. George Allen & Co. Ltd., London.
11. McFadden, William D. 1967. *Wool Science,* pp. 12–17. Pruett Press, Boulder, Colo.
12. Onions, W. J. 1962. *Wool,* pp. 17, 48, 64. Ernest Benn Ltd., London.
13. Pexton, Ed, LeRoy Johnson, and Frank Craddock. 1980. 4-H Wool Judging. University of Wyoming Agricultural Extension Service, Laramie, No. 12500h-80.
14. Reed, C. A. 1959. Animal domestication in the prehistoric Near East. *Science* 130, No. 3389:1629.
15. Ryder, M. L., and S. K. Stephenson, 1968. *Wool Growth,* pp. 21, 241, 246, 320, 343. Academic Press, New York.
16. Von Bergen, Werner. 1963. *Wool Handbook,* Vol. 1, pp. 142, 146, Wiley-Interscience, New York.
17. Wildman, A. B. 1932. Coat and fiber development in some British sheep. *Proc. Zoo. Soc. Lond.,* Part I, 257.
18. Yeates, N. T. M., T. N. Edey, and M. K. Hill. 1975. *Animal Science,* pp. 274, 276, 290. Pergamon Press, Elmsford, N.Y.

SUGGESTED READING

Lydekker, R. 1912. *The Sheep and Its Cousins.* Geo Allen & Co. Ltd., London.
Onions, W. J. 1962. *Wool.* Ernest Benn Ltd., London.
Ryder, M. L., and S. K. Stephenson, 1968. *Wool Growth.* Academic Press, New York.
Von Bergen, Werner, 1963. *Wool Handbook,* Vol. 1. Wiley-Interscience, New York.

Wentworth, E. N. 1948. *America's Sheep Trails.* Iowa State College Press, Ames.

Wool Science Review. Volumes 4, 5, 6, 21, 22, and 23. International Wool Secretariate, Dorland House, London.

Yeates, N. T. M., T. N. Edey, and M. K. Hill. 1975. *Animal Science.* Pergamon Press, Elmsford, N.Y.

20 Wool Measurement

Wool is a *natural* fiber; as such, it is understood that individual fibers are seldom alike in most respects so they are usually described as "average" for specific characteristics. Studies of the chemical and cellular structure of wool (Chapter 19) have led to descriptions such as "hygroscopic" and "elastic." The importance of understanding and describing the microstructure of wool fibers should not be minimized because it is related to practical characteristics such as the dyeability, durability, and insulative value of materials made from wool.

For producers, processors, and consumers, the measures of the characteristics of wool discussed in this chapter are especially important. They are used to describe the quantity and quality of wool grown and represent the primary limiting factors involved in the conversion of grease wool to consumer products.

CHOICE OF METHOD

Decisions concerning which characteristics and when and how to measure them are required in the general areas of breeding, marketing, and manufacturing of wool. In each situation, the purpose for measurement, the degree of accuracy required, the methods available, and their relative costs need to be considered.

For example, the time and cost required for objective measurement of individual fleeces at shearing floors, warehouses, and scouring plants cannot be justified when sorting large numbers of fleeces into *graded lots*. In these situations, the less accurate subjective methods are the most efficient and the trained eye is indispensable. Monitoring the relative accuracy of visual sorts is accomplished by objective measure of representative samples.

However, *marketing situations require special attention to measurement.* It was noted in Chapter 18 that correct pricing of raw wool requires correct description. This concept also applies to semiprocessed and processed wool such as tops, yarns, and fabric. Although trade-offs between accuracy and time and cost of measurement may be justified in certain cases associated with breeding (Chapter 21) and processing, objective methods should be used whenever possible in wool marketing.

SUBJECTIVE MEASURES

Subjective estimates, those obtained by sensory (visual and tactile) observation and without the aid of measuring devices, of the value-determining characteristics of wool are the most frequently used in breeding programs and at all stages in the movement of wool from producer to consumer. Although the skill, developed through training and practice, of the evaluator is closely associated with the degree of accuracy of subjective estimates, it is understood that it is seldom as accurate as methods that employ measuring devices. However, cost–benefit ratios indicate that for certain characteristics (particularly those that are considered less important than diameter and clean wool content) such as strength, handle, and perhaps color, subjective estimates may be optimum.

OBJECTIVE MEASURES

The term *objective* when used to describe measurement of the value-determining characteristics of wool implies the use of measuring devices. Much time and money has been expended worldwide by universities, government organizations, and marketing and manufacturing groups in the development of instruments for measuring the many factors associated with wool quality and quantity. Some have been accepted nationally and internationally, while others have been abandoned.

One of the most promising aspects of current research is the inclusion of sophisticated electronic systems in wool measurement methods. Decreasing costs by increasing rapidity, without decreasing the accuracy and precision of conducting measurement procedures, with the aid of electronic equipment is expected. When this happens, objective measurement of wool characteristics will be more practical in many situations. Producers and processors need to be alert to these advances in technology as they occur.

Detailed procedures for measuring both quantitative and qualitative aspects of wool objectively, which also describe the various instruments involved and their correct use, have been provided by the American Society for Testing and Materials (ASTM).[3] This reference, updated and published annually, is considered the single most important item in an American wool testing laboratory. For laboratories involved with measures of wool traded internationally, the International Wool Textile Organization (IWTO) Specifications of Test Methods[13] can be especially helpful.

Sampling

Most objective measurements of wool characteristics are made from samples of larger quantities. Therefore, even objective measurement procedures are at best *estimates*. The accuracy of these estimates is limited by the degree to which the sample represents the fleece, group of fleeces, or processed lot.

Large lots are typically sampled, subsampled, and subsampled again, so the *most important part* of any procedure for measuring wool objectively is the drawing of a representative sample. For example, when estimating the average grade and yield of a truckload of wool, approximately 40,000 lb (18,240 kg) is reduced to a few hundred grams for evaluation in laboratory settings. Good procedures for sampling wool in several situations have been developed, tested, and described. They must be followed carefully.

Variation

A major problem associated with the conversion of raw wool into a wide variety of products is the *variability* of both diameter and length of fibers. Without discussing all the technical aspects of manufacturing problems associated with the variability of wool fibers, it is appropriate to assume that the greater the uniformity for most of the manufacturing processes, the more desirable and therefore more valuable the wool.

The challenge at all stages from the production of wool to finished product is when and how to measure variability and particularly how to interpret it. Estimates of variation are included in several measurement methods and are expressed as standard deviation or coefficient of variation for comparison. The sources of variation represented in the sample measured[10,28] should always be considered in the interpretation of these estimates and choice of appropriate remedial actions (Chapters 18, 19, and 21).

QUANTITATIVE MEASURES

The procedure for pricing wool (Chapter 18) indicates that the *amount* of clean wool in a lot is a key factor in determining its value. The proportions of specific nonwool components such as vegetable matter are also important, particularly when excessive.

Yield

"Clean wool content," "clean fibers present," or simply "clean wool" are expressions used interchangeably in wool marketing and selection to describe the amount of usable wool in a quantity of raw wool. Because an estimate of yield is required to determine clean wool content (weight of raw wool × percent yield/100 = weight

Quantitative Measures

of clean wool at standard conditions), the following discussion is limited to yield determination and clean content is implied.

In American marketing systems, yield is used to describe the proportion (percent) of "clean wool fibers present" (ASTM-D584)[3] in any samplable amount of grease wool. It is the weight of wool base present in raw wool adjusted to a standard condition of 12% moisture, 1.5% alcohol extractives, 0.5% mineral matter, and no vegetable matter, where "wool base" is oven-dry scoured wool, free from alcohol-extractable matter, mineral matter, vegetable matter, and all impurities. Wool at standard conditions is 86% wool base and 14% allowable nonwool components.

In practice, wool seldom reaches these standard conditions. However, because the nonwool components of grease wool vary greatly, this *theoretical* state is used to standardize the *condition* for describing the clean wool content of wool. In international marketing and for wool in semiprocessed forms, standard conditions are sometimes slightly different and appropriate conversion factors are provided (ASTM-D2720).[3]

The two words describing the amount of moisture in wool at standard condition can be confusing. For international communication, the most frequently used term is *regain,* which expresses moisture as a proportion of dry wool. A more familiar term in the United States is *moisture content,* which describes it as a proportion of undried wool. Therefore, a wool sample that has a moisture content of 12% (12 units/100 units) would have a regain of 13.64% (12 units/100 − 12 or 88 units).

The method described by the ASTM for determining yield of raw wool requires laboratory facilities. By steps and briefly, yield determination from a sample of wool in a testing laboratory is as follows:

1. Weigh sample.
2. Scour.
3. Oven dry.
4. Weigh oven-dry sample.
5. Subsample and determine residual (portions not removed in scouring) vegetable matter, mineral matter, and alcohol extractables and express as percents.
6. Calculate wool base: oven dry weight − (% residuals/100 × oven-dry weight).
7. Correct to standard condition weight (wool base/0.86).
8. Calculate percent of yield (standard condition weight/sample weight × 100).

Yields reported by laboratories are sometimes challenged by either buyers or sellers. It is understood that errors can occur in the many laboratory procedures involved, but they are rare as samples are processed in duplicate and reevaluated if errors are suggested.

Errors are more likely associated with sampling and the handling of samples. Because fleeces are usually stratified in bags when sacked on farms and ranches,

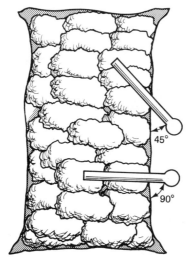

Figure 20-1 Stratification of fleeces in a bag of wool. A core taken perpendicular (90°) to the bag may sample only one or two fleeces. A core taken at 45° may sample three or more fleeces depending on the density of packing.

cores are always taken at an angle (Figure 20-1) in order to subsample as many fleeces as possible. Bales are sampled from the compressed surface.

The most probable source of error in sampling for yield is moisture as it moves in and out of bags of wool depending on the humidity of the air in which it is stored. The rate of moisture movement is related to the degree to which fleeces have been compressed in the bags and is slower in more tightly packed bags. However, moisture *zones* are typical (Figure 20-2). For example, if wool has been shorn and bagged

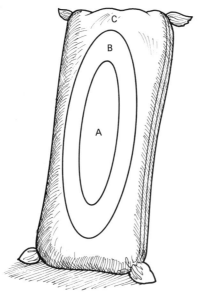

Figure 20-2 Moisture zones in a bag of wool.

on the damp side and stored where humidity is low, the wool in the outer sides of the bag starts to dry. After a period of time, the mid-zone (A) will be damper than the outer zones (B and C). A reverse situation is often observed. These differences in moisture zones do not present a problem if they are represented proportionately in the sample. Another source of moisture-related error in yield determination is the handling of the sample. Samples *must* be placed in moisture-tight containers (heavy plastic sacks are good) as soon as they are drawn and sealed tightly for shipping to a laboratory.

Also, because of the hygroscopic nature of wool, it is essential for the lot or fleece to be weighed at the time it is sampled. Estimates of clean content derived from estimates of yield are valid only for the weight of the raw wool *at the time of sampling*. Table 20-1 demonstrates that, although the moisture content, grease weight, and yield of bagged wool can change with environmental humidity, its clean content does not. Therefore, value remains constant provided the grease weight *at sampling* is used as pay weight.

Detailed procedures for sampling *lots* (groups of fleeces) of wool by coring have been outlined by the ASTM (D-1060).[3] Buyers and sellers of raw wool find it mutually beneficial to do everything possible to ensure these procedures are conducted correctly. At the time of coring, free access to all bags or bales must be provided. Adequate personnel or equipment should be available to weigh and move bags, place core samples in plastic sacks, and so on.

Current laboratory procedures for estimating yield are relatively slow and expensive. Because there is a definite increase in the number of lots traded based on objective measures of clean content, scientists worldwide continue to search for more efficient methods. The use of infrared light rays is an example.[17,25,27]

Estimates of yield of *individual fleeces* are useful for selection purposes (Chapter 21). Clean weight obtained by scouring entire fleeces is the best measure avail-

TABLE 20-1 EFFECT OF FLUCTUATING MOISTURE CONTENT ON YIELD, PRICE AND VALUE OF WOOL IN A 300-LB (136.2 KG) BAG OF GREASE WOOL

Net Grease Weight of Wool in Bag		Amount of Moisture in Bag		Moisture (%)	Yield[a] (%)	Grease Price/Lb[b] ($)	Value of Wool in Bag ($)
(lb)	(kg)	(lb)	(kg)				
315	143.0	51	23.15	16.19	47.62	0.9524	300.00
310	140.7	46	20.88	14.84	48.39	0.9678	300.02[c]
305	138.5	41	18.61	13.44	49.18	0.9836	300.00
300	136.2	36	16.34	12.00	50.00	1.00	300.00
295	133.9	31	14.07	10.51	50.85	1.017	300.02[c]
290	131.7	26	11.80	8.97	51.72	1.0344	299.98[c]
285	129.4	21	9.53	7.37	52.63	1.0526	299.99[c]

[a]Clean fibers present in bags at each weight remain constant at 150 lb (68.2 kg).
[b]Clean basis price = $2.00/lb or $4.40/kg.
[c]Differences due to rounding.

able, but it is an impractical method due to relative cost and a lack of scouring operations that can or will handle individual fleeces.

A good procedure for core sampling single fleeces has been described (Figure 20-3).[14] However, laboratory facilities are required for a complete analysis of the samples, so this method is available only through commercial testing laboratories and a few research centers in the United States. Although coring or scouring is considered best for determining the clean content of single fleeces, the costs involved and availability of facilities represent a practical limit to the number of fleeces that can be tested in this manner.

Several researchers have attempted to estimate the clean content of fleeces by scouring small samples (usually from the side of the sheep). Estimates based on shoulder, side, or thigh samples are biased upward because the lower yielding portions (back, belly, and tags) are not included. Relatively high positive correlations have been reported between these small (1/4 lb, 100 g) scoured samples and scoured whole fleeces, which suggests this procedure is a reasonable alternate to coring.[22,26] Although overestimated, clean content of fleeces estimated from the yield of side samples can be considered comparative for selection purposes. It is a method used extensively, particularly for ram selection, in some of the major wool-producing countries.

A procedure has been described for scouring side samples on farms.[24] It requires only tubs and wringers, along with cleaning agents and a source of hot water. Samples may be air dried or dried in an oven. In either case, samples of fleeces to be compared should be treated at the same time. A small kitchen scale, capable of weighing samples to the nearest 1/4 oz (7 g) can be used to weigh samples before scouring and after they are dried. The fleece must be weighed when sheared and the sample kept in a closed plastic bag until scoured to avoid weight change due to environmental humidity.

Figure 20-3 Hydraulic machine for coring individual fleeces. (University of Wyoming photo)

Quantitative Measures

A mechanical method (Neale's Squeeze Machine) has been developed to estimate the clean content of fleeces.[19,20] Its design is based on a theory that resistance to compression is a function of the fibers in fleeces rather than the nonwool components. With the aid of this machine, estimates of the clean content of individual fleeces can be had in minutes. Fleeces are compressed at a constant pressure by a belt inside a round chamber 30 in. (76.2 cm) in diameter and 8 in. (20.3 cm) deep with a hinged cover (Figure 20-4). Correlations reported between machine readings and actual scoured clean weights were positive and relatively high so a table converting squeeze reading to clean weight was prepared. Clean weights of fleeces grown under different conditions and representing different breeds and average fiber diameters may not always agree with tabular values provided for scoured weights. How-

Figure 20-4 Neale's squeeze machine: (A) mechanical; (B) hydraulic. (University of Wyoming photos)

ever, the machine has been shown extremely useful in ranking fleeces from similar management systems (periods of growth, age, and production states) within flocks. Reasons why this machine has not gained enough popularity among producers to warrant its manufacture and distribution through normal livestock equipment channels are unclear.

Multiple regression equations that include grease weight, staple length, squeeze machine reading, body, weight, and/or fiber diameter to predict clean fleece weight have been developed[4,22,26] with correlations between predicted and actual clean fleece weights ranging from 0.87 to 0.96.

Although it does include differences in the nonwool components of fleeces and can lead to errors in selection for increasing clean fleece weight, *grease fleece weight* should not be overlooked as a means of comparing wool growth of individual sheep.[22] Yields of fleeces measured in ram tests that are grown for a constant period of time on a uniform diet and within breed normally vary 20%, probably due largely to differences in moisture. In spite of potential errors due to yield differences when using grease weight as an estimator of clean weight in individual fleeces, it is the most practical approach in many situations. Grease weight can be obtained quickly and inexpensively with a simple dial scale and tray (Figure 20-5) and is certainly superior to no weight.

Figure 20-5 Dial scale and tray for weighing individual fleeces. (University of Wyoming photo)

Moisture

If correctly sampled, relative amounts of moisture in grease wool are accounted for in laboratory procedures for determining yield. Moisture is also standardized when measuring fiber diameter. However, in special situations, the specific amount of moisture in wool needs to be determined. It becomes a particular concern in the storage of raw wool that has been shorn damp. Wools stored with moisture contents above 15% can suffer fiber damage due to mold and/or heat buildup. Provided representative samples can be taken, moisture content can be determined quite quickly and accurately by oven drying. Electronic moisture meters can be used to estimate moisture in bagged or baled lots.

Tactile evaluation of moisture in wool is extremely difficult because wool does not reach saturation (feel wet) until it has a moisture content of 25% (33.3% regain). Fleeces that feel damp at shearing probably contain more than 15% moisture. Wool buyers *twist* locks of wool tightly and look for *beads* of moisture to detect wools with relatively high moisture content.

Vegetable Matter

Vegetable matter (VM) in wool is readily determined by placing samples in boiling solutions of 3% or 10% sodium hydroxide (ASTM-D1113).[3] The wool is solubilized in this test, but the VM is alkali insoluble, so it is strained from the solution, dried, and weighed. VM is normally included in test reports of clean wool fibers present as a single quantitative measure (vegetable matter present, ASTM-D584).[3] These reports may also include estimates of the proportion of various types of VM present.

Mineral Matter

The relative amount of mineral matter in raw wool, which may vary from near none to over 25%, has a profound effect on yield. However, it is removed readily during scouring and is not normally included as a separate quantity in test reports of the analysis of raw wool samples. In laboratory procedures used to estimate yield, residual mineral matter (small amounts not removed during scouring) is measured by ashing (ASTM-D584).[3]

Density

This is a quantitative measure applicable to wool growing on sheep. An objective measure could be used to advantage in predicting fleece weight. Several instruments such as the Wyedesa and the Wool Industries Research Association (WIRA) Fleece Calipers[6,32] have been devised but have been shown similar in accuracy to subjective estimates by skilled evaluators. Accurate estimates of fleece density are best achieved

Figure 20-6 The "touch method" for evaluating the density of a fleece. (University of Wyoming photo)

by counting follicle populations in skin sections,[7] a procedure that requires laboratory facilities and trained personnel. The *touch* method of evaluating differences in density by squeezing a portion of the fleece between the thumb and fingers (Figure 20-6) is the most practical method available to producers.

QUALITATIVE MEASURES

The relative quality of raw wool is directly related to the use for which it is intended. Measurement of the many factors that determine quality is usually associated with groups of fleeces (lots) for marketing and manufacture and with individual fleeces for selection purposes.

Fiber Diameter

Because diameter is the single most important dimensional measure of wool fibers, it has received the most attention by researchers. Considerable effort has been put forth in the search for optimum methods of measuring diameter for research, selective breeding, marketing, and quality control during processing. Many instruments have been developed and tested. Some have persisted as useful tools for measuring diameter, while many others have not.

A major problem associated with measuring the diameter of wool fibers is that they are seldom round (Chapter 19), so procedures must be used that minimize bias due to measuring a predominance of either the major or minor axes. An equally important problem is the hygroscopic behavior of wool. Wool fibers swell when they absorb moisture and become finer as they dry. A change of 10% in relative humidity (r.h.) can cause a 1% change in mean diameter.[24] It is therefore essential that samples be measured at a constant humidity. The standard atmosphere for measuring fiber diameter is 65% r.h. at 21°C. Dry fibers, preconditioned at a r.h. of 10% to 25% and a temperature of not over 50°C, are conditioned for at least 4

Qualitative Measures

hours in this standard atmosphere. When slides are made, a mounting medium is used that maintains the moisture content of the fibers.

Objective methods for measuring wool fiber diameter are classified broadly as precision and rapid. The most widely used *precision* methods include a microscope and a means of projecting fiber images for measurement at a magnification of 500 times.

Projected images are either measured with a rule as described by the IWTO[13] or by grouping them in 2.5-μ intervals as described by the ASTM (D2130).[3] The ASTM method lists the number of fibers required to be measured from a sample for confidence limits of ± 0.2, 0.4, or 0.5 μ. It is considered to have no known bias and is generally accepted as the *referee method* for measuring the diameter of wool.

An example of the instruments used in precision methods is a Bausch and Lomb microprojector (Figure 20-7). It is shown equipped with a Ladd digitizer that uses a sensitized pad to measure the distance between pressure points applied to

Figure 20-7 Microprojector for measuring wool fiber diameter. (University of Wyoming photo)

the outside edges of fiber images. This pad is attached to a unit that records each measurement and provides averages and standard deviations. The microprojector shown in Figure 20-7 is normally used with a plain white field for focusing fiber images that are measured with a standard *wedge* (ASTM-D2130). Other instruments such as the Lanameter and WIRA projection microscope are slightly different in design but accomplish the same purpose of magnifying and focusing fiber images for measuring their diameter.

Because of the time required for moving the slide and focusing, precision methods currently used are relatively slow. However, current research reports indicate that instruments that can measure fiber diameter as accurately and faster are forthcoming. The Fiber Fineness Distribution Analyzer developed by the Commonwealth Scientific Industries Research Association (CSIRO) of Australia is an example. Fibers are measured as they pass through laser beams, which removes the potential for operator error in focusing.

Serious errors in the measurement of projected images of fibers can occur due to focusing. The ellipticity of fibers usually prevents both sides of the fiber being brought into focus. The correct way is to focus so one side is in sharp focus while the other shows a white *Becke line*. Measurement is made to the inside of the Becke line. As a guide to accuracy and checking equipment, reference wools are available from the IWTO[13] and USDA.[29]

An important feature of precision methods is that, because they are based on measuring individual fibers, they provide an estimate of the variability of the fibers in a sample.

Rapid methods are most useful when estimates of variations are not considered as important as average diameter. The most widely used rapid methods are based on a relationship between the number of fibers in a known sample weight at a constant compression and their resistance to the flow of air through them. The specific surface of fibers and therefore resistance to airflow increases proportionately with decreasing fiber thickness. The relative resistance to airflow at a constant pressure measured by these instruments is then converted to an estimate of average diameter.

Frequently used airflow instruments include the WIRA Fiber Fineness Meter,[33] which uses a 2.5-g sample, the Micronaire,[18] which uses 4.8- or 5.9-g samples, and the Port-Ar[21] which uses a 12.5-g sample. Detailed procedures (D1282)[3] have been described by the ASTM for these three airflow methods. Average biases as compared to the microprojection method (ASTM-D2130) are reported as:

Micronaire	-0.59μ
Port-Ar	$+0.68\mu$
WIRA	-0.41μ

Therefore, a 26.0μ diameter estimated by microprojection would tend to be reported as 26.68μ when using the Port-Ar procedure. Although testing wool that has been *conditioned* at 65% r.h. and 20° to 21°C is desirable, tables for correcting the effects of temperature and relative humidity under field conditions are provided for use with the WIRA.

The CSIRO Sonic Fineness Tester[8] (two models) is similar in principal to the

Qualitative Measures

airflow instruments except that a low-frequency audio signal is used instead of air. No ASTM procedure is described for the Sonic Tester but the Standards Association of Australia has provided a test method (AS 1404-1973, Sonic Fineness Testing of Raw Wool). The manufacturer provides adjustment factors for differences in relative humidity, ambient temperature, and elevation when used in field conditions. Airflow and sonic instruments, such as those shown in Figure 20-8, do not provide estimates of variation.

An apparent tendency for fiber diameter to be inversely related to crimp frequency (crimp/unit length) has led to the development of crimp scales (Figure 20-9).[9] Although a measuring device is used, this method is not considered much more accurate than visual evaluation for estimating fiber fineness.

The relationship between crimp and diameter may be true in the broad sense but it is inconsistent. Traditionally, crimps per unit of length have been considered a good visual indicator of relative fineness (the more crimps per unit of fiber length, the finer the wool). However, results of a New South Wales, Australia, survey[1] showed that only 37 of 61 clips evaluated had this relationship. Nineteen clips showed no relationship at all, and 5 showed a reverse relationship. The same report noted an auction in Geelong, Victoria, where only 8 of 23 clips examined showed a

Figure 20-8 Two instruments used for rapid estimates of the average fiber diameter in samples of wool fibers: (A) sonic; (B) air flow. (University of Wyoming photo)

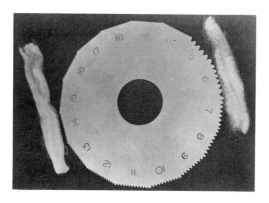

Figure 20-9 Scale for estimating wool fiber diameter based on crimp frequency. (University of Wyoming photo)

significant relationship between quality number and fiber diameter. Several other rapid methods involving instrumentation have been tried,[11,24] but have not been widely accepted by the industry for various reasons.

Although its accuracy has been shown inconsistent both within and between evaluators,[5,30,31] the most rapid method of describing wool fiber diameter is by sensory judgment. It is a strictly subjective approach based largely on crimp frequency and softness with estimates expressed in terms of spinning counts (quality number).

Length

The length of wool fibers is described as "average fiber length" or "staple length." Measurement of individual fibers is often useful in studies of wool growth, in quality control, and as a basis for machine settings during processing. There is considerable variation in length among fibers in a given lock of wool, so true average length is determined by measuring each fiber. Individual fibers can be measured to the nearest 0.5 cm by gripping both ends of a fiber with forceps and stretching it against a black scale. However, it is difficult to ensure constant tension. A machine has been developed by the WIRA that permits measurement of individual fibers under controlled tension.[34]

Staple length (also sometimes called bundle length or lock length) in grease wool can be obtained quite accurately by simply measuring an unstretched lock of wool with a ruler (Figure 20-10). A low-cost ruler can be used effectively for measuring staple length of fleeces growing on sheep or of samples taken from individual classes or lots. Wool buyers use a premeasured finger or thumb as described in Chapter 22 to estimate average staple length. Tippy wools can lead to some inaccuracy as the tendency is to measure to the longest fibers present in a lock. The ASTM[3] has described procedures for measuring staple length in grease wool (D1234) and fiber length in scoured wool and card sliver (D1575) and in wool top (D519).

Attempts to develop equipment to measure locks of wool rapidly at constant

Qualitative Measures

Figure 20-10 Using a ruler to measure staple length on a sheep. (University of Wyoming photo)

tension[15,23] have not been widely accepted in the United States. However, Australian presale test certificates are beginning to include mean staple length measured on their CSIRO designed ATLAS instrument.

Strength

Several machines have been developed for testing the tensile strength of wool fibers.[24,35] The general principles and procedures for using some of them are described by the ASTM (D76).[3] Procedures for conducting objective measures of the average tensile strength of bundles of wool fibers are also described by the ASTM (D1294 and D2524).[3] Although possible, testing the strength of individual fibers with sophisticated machines such as the Instron Tester[12] is labor intensive and usually used only in basic research settings.

Besides measuring staple length as reported above, the ATLAS can also measure mean staple strength and the position, as well as distribution of position of weakest points along sample staples taken from grease wool. These objective measures of staple strength and location of weakest points are not currently included in tests of U.S. grease wool for marketing purposes or as standard data in Australian presale test reports. However, they are expected to be used much more frequently as a factor in the evaluation of grease wool around the world.

Handle

The relative softness of wool may be a function of crimpiness, fiber diameter, and/or chemical composition. Wools with less sulfur that are finer and with greater crimp frequencies have been described as "softer." Also, the proportions of tyrosine or orthocortex in wool tends to be related to softness (Chapter 19). Measures of chemical composition, particularly the alkali-solubility test (ASTM-D1283),[3] along with diameter and crimp frequency, have been utilized to evaluate degrees of softness.

Another approach is to describe softness in terms of its plasticity as measured on tensile strength machines and expressed as load-extension curves. An objective mechanical method of measuring the relative softness of wool was developed in a study that also demonstrated that tactile evaluation of softness is extremely difficult.[16] These mechanical methods have not received much attention from wool processors or marketing groups, but continuing research will probably lead to the development of a practical instrument for evaluating the softness of wool objectively.

Medulation

The proportion of medulated fibers in a sample can be obtained accurately at the same time fibers are measured for diameter by microprojection because they are easily identified (Figure 20-11). Also, the types of medulation and diameters of the medullae are readily apparent.

A rapid procedure is to immerse fibers in a fluid having a refractive index similar to wool keratin. Nonmedulated fibers disappear while medulated fibers remain visible. Benzene is a typical fluid, so this method is referred to as the benzene test. It is based on a visual judgment, so it is more subjective than objective and the proportion of medulated fibers cannot be determined.

Color

Because of increased demand by consumers for very light colored fabrics, degree of coloration is becoming more and more important as a quality factor in wool. Although normally evaluated visually, a method has been developed[2] for measuring color differences in wool objectively. Other methods for measuring degree of color in wool ranging from near white to very yellow are anticipated.

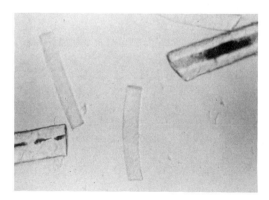

Figure 20-11 Medulated and nonmedulated images of wool fibers from a projection microscope. (University of Wyoming photo)

REFERENCES CITED

1. Anonymous. 1973. Classing the clip to suit the textile maker. *CSIRO Rural Research* 80:22.
2. Anonymous. 1983. Dark fiber contamination of wool. *CSIRO Rural Research* 121:13.
3. American Society for Testing and Materials. *Annual Book of ASTM Standards,* Parts 32 and 33. Philadelphia.
4. Bassett, J. W., and others. 1973. Estimates of clean fleece weights for ram performance tests. *Proc. Western Section Amer. Soc. Animal Sci.* 24:9.
5. Burns, R. H. 1962. Sensory judgments of greasy wool. *J. Textile Inst.* 53, 11:P756.
6. Burns, R. H., and W. C. Miller. 1931. Sampling instruments to determine fleece density in sheep. *J. Textile Inst.* 22:T547.
7. Chapman, R. E. and S.S.Y. Young. 1957. A study of wool production per unit area of skin in Merino sheep. *Aust. J. Agr. Research* 8:723.
8. CSIRO Sonic Fineness Tester. Paton Industries Pty. Ltd., 35 Henry St., Stepney, South Australia.
9. Duerden, J. E., 1929. Standards of thickness and crimps in Merino grease wools. *J. Textile Inst.* 20:T93.
10. Dunlop, A. D., and P. R. McMahon, 1974. The relative importance of source of variation in fiber diameter for Australian Merino sheep. *Aust. J. Agr. Research* 25:167.
11. Hardy, J. I., and H. W. Wolf. 1939. Two rapid methods for estimating fineness and cross-sectional variability of wool. *USDA Circular no. 543.*
12. Instron Corp., Canton, Mass.
13. International Wool Textile Organizations. Specifications of test methods. International Wool Secretariate, Carleton Gardens, London, SWIY 5AE, United Kingdom.
14. Johnson, C. LeRoy, and S. A. Larsen. 1978. Clean wool determination of individual fleeces. *J. Animal Sci.* 47, 1:41.
15. Larsen, S. A. 1969. Staple length measurer for wool. *J. Textile Inst.* 60, 8:39.
16. Larsen, S. A., K. E. Hoke, and A. W. Kotula. 1976. The interrelationships of wool fineness, softness, quality and market evaluations of U.S. domestic wool. II. Determination of softness and its relationship to physical characteristics and geographic origin. *Science Monograph 33.* Agr. Expt. Sta., Laramie, WY.
17. Larsen, S. A., and J. L. Kinnison. 1982. Estimating quality components of natural fibers by near-infrared reflectance. II. Clean wool base and average fiber diameter. *Textile Research J.* 52:1:25.
18. Micronaire, Sheffield Corp., Dayton, Ohio.
19. Neale, P. E., G. Sidwell, and W. McFadden. 1956. Clean fleece yield in two minutes. *Proc. Western Section Amer. Soc. Animal Prod.* 7:74.
20. Neale, P. E., G. M. Sidwell, and J. L. Ruttle. 1958. A mechanical method for estimating clean fleece weight. *Agr. Exp. Sta., New Mexico* College of Agriculture and Mechanic Arts, Las Cruces, Bulletin 417.
21. Port-Ar. Special Instruments Laboratory, Inc. (Spinlab), 312 Vine Ave. S. W., Knoxville, Tenn.

22. Price, D. A., S. K. Ercanbrack, and L. O. Wilson. 1960. An evaluation of the "squeeze machine" and other methods of estimating clean fleece weight. *Proc. Western Section Amer. Soc. Animal Prod.* 11, 14:1.
23. Ray, H. D., and others. 1964. Measuring wool by staple length recorder. *USDA Marketing Research Report 668.*
24. Ryder, M. L., and S. K. Stephenson. 1968. *Wool Growth,* pp. 636, 638, 688, 725. Academic Press, New York.
25. Sabbagh, H., and S. A. Larsen. 1978. Wool yield by infrared reflectance analysis. *Proc. Western Section Amer. Soc. Animal Sci.* 29:108.
26. Sidwell, G. M., P. E. Neale, and G. L. Jessup. 1958. A comparison of five methods of estimating clean fleece weight. *J. Animal Sci.* 17:593.
27. Scott, R. F., and E. M. Roberts. 1978. Objective measure of clean fleece weight by infrared reflectance spectroscopy. *Wool Technology and Sheep Breeding* 24, 2:27.
28. Stobart, R. H., and others. 1986. Sources of variation in wool fiber diameter. *J. Animal Sci.* 62:1181.
29. U.S. Department of Agriculture, Agricultural Marketing Service, Livestock Division Wool Laboratories, 711 "0" Street, Greeley, Colo. 80631.
30. Whan, R. B. 1968. Is wool classing worthwhile? *Wool Technology and Sheep Breeding* 15, 1:87.
31. Whan, R. B. 1969. The role of objective measurement in Australian wool marketing. *Textile Inst. and Industry* 7, 6:154.
32. Wildman, A. B. 1936. An improved fleece sampling instrument. *J. Textile Inst.* 27:T177.
33. WIRA Fiber Fineness Meter. Precision Specialities International, P.O. Box 5614, Greenville, S.C. 29606.
34. Wool Industries Research Association, Wira House, West Park Ring Road, Leeds, LS16 6QL, United Kingdom.
35. Yeates, N. T. M., T. N. Edey and M. K. Hill. 1975. *Animal Science,* p. 328. Pergamon Press, Elmsford, N.Y.

SUGGESTED READING

Ryder, M. L., and S. K. Stephenson. 1968. *Wool Growth,* Academic Press, New York.

Von Bergen, Werner. 1963. *Wool Handbook,* Vol. 1. Wiley-Interscience, New York.

Yeates, N. T. M., T. N. Edey and M. K. Hill. 1975. *Animal Science.* Pergamon Press, Elmsford, N.Y.

21 Inheritance and Environment in Wool Production

One of the goals of the ewe flock owner is to produce heavy fleeces that will command high prices. Wool growers can achieve this goal by influencing environmental factors such as nutrition and by changing the genetic potentials of desired wool traits in their flocks.

Differences in any measure of wool production are due to a combination of genetic and environmental influences. Sheep flock managers must decide if, when, and how to alter either or both of these influences in a manner that is economically advantageous to their respective operations.

A basic concept in wool production is that the amount of *clean wool* **produced per sheep is a function of** *fiber diameter, staple length, density* **and** *body surface.*

Regardless of whether they are expressed as a result of environmental or genetic influences, observed differences in clean wool production among breeds and among sheep within breeds are due to differences in diameter, length, density, and body surface, independently or collectively. When these measures are considered independently, the amount of clean wool produced by a sheep varies directly and proportionately with changes in density, length, or body surface within each grade. For example, from a theoretical standpoint, doubling density (with grade, length, and body surface remaining constant) results in a doubling of clean wool produced during a given period of time. The same is true for doubling length and body surface with the other factors constant. However, when the diameter of fibers is doubled, with the other three factors constant, the volume, and therefore weight, increases four times, so clean wool production can be increased most rapidly by increasing fiber diameter. Although it is possible to increase the clean wool production of sheep quite rapidly by selection for coarser grades, it may not be a practical approach when selection goals established by breed standards, price structure, or other factors such as environment indicate specific grades as optimum.

BREEDING FOR IMPROVED WOOL PRODUCTION

Because sheep production is normally a two-crop enterprise, flock managers need to consider both lamb and wool when developing their breeding plans. Relative amounts of selection pressure placed on wool and lamb production varies with management system and breed: Wool production is relatively more important in range than farm flocks (Chapter 7). Also, most associations representing the breeds classified broadly as wool breeds recommend a 60% emphasis for growth and carcass traits and 40% for wool traits. On the other hand, wool traits usually receive little or no attention in those classified as meat breeds. These breed differences suggest that the responsibility for changing genetic potentials of wool traits in the national sheep flock is largely in the hands of seedstock producers of the wool breeds.

When breeding for improved wool production, factors that need to be considered are heritability, correlation among traits, methods of measurement, whether to apply selection pressure on the male or female side or both, age at selection, average potentials of different breeds, systems for identifying animals, and the relative importance of specific traits.

Inheritance

The heritability of most measures of wool production is considered to be relatively high (Table 7-1). When combining these heritability estimates with realistic estimates of possible selection differentials, it is apparent that positive changes in the characteristics used to describe the quantity or quality of wool growth can be made quite rapidly through selection.

Relationships among Traits

The genetic and phenotypic correlations of wool traits with each other and with other important sheep production traits have been documented by several researchers. These relationships among the five basic measures of wool production are usually expressed as diagrammed in Table 21-1. They suggest that continued selection for a decrease in fiber diameter to meet a local demand for finer wool, without attention to the other measures, can be expected to increase density and decrease length, body surface, and average clean fleece weight in a flock. Also, selection for increased clean fleece weight as a single trait usually results in coarser, longer, more open fleeces growing on larger sheep.

More than one trait can be included in a selection program with somewhat slower progress for each (Chapter 7). It is possible to select for increases in staple

TABLE 21-1 DIRECTION OF EXPECTED CHANGE IN ALTERNATE MEASURES OF WOOL PRODUCTION WITH SINGLE TRAIT SELECTION

Trait Selected for		Expected Results		
Staple length →	↑ diameter,	↑ surface,	↓ density,	↑ clean weight
Density →	↓ diameter,	↓ length,	↓ surface,	↓ clean weight
Diameter →	↓ density,	↑ length,	↑ surface,	↑ clean weight
Body surface →	↑ diameter,	↑ length,	↓ density,	↑ clean weight
Clean weight →	↑ diameter,	↑ length,	↑ surface,	↓ density

length, density, and body size (surface) while attempting to maintain a desired grade and thereby increase clean wool production within that grade, but progress will be slower.

Other important production traits such as reproduction, growth rate, and desirable carcass traits have been shown negatively associated with most wool traits in varying degrees. These relationships are demonstrated most dramatically when comparing meat breeds with wool breeds. After several hundred years of selection pressure emphasizing carcass merit and growth, the meat breeds tend to produce relatively little wool of mixed qualities. The reverse situation is apparent in wool breeds.

Research reports indicating a negative relationship between wool traits and other desirable characteristics such as reproduction have often led producers to avoid selection programs that include both aspects of production. Although usually negative, most correlations of wool and other production traits are relatively small, so selection for both is considered practical. Australia's Young and Turner[18] demonstrated that increasing reproductive performance and clean fleece weight by simultaneous selection can be accomplished. They estimated that, by selection for clean fleece weight among twin born animals for a period of 10 years, an increase in reproduction rate of 40% along with an increase of 1.2 lb (0.55 kg) of clean wool per head can be expected.

Measurement of Traits

Progress through selective breeding is directly related to the accuracy of identifying animals that are superior or inferior for desired traits. Conservative estimates of potential selection differentials (the difference between the average of those chosen for replacements and the average of the flock from which they came) for fleece weight combined with its relatively high heritability (Table 7-1) suggest that the average weight of fleeces shorn in the United States should have increased significantly during the past 25 to 50 years. This has not happened, probably because individual fleece weight (either grease or clean) has not been included as a selection factor in most flocks. Both heavy and light shearing individuals have been retained for breeding, resulting in random selection for fleece weight.

Performance and/or progeny tests (Chapter 7) combined with objective measures are considered ideal for comparing wool traits of individuals in similar environments. Methods for measuring specific traits are discussed in Chapter 20.

For selection purposes, objective measures are preferable to subjective measures (Figure 21-1), but trade-offs between relative accuracy, cost, and rapidity of measurement must always be considered. Some examples of these trade-offs are as follows:

1. When marketing wool, fiber diameter must be measured by microprojection to provide a confidence limit of $\pm 0.4~\mu$.[3] However, a confidence limit of $\pm 1.0~\mu$, which can be achieved faster by measuring fewer fibers from each sample, is usually considered acceptable by most breeders and is less expensive. Unless uniformity is considered a factor, other instruments that do not provide estimates of variation in diameter may be appropriate for selection but not for marketing.
2. Accurate estimates of clean content (yield) are obtained by processing *cores* of clips for marketing. Individual fleeces can also be cored and processed for yield, a procedure that provides excellent accuracy but is relatively expensive and is not readily available to producers in all areas of the United States. For selection purposes, grease weight or machines or instruments designed for quicker but less accurate estimates of yield (Chapter 20) may be more practical.
3. In most flocks, objective measures of desired traits may be considered an *investment* with good profit potential when selecting either commercial or purebred rams.[11] On the other hand, although subjective estimates are less accurate, they may be best for ewe selection because they can be obtained quickly and inexpensively.

Sex

There are two basic reasons why special attention should be given to the *sire* side of selection programs designed to change wool production traits in a flock: First, meaningful selection differentials can be developed most easily by careful sire selection (Chapter 7). Second, and perhaps most important, the cost of objective measures of desired wool traits usually limits their use to sire selection only.

Age

The best time to measure traits is usually associated with other management practices: In commercial flocks, replacement ewe lambs are typically selected and ewe flocks culled at or near weaning. At this point, the fleeces of ewe lambs must be evaluated "on the sheep," so selection for wool traits is limited to staple length, density, and a visual grade. In some situations, when there is a good market for yearling ewes, all ewe lambs are kept until first shearing, which also allows selection

Breeding for Improved Wool Production

Figure 21-1a Which ram grows more wool? Data and photos from University of Wyoming Ram Performance Test.

Figure 21-1b

	RAM A	RAM B
Weight	232 lb.	235 lb.
Weight/day age	.70 lb.	.62 lb.
Fleece grade	64's	64's
Staple length	4.67 in.	4.62 in.
Annual clean fleece	13.37 lb.	9.45 lb.

The only "real" difference in performance of the two rams is clean fleece weight, which is not detectable except by objective measure.

for fleece weight. When culling mature ewes, fleece weights may be considered if they were recorded at shearing. In purebred flocks of the wool breeds, there is normally greater flexibility as to age when ewe selection and culling are conducted, as well as measurement method, because individuals are identified and records kept, and management and marketing systems can be adjusted more readily than in commercial flocks.

Ram selection in the wool breeds usually occurs at several times from birth to 12 to 14 months of age. Lambs with structural defects, faulty mouth formation, entropion, hairiness, and the like, should be castrated shortly after birth.

The next logical time for culling ram lambs is near weaning because undesirable individuals can be sold as feeders or for slaughter with relatively small price discounts as compared to wethers (Chapter 14). However, at this age, maternal influences on type of birth and rearing, milking ability of ewes, as well as age itself, tend to affect the phenotypic expression of wool traits more strongly than when they near one year. Table 21-2 indicates that differences due to type of birth and rearing may be small, and the growth rate of wool increases with age until lambs reach about 10 months when it appears to level off. Table 21-3 shows fiber diameter following the same pattern.

In a typical flock, age at weaning may vary as much as 2 months. At this point, the data in Tables 21-2 and 21-3 suggest serious errors in evaluating wool traits may occur: If selection for finer fleeces is considered important, a 4-month-old lamb that grades 64's (but can be expected to grade 62's at 6 months) may be chosen over a 6-month-old lamb that grades 62's that is probably genetically similar. Likewise, if shorn at weaning, ram lambs with 6-month fleeces can be expected to have grown about twice as much wool as those with 4-month fleeces. The growth pattern for wool shown in Table 21-2 suggests that attempts to correct for age by expressing actual fleece weight as weight per day of age for comparison would still show the younger lamb as less desirable at weaning.

It is apparent that the ideal time to evaluate fleece traits of rams in a flock is between 10 and 12 months of age when maternal influences and the effects of age tend to be reduced (Tables 21-2 and 21-3). However, the extra feed required and low prices received for yearling rams sold for slaughter have discouraged meaningful culling at that age. Consequently, very few rams are evaluated for wool traits and are sold for breeding without estimations of their relative wool production merit. Until ram buyers indicate a willingness to pay enough premium for individuals with superior wool traits (identified by objective measure) to compensate for those that should be culled and slaughtered, little progress by way of genetic change can be expected.

Breed

Breeds are usually chosen for specific management situations based on their adaptability and overall potential for meeting specific production goals (Chapter 5). In terms of wool traits, breeds represent either a limit or opportunity for change by

TABLE 21-2 AVERAGE PERIODIC CLEAN FLEECE WEIGHT OF LAMBS DURING FIRST YEAR OF LIFE

Age (months)		2	4	6	8	10	12
Type of Birth and Rearing				Clean Fleece Weight, lb (kg)			
Single/single (n = 11)	Periodic	0.52 (0.24)	0.87 (0.39)	1.21 (0.55)	1.68 (0.76)	1.92 (0.87)	1.81 (0.82)
	Cumulative		1.39 (0.63)	2.60 (1.18)	4.28 (1.94)	6.20 (2.81)	8.01 (3.63)
Twin/single (n = 9)	Periodic	0.50 (0.23)	0.89 (0.40)	1.24 (0.56)	1.57 (0.71)	1.71 (0.78)	1.53 (0.69)
	Cumulative		1.39 (0.63)	2.63 (1.19)	4.20 (1.90)	5.91 (2.68)	7.44 (3.37)
Twin/twin (n = 11)	Periodic	0.47 (0.21)	0.70 (0.32)	1.13 (0.51)	1.69 (0.77)	1.87 (0.84)	1.76 (0.80)
	Cumulative		1.17 (0.53)	2.30 (1.04)	3.99 (1.81)	5.86 (2.65)	7.62 (3.45)
Combined (n = 31)	Periodic	0.50 (0.23)	0.82 (0.37)	1.19 (0.54)	1.65 (0.75)	1.84 (0.84)	1.71 (0.78)
	Cumulative		1.32 (0.60)	2.51 (1.14)	4.16 (1.89)	6.00 (2.73)	7.71 (3.51)

Adapted from Harvey.[8]

TABLE 21-3 AVERAGE PERIODIC FIBER DIAMETER OF LAMB FLEECES DURING FIRST YEAR OF LIFE

Age (months)	2	4	6	8	10	12
Type of Birth and Rearing			Fiber Diameter[a]			
Single/single	19.04	20.75	22.68	25.29	26.87	27.38
($n = 11$)	(80's)	(64's)	(62's)	(58's)	(56's)	(56's)
Twin/single	19.38	22.32	23.75	25.91	26.91	27.11
($n = 9$)	(70's)	(62's)	(60's)	(58's)	(56's)	(56's)
Twin/twin	18.87	20.51	22.68	25.71	26.88	27.07
($n = 11$)	(80's)	(70's)	(62's)	(58's)	(56's)	(56's)
Combined	19.08	21.12	22.99	25.63	26.89	27.20
($n = 31$)	(80's)	(64's)	(62's)	(58's)	(56's)	(56's)

Adapted from Harvey.[8]

[a] Upper value is micrometers (microns); lower value in parentheses is spinning counts.

selection. Consideration of wool traits in breeding plans involving the meat-type breeds should be minimal if included at all. In purebred flocks of the wool breeds, selection for qualities not typical of the breed, for example, a 64's grade in Columbias or Corriedales, is difficult and should be discouraged. On the other hand, average breed differences in fleece traits can be used to develop large selection differentials for specific traits in commercial flocks.

Identification of Animals

Identification of animals is essential to a well-controlled breeding plan. In purebred flocks, each individual is identified by ear tag and/or tattoo. Matings are controlled and the parents of offspring are identified and recorded at birth, so good records of production and relationships can be maintained. Some farm flocks use the same procedure. In these situations, any measurable trait can be included in a breeding plan and selection pressure applied to both rams and ewes.

In large, commercial flocks, particularly where lambing is on pastures or ranges, individuals are normally not identified and parents of individuals are unknown. Selection in these management systems should be conducted on a *group* rather than individual basis, with groups identified by ear notch or brand and maintained separately during breeding and lambing. A typical plan is a three-stage arrangement of A (super), B (above average or multiplier), and C (average or commercial) groups of ewes, with primary selection pressure for given traits applied to the male side, particularly in Group A (see Stratification, Chapter 7).

The top end (normally established by visual evaluation of fleece density, length, and grade along with body size and conformation, face cover, and skin folds) of ewe lambs born in group A remains there as replacements, with the balance providing the majority of replacements for group B. Except for faults such as coarse britches, extremely open or short fleeces, and colored spots, little attention is given to wool traits when selecting ewe lambs in group B. The top end (evaluated as in or for group A) stays in this group if needed to maintain numbers, and the rest are used as replacements for group C. Ewe lambs born in group C are normally marketed, so selection factors are not a concern.

Choice of Traits

Designers of sheep breeding plans need to decide which traits should be included for maximum total production in a particular management system. If wool production is considered important, each of the following traits should be evaluated with particular attention to method of measurement.

Fleece weight. Because it contributes more to the *value* of clips than other traits, fleece weight must be included in all breeding plans designed to increase returns from wool production. *Clean* fleece weight is considered the ideal quantitative measure of wool production. It measures the combined effects of all four of the major variables (length, grade, density, and body surface) and removes the possibility of error in evaluation due to variation in nonwool components. However, determination of the clean content of fleeces requires facilities and expertise that are currently either not available to most American producers or too expensive for evaluating large numbers of fleeces.

Alternative methods for estimating clean wool production and their associated limitations are discussed in Chapter 20. For ewe selection, grease weight is usually considered the optimum measure, but producers of purebred wool breeds are encouraged to use the more accurate methods if possible.

For ram selection, many producers may be limited to grease weight as the only practical measure available when selecting for heavier fleeces. However, an objective measure of clean weight is always encouraged, regardless of management system. If either clean or grease fleece weights representing similar periods of growth are not available when selecting for increased wool production, the incidence of belly wool (which is shorter and less dense than the rest of the fleece (Chapter 7) can be combined with staple length and density as indicators of potential fleece weight. Differences in these indicators can be identified with reasonable accuracy by experienced people.

Average clean fleece weight in a flock can be expected to increase by about 1.1 lb (0.5 kg) over a period of 8 to 10 years (two generations) as a result of selection. However, some attention must be given to diameter, because a weight increase of 1.1 lb (0.5 kg) by selection increases diameter by about 0.5 μ unless something is done to prevent it.[2]

Fiber diameter. Diameter is the most important *price* determining measure of wool. However, its direct relationship to fleece weight should be evaluated carefully. Clean wool production can be increased rapidly by selecting for coarser fleeces. As fiber diameter increases, staple length and, in most cases, body surface also increases, offsetting an associated decrease in density (Table 21-1). However in the current market structure, selection for increased fiber diameter has resulted in a price reduction, so this approach has not been widely accepted.

Some producers have found it financially rewarding to sacrifice quantity for fineness when choosing between increasing fleece weight by increasing diameter or maintaining a more desirable grade for the current market. However, progress is slow. Within-breed selection for finer wool can reduce average fiber diameter by about 2 μ (approximately one spinning count) over a period of 8 to 10 years.[2]

Producers should be encouraged to select for a grade typical of the breed that is most suitable in their overall management system rather than attempt to "follow a market trend" for grade. Objective measures of diameter (Chapter 20) in ram fleeces and visual evaluation of ewe fleeces is the most practical approach when selecting for grade. Both methods can be used at almost any time during the year to compare contemporaries.

Producers are cautioned to be very careful in the interpretation of objective measures of fiber diameter. Average diameter in fleeces can be described as "ever changing" due to nongenetic factors[10] (Figure 19-8). A fleece measured at different times of the year or in different years can vary as much as 6 μ (56's to 64's) or more. Therefore, the diameter of fleeces should not be compared unless grown on sheep of similar ages, production states, and nutritional status.

Uniformity of diameter. Typical histograms showing the distribution of fibers by diameter in very uniform and variable samples of wool taken from individual clips are shown in Figure 21-2. Methods for increasing the uniformity and consequently relative value of clips are described in Chapter 18. The most important source of variation is between fibers within locks,[6,13] a problem that can only be corrected by selective breeding.

The degree of fiber-to-fiber variation in individual fleeces can also be extreme (Figure 21-3). Although heritability estimates for fiber-to-fiber variation have not been reported, it is assumed that it is related to S/P ratio, which is considered quite highly heritable.[1] Therefore, good progress through selection for uniformity of diameter is feasible provided differences in variation can be correctly identified. Accurate evaluation of variation requires a method that measures relatively large numbers of individual fibers in a sample rather than one of the more rapid (less expensive) methods that provides average diameter with no estimate of variation.

Because variation in average diameter of fleeces from shoulder to side to thigh is often visually apparent, some breeders have expressed interest in comparing side-to-thigh variation, particularly in potential sires for purebred flocks. However, this has been shown as a minor source of total variation in fleeces.[6,13] If diameter variation is to be included as an important trait in a selection program, the best procedure is to compare estimates of variation from side samples. This trait should be considered seriously as a selection factor in purebred flocks that are recognized as leaders in their respective breeds and consequently provide the majority of the foundation sires for their breeds.

Staple length. All breeding plans for increasing wool production should include staple length. It can be measured easily and economically and is directly related to fleece weight within grade. When selecting ewe lambs at weaning, it is the only objective measure of wool production available unless they are shorn or side-sampled for diameter, it is also an important price-determining factor.

Fiber density. Except for histological evaluations of skin sections, which is considered impractical, particularly for selecting ewes, a good objective measure of the density of fibers growing on a sheep has not been developed (Chapter 20). However, the subjective "touch method" has been shown useful when selecting for density. Average fleece weight in a large flock was increased by 3.5 lb (1.6 kg) over a period of 20 years.[15] Breeding plans designed to increase the genetic potential for fleece weight in a flock should always include selection for increased density within the desired grade.

Body surface. This trait should not be included in breeding plans related to wool production because it is difficult to measure and the potential for increasing fleece weight by using other measures is considered more effective. Because of its relationship to growth and carcass traits, body weight and/or frame size (and coincident smooth skin area) are usually included in breeding plans for increasing lamb production, which has a positive effect on clean wool production.

Grease content. It has been observed that, within grade, certain fleeces in a flock managed in a common environment contain less grease than others. This implies that the difference may be due to genetic variation. However, few producers have attempted to select for fleeces with low grease content. Other wool production traits such as grade, length, density, and clean wool content are usually considered more important for inclusion in selection systems than relative greasiness. For an effective selection program designed to decrease the grease content of fleeces, this component must be evaluated correctly, which requires relatively expensive laboratory procedures.

Face cover and skin folds. These are frequently considered as traits associated with wool growth. For a period of time in the history of sheep production, it was believed that increasing the amount of face cover and the degree of skin folds (wrinkles), particularly on the neck area, would also increase fleece weight. This concept has been abandoned and these traits are considered defects in the United States (Chapter 7).

Although individual fleece weights may be slightly heavier in woolly faced sheep, there are distinct disadvantages. Besides requiring periodic "facing" to avoid

```
10.00
11.25#
12.50
13.75###
15.00
16.25##################
17.50
18.75#######################################
20.00
21.25#####################################
22.50
23.75###################################
25.00
26.25##################
27.50
28.75####
30.00
31.25###
32.50
33.75#
35.00
36.25#
37.50
38.75#
40.00


GRADE.........................................64'S

AVERAGE FIBER DIAMETER........................21.56 MICRONS

STANDARD DEVIATION............................ 3.82 MICRONS

COEFFICIENT OF VARIATION......................17.72 PERCENT
```

Figure 21-2 Typical distribution of fiber diameters by 2.5-μ increments of composite samples of (A) uniform and (B) variable *clips*. (Courtesy of Angus McColl[9])

wool blindness, woolly faced ewes tend to produce fewer lambs that are lighter at weaning than open-faced ewes.

Smooth-bodied sheep have been shown to produce as much clean wool as those with wrinkles. The more wrinkled sheep tend to produce shorter stapled and heavier but lower-yielding fleeces. Also, there is an apparent negative association between degrees of wrinkling and number of lambs born and weaned.[14]

Balancing trait emphasis. Because they do not function independently, striking a balance of emphasis given to different wool traits is challenging but can be rewarding. A recommended procedure is to use the most accurate measurement

```
10.00
11.25#
12.50
13.75##########
15.00
16.25######################
17.50
18.75##################################
20.00
21.25#######################################
22.50
23.75############################
25.00
26.25####################
27.50
28.75########
30.00
31.25#####
32.50
33.75###
35.00
36.25###
37.50
38.75#
40.00
41.25#
42.50


    GRADE.........................................64'S

    AVERAGE FIBER DIAMETER.........................21.92 MICRONS

    STANDARD DEVIATION............................ 5.10 MICRONS

    COEFFICIENT OF VARIATION......................23.27 PERCENT
```

Figure 21-2 (*Continued*)

methods available and *select sheep with the heaviest, longest-stapled, densest fleeces of the most uniform possible grade that is typical for the breeds best adapted to individual management systems.* In other words, *go for weight within grade.*

Design of Breeding Plans

Realistic goals for changing wool production traits by selection are not reached in many flocks. Problems involving relationships among traits, measurement of traits, age when selections are made, and animal identification (discussed elsewhere in this

```
12.50
13.75#
15.00
16.25#############
17.50
18.75##################################################
20.00
21.25##################################################
22.50
23.75###########################
25.00
26.25###
27.50
28.75#
30.00
31.25#
32.50


Animal Number..................................381
Animal Sex.....................................RAM
Sample Location................................Side
GRADE..........................................64'S
Average Fiber Diameter.........................20.67 Microns
Standard Deviation............................. 2.60 Microns
Coefficient of Variation.......................12.58 Percent
```

Figure 21-3 Typical distribution of fiber diameters by 2.5-μ increments in side samples of (A) uniform and (B) variable *ram fleeces*. (Courtesy of Angus McColl[9])

chapter) are limiting factors. However, in most situations, the lack of a well-designed breeding plan is the primary reason why desired genetic changes are not accomplished. Too often, there is either no plan or plans are changed so frequently that they lose direction.

The following schemes are presented as "places to start" rather than ideals for all flocks in developing plans suitable for particular management systems. They cannot be applied directly as shown for purebred flocks where crossbreeding should not be considered (see Systems of Breeding, Chapter 7). However, they may be considered for controlling family lines in purebred flocks and can certainly be *adapted* to develop plans for reaching specific genetic goals in most commercial systems.

Figure 21-4 is an example of a breeding plan that can be used to reach many commercial sheep production goals. Depending on the management system and potential market for both wool and lamb, it offers excellent opportunities for maximizing production by choosing breeds based on careful evaluation of their typical production patterns and adaptability. For example, if a breeder wants to emphasize

```
10.00
11.25#
12.50
13.75#######
15.00
16.25##############
17.50
18.75#########################
20.00
21.25##############################
22.50
23.75################################
25.00
26.25####################
27.50
28.75#####
30.00
31.25###
32.50
33.75#
35.00
36.25#
37.50
```

```
Animal Number..................................5-24
Animal Sex......................................RAM
Sample Location................................Side
GRADE..........................................64'S
Average Fiber Diameter.................21.79 Microns
Standard Deviation......................4.36 Microns
Coefficient of Variation...............20.01 Percent
```

Figure 21-3 (*Continued*)

wool production, typical American ewe breeds (Chapter 5) would probably be Rambouillet, Targhee, Columbia, or Corriedale or even the "western white-faced" crossbreds because they are available in large numbers. If a high proportion of multiple birth is considered a major goal with some trade-off in wool production, the Finnsheep, Polypay, or their crosses may be used as ewe breed A. In some areas, where the price of wool is low compared to the price of lamb and primary emphasis is on lamb production, one of the meat breeds might be considered as a ewe breed and included as either A or B. This plan assumes that replacements for breed B ewes (approximately one-third of the flock) would be purchased.

A simple crossbreeding plan that is used widely by American producers is shown in Figure 21-5. It usually involves only two breeds and eliminates the need for purchasing replacements. The numbers shown represent the number of ewes

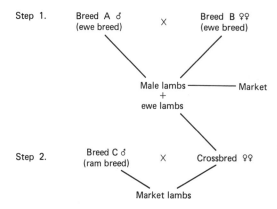

Breed A: adapted, excel as ewe breed, not necessarily in large numbers

Breed B: adapted, excel as ewe breed, necessary in large numbers, replacements must be available

Breed C: sire growthy lambs with desirable carcasses

Figure 21-4 Two-step plan to produce crossbred ewes and lambs. (From Whiteman,[16] with permission)

typically mated to one ram, so flock A becomes a "one-ram flock" and flock B is a "two-ram flock." This plan can be used most efficiently in three, or any multiple of three, *ram flocks*. In estimating numbers, the assumptions given for weaning rate may need to be adjusted for specific situations. Ewe lambs sired by the base breed (I) ram are retained for replacements, and all the crossbred lambs are marketed. In

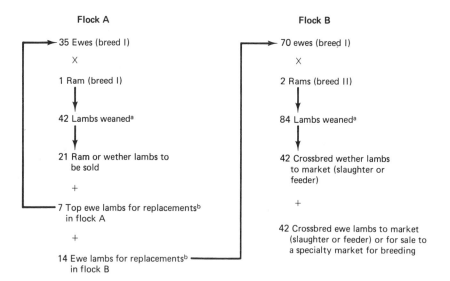

Figure 21-5 Crossbreeding plan using two breeds in a three-ram flock. [a]Assume 120% weaned. [b]Assume 20% replacement rate.

Breeding for Improved Wool Production

this example, the producer can expect an increase in the survival rate and weaning weight of his crossbred lambs, but very little benefit in fertility and prolificacy due to heterosis (Table 7-4).

If separate breeding pastures are available (with the flock managed as a single unit except at breeding), many options are available as suggested in the discussion of Figure 21-4. A great deal of selection pressure for wool traits could be applied to both ewes and rams in flock A. This plan could even be specialized to the point that flock A would be purebred (probably a wool breed). It could also be adapted to produce ewe lambs of any desired cross for sale to breeders who prefer purchasing rather than raising them as suggested in Figure 21-4.

A typical adaptation of this plan, where separate breeding pastures are *not* available, is to turn out one breed I ram and two breed II rams in a three-ram flock. With a white-faced base flock (usually one or a combination of the wool breeds), a meat-type ram is usually used as breed II. If the meat type rams are black-faced and the base flock breed is white-faced, the crossbred lambs will be speckled and the straight-breds white-faced, so their identification by tags or brands at birth is not required. With approximately one-third of the flock mated to the base breed ram, enough ewe lambs can be expected for use as replacements to maintain flock size. With this system, mating is random, so no selection pressure is applied to the ewe side of the picture.

An expansion of Figure 21-5 is presented in Figure 21-6. It is limited to flocks requiring nine, or multiples of nine, rams. The assumption for weaning rate in flock C has been estimated as 20% greater than for flocks A and B because of the ex-

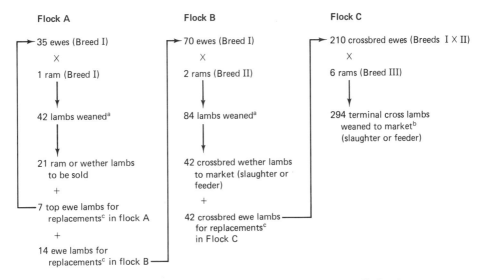

Figure 21-6 Crossbreeding plan using three breeds in a nine-ram flock. [a]Assume 120% weaned. [b]Assume 140% weaned. [c]Assume 20% replacement rate.

pected increase in reproductive performance due to probable cumulative effects of individual and maternal heterosis for lambs reared per ewe mated (Table 7-4). Actual responses may vary widely from one cross to another.

The opportunities for combining the average production potentials of breeds are many. If wool production is considered worthy of emphasis, breeds I and II would be *wool* breeds. On the other hand, if year-long fertility and/or multiple birth are considered primary traits, combinations of Finnsheep and Rambouillets or Dorsets would be used. Many successful producers of market lambs believe flock C ewes should be a Suffolk × Finnsheep or Rambouillet combination.

Sires used in flock C are usually considered "terminal" and the meat breeds are most popular. Careful evaluation of this plan suggests that it is possible to select for the highly heritable wool traits and include desirable traits of other breeds (Chapter 5) and heterosis (Chapter 7) in about two-thirds of the flock.

ENVIRONMENTAL EFFECTS ON WOOL PRODUCTION

All sheep have genetically determined limits (potentials) for the four basic measures of wool production. The expression of these potentials is directly related to many environmental influences.

The genetic potential for wool growth in sheep becomes truly limiting only when all environmental effects are ideal. It is for this reason that sheep flock managers often express more interest in manipulating environmental factors than genetic potentials. Changes in wool production related to specific changes in environmental factors are observed more quickly and tend to be larger than changes due to selection. Control of environmental factors in many situations is, or appears to be, easier to manage than breeding programs.

Although they seldom function independently, several specific environmental factors need to be considered in the development of management systems that include emphasis on wool production.

Nutrition

Because it is directly or indirectly related to all other environmental influences, nutrition is considered the most important environmental factor affecting wool traits. The extent to which sheep express their genetic potentials for wool production is directly related to the supply of nutrients to the follicles (skin).

All nutrients consumed by a sheep are divided between wool growth and other body functions. The biological mechanisms that control this separation of nutrients have been studied extensively but have not as yet been completely clarified.

The nutritional requirements for sheep at various stages of production and body weight are discussed in Chapter 8. However, the ways specific dietary components influence wool growth has been, and continues to be, an area of concern for both researchers and producers.

Minerals and vitamins. Because the sulfur-containing amino acids (particularly cystine) are especially important in the chemical structure of wool (Chapter 19), it has been suggested that feeding levels of sulfur in excess of recommended amounts and balances could increase wool growth. This has not been demonstrated. Copper is directly involved in the formation of wool fibers, and deficiencies result in a harsh, nearly crimpless "steely" wool. Copper metabolism is closely related to dietary levels of molybdenum and sulfate, so the balance of these factors may be more important than absolute levels. Zinc deficiencies have also been implicated in the growth of steely wool.

Protein. Wool is considered pure protein with a high cystine content. This analysis suggests that wool growth might be increased by increasing the kind and/or quantity of protein fed to sheep.

When related to wool growth, the optimum level of crude protein in sheep rations is apparently close to 10%, provided other nutrients are not limiting. Clean fleece weight has been shown to increase directly with increases in dietary protein up to 10%, but a similar response at higher levels has not been evident. Further research, particularly in the area of energy/protein ratios, may lead to adjustments in the level of dietary protein recommended for optimum wool production.

The sulfur-containing amino acids have attracted special attention by nutritionists interested in wool production. Although abomasal (postruminal) infusions of methionine and casein have been shown to increase wool growth dramatically, the conversion of dietary to microbial protein in the rumen limits the potential for increasing wool growth by *feeding* high levels of sulfur-containing amino acids.

Current research relating protein nutrition to wool growth in sheep suggests that a realistic goal is the development of a practical method for avoiding ruminal degradation of these amino acids, permitting their subsequent absorption and utilization by wool follicles.

Energy. The effects of changes in feed intake are evident when observing both within and between year changes in fiber diameter and length (Chapter 19). The relationship between energy level and wool growth in diets containing a minimum of 8% to 10% protein is apparently linear and positive (within genetic limits) in mature sheep. In most diets, wool growth is probably more closely related to energy than protein levels. However, because of the relative prices of wool and high-energy feeds, the addition of high-energy supplements to normally adequate sheep diets may not be a profitable procedure for increasing fleece weight.

Physiological State

The degree of competition for nutrients between wool growth and other body functions is directly related to the *production states* of sheep, which vary greatly in ewes during a typical yearly cycle. Nutritionally, the most critical period of a ewe's year is during the last 6 to 7 weeks of gestation and first 6 to 8 weeks of lactation (Chapter 8). Demands for nutrients during this period must be met from body reserves and/or increased feed intake.

Competition for nutrients by the lamb during late gestation and early lactation may reduce wool growth two ways. First, the fleeces of ewes raising single lambs can be expected to weigh 10% to 20% lighter than those from dry ewes. A further reduction of up to 4% is typical for ewes raising twins.[5] Second, the plane of nutrition of the ewe during these periods has been suggested as an important factor determining whether or not their lambs reach their genetic potentials for density (S/P ratio). The supply of nutrients to the fetus during late gestation has been shown to affect the extent of *initiation* and probable *degree of branching* of secondary follicles. The degree to which these follicles *mature* to produce fibers is affected by the level of nutrition of the lamb during early life. A reduction in the number of follicles initiated due to limited nutrients during prenatal life is considered permanent. However, there is some disagreement among researchers as to whether limited nutrition during early postnatal life inhibits follicle maturation permanently or only delays the process.

The net effect of the relationship of nutrient availability to follicle development can be observed when comparing single lambs with those born and raised as twins. The singles tend to produce more wool than twins. The typical lower density and smaller size of twin-born ewes combined with their likelihood of producing twins themselves has led to estimates that they produce 2% to 5% less wool during their lifetime than those born single.[7]

Most producers consider this decrease in wool production fair trade for more lambs. However, the effect of these maternal handicaps suggests that type of birth and rearing must be considered when comparing wool growth in the selection of young sheep for replacements. Selection for increased fleece weight as a single trait may tend to discriminate against twins unless a proper increase in dietary intake is provided for ewes giving birth to and raising twins.

Variations in wool production, apparently due to age, have been well documented. On the average, maximum fleece weights have been observed in 2- to 4-year-old sheep, with an approximate 4% per year decline thereafter (Figure 6–11). It is difficult to separate the true effects of age from other factors such as changes over time in grazing ability, reproductive performance, ability to withstand stress or changes in number or efficiency of active wool follicles[17] From a practical standpoint, the effects of age on wool production should probably be ignored except when comparing fleeces in the selection of lambs for replacement (Table 21-2 and 21-3).

Awareness of the direct relationship between level of nutrition and wool production during various physiological states suggests that the best way to increase wool production is to increase the intake of well-balanced rations, particularly during late gestation and early lactation. Although this suggestion is true, the relative prices of wool and sheep feeds in most production situations indicate there is a point of diminishing return for this practice. The encouraging aspect is that adequate nutrition *also* enhances lamb production. Well-fed ewes produce lambs that are

Environmental Effects on Wool Production

heavier at birth and grow faster with fewer death losses than poorly fed ewes. At the same time, well-fed ewes produce more wool and their lambs have a potential for producing more wool.

A valid assumption is that if ewes are kept on diets balanced to meet suggested requirements for all production states, they produce *both* lamb and wool at levels that may not be maximum for their genetic potentials, but are close to an optimum for profit.

Disease and Parasites

Parasites, particularly internal parasites, represent a source of competition for nutrients available for both wool and lamb production. Wool growth is particularly sensitive to these competitions. It has been shown[4] that, although differences in weight gains could not be associated with varying loads of certain internal parasites in lambs, fiber diameter was reduced as parasite population increased. In cases of disease situations where high body temperatures are experienced or where extremely heavy loads of internal parasites are involved, fiber diameter is reduced dramatically and "breaks" are typical. The use of available anthelmintics, pesticides, vaccines, and antibiotics is considered cost-effective when included in well-designed health programs.

Soil Type

An awareness of soil type and composition can lead to dietary adjustments beneficial to wool growth. Calcium–phosphorus imbalances and/or deficiencies and trace mineral deficiencies (and in some cases excesses) in the feeds provided for sheep are usually related to the types of soil where they are grown. These problems can usually be rectified easily by supplemental feeding. In some areas, the alkalinity of soils represents a unique problem by causing abnormal "weathering" of the tips of fleeces (Chapter 19). In other areas, soils tend to discolor fleeces. If other arrangements can be made, grazing of these areas where special soil types are detrimental to wool production should be avoided.

Season

It is generally agreed that there is a definite seasonal rhythm in the rate of wool growth. Sheep tend to grow coarser, longer fibers and therefore more wool in summer than winter. Many attempts have been made to identify the major cause(s) of this trend in search of practical ways to adjust seasonal environmental effects to increase annual fleece weight. Figure 21-7 diagrams an example of seasonal changes associated with changes in nutrition under grazing conditions.

From a research standpoint, separation of the many different environmental effects that have been shown to coincide with changes in season is extremely difficult. For example, controlled lighting has been used to study the effects of photope-

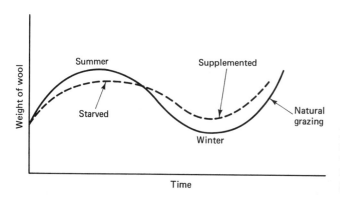

Figure 21-7 Diagrammatic representation of the seasonal rhythm of wool growth based on changes observed in British breeds. (From Ryder and Stephenson,[14] with permission)

riod, but the relative intensities of natural and artificial light make interpretation of results difficult.

Several factors shown to affect wool growth are included when discussing seasonal changes in the broad sense. They are photoperiod (day length, including light intensity, change and direction of change); physiological state (discussed previously); adaptability of breeds (genetic-environmental interactions in different climates and weather situations); quantity and quality of feeds (particularly important in grazing situations); and temperature (actual or adjusted as in the case of wind-chill estimates), which has a major effect on competition for nutrients and degree of circulation of blood, and therefore nutrients, to the skin. Some researchers[14] suggest that although there is probably no single cause of seasonal variation in wool growth, day length is most important. Others[12] do not completely refute this but suggest that the effects of photoperiod force wool growth to oscillate about a mean value, so *annual* wool production is not affected by changes in photoperiod. It is, therefore, of little commercial interest in most situations.

For producers, management decisions as to whether and how attempts should be made to adjust special season-related effects need to be considered. Providing shelter or shade for extremes in weather is frequently suggested but may not be economically justifiable, particularly for large numbers.

A practical approach is to do as much as possible to help sheep *cope* with extremes in weather and feed sources as well as the demands of various physiological states. Decisions as to whether or not and how to adjust season-related factors should be based on optimum lamb production, because it is the major source of income for most American producers. Wool production will probably benefit coincidentally.

REFERENCES CITED

1. Abouheif, M. A., C. LeRoy Johnson, and M. P. Botkin. 1984. Heritability estimates of wool follicle traits in sheep skin. *Animal Prod.* 39:399.

2. Anonymous. 1973. New aims for the sheep breeder. *CSIRO Rural Research* 80:8.
3. American Society for Testing and Materials. 1982. D-2130 diameter of wool and other animal fibers by microprojection. *Annual Book of ASTM Standards,* Part 33. The Society, Philadelphia.
4. Bergstrom, R. C., J. L. Kinnison, and B. A. Werner. 1977. Parasitism in sheep: Relationship between wool fiber diameter changes and feed conversion efficiency. *Amer. J. Vet. Research* 38, 6:888.
5. Corbett, J. L. 1979. Variation in wool growth with physiological state. In J. L. Black and P. J. Reis, eds., *Physiological and Environmental Limitations to Wool Growth.* University of New England Publishing Unit, New South Wales, Australia.
6. Dunlop, A. A., and P. R. McMahon. 1974. The relative importance of sources of variation in fiber diameter for Australian Merino Sheep. *Aust. J. Agr. Research* 25:167.
7. File, G. C. 1981. Highly fertile Merinos and their nutritional management through pregnancy and lactation. *Wool Technology and Sheep Breeding* 29, 1:7.
8. Harvey, Jean. 1986. Evaluation of wool production parameters in selection programs. M.S. thesis in preparation. University of Wyoming, Laramie.
9. McColl, Angus. 1985. YOCOM-McCOLL Testing Laboratories, Inc., 540 West Elk Place, Denver, Colo. 80216.
10. McFadden, William D. 1967. *Wool Science,* p. 13. Pruett Press, Boulder, Colo.
11. McGuirk, Brian. 1978. Objective measurement of flock rams. *Wool Technology and Sheep Breeding* 26, 1:17.
12. Nagorcka, B. N. 1979. The effect of photoperiod on wool growth, p. 127. In J. L. Black and P. J. Reis, eds. *Physiological and Environmental Limitations to Wool Growth.* University of New England Publishing Unit, New South Wales, Australia.
13. Stobart, Robert H., and others 1986. Sources of variation in wool fiber diameter. *J. Animal Sci.* 62:1181.
14. Ryder, M. L., and S. K. Stephenson. 1968. *Wool Growth,* pp. 610–617. Academic Press, New York.
15. Warren Livestock Co., Cheyenne, Wy. 1947. Field day report. University of Wyoming Wool Library.
16. Whiteman, Joe V. 1972. Types of sheep for various environments, p. 3. *Intensive Sheep Management Symposium,* Paducah, Ky., and Dixon Springs, Ill. SID—Sheep Industry Development Program, American Sheep Producers Council, Denver, Colo.
17. Yeates, N. Y. M., T. N. Edey and M. K. Hill. 1975. *Animal Science.* Pergamon Press, Elmsford, N.Y.
18. Young, S. S. Y., and Helen N. Turner. 1965. Selection schemes for improving reproduction rate and clean wool weight in the Australian merino under field conditions. *Aust. J. Agr. Research* 16:863.

SUGGESTED READING

Black J. L., and P. J. Reis. 1978. *Physiological and Environmental Limitations to Wool Growth.* University of New England Publishing Unit, New South Wales, Australia.

Coop, I. E. 1982. *World Animal Science, C1, Sheep and Goat Production.* Elsevier Science Publishing Co., New York.

Ryder, M. L., and S. K. Stephenson. 1968. *Wool Growth.* Academic Press, New York.

Tomes, G. L., D. E. Robertson, and R. J. Lightfoot (eds.), revised by W. Haresign. 1979. *Sheep Breeding,* 2nd ed. Butterworths, Woburn, Mass.

Yeates, N. T. M., T. N. Edey, and M. K. Hill. 1975. *Animal Science.* Pergamon Press, Elmsford, N.Y.

22 Wool Shows and Wool Judging

WOOL SHOWS

Wool shows play a minor but important role in agricultural shows across the United States. They range in size from less than a dozen to several hundred fleeces at county, state, regional, and national levels. Regardless of size, their ultimate purpose is twofold: *education* and *promotion*.

Educational Aspect

Learning experiences are frequently reported by producers involved in wool shows either as exhibitors or "just lookers." Wool shows usually demonstrate clearly that there are dramatic differences between fleeces, which imply a real potential for increasing both the quality and quantity of wool grown by individual sheep. Observation of this potential demonstrated at wool shows can, and often does, motivate sheep flock managers to intensify their efforts to produce more and higher quality wool in their flocks.

Also, many consumers recognize wool fibers only as a component of the garments, blankets, and the like, that they use. Wool shows introduce them to wool as it is grown and helps them develop an appreciation for the changes that take place in the conversion of wool fibers from fleece to fabric.

Promotional Aspect

The promotional aspect of wool shows is, like the educational aspect, aimed at both producers and consumers. Owners of the superior fleeces in a show are in reality

Figure 22-1 Hand-spinning demonstrations at wool shows attract consumers. (University of Wyoming photo)

urging other producers to buy their sheep for breeding purposes. The message implied is "you can produce this kind of wool if you buy rams or ewes bred like those that grew these fleeces." Wool shows attract the attention of consumers, as well as wool growers, by reminding them that these natural fibers should be considered for their use as manufactured items.

Consumer oriented promotional efforts such as shearing and hand-spinning demonstrations are often held in areas close to wool shows (Figure 22-1). Organizations such as the American Sheep Producers Council[1] and local Wool Growers Auxiliaries prepare booths with colorful pictures and brochures encouraging passers-by to include both wool and lamb in their budgets.

Wool Show Management

To be successful, a wool show must be prepared for and conducted in an orderly manner. Too often, many important details are overlooked, particularly at the county and state levels, and wool shows are dropped because of lack of exhibitor participation. The following outline of management concerns is presented as an aid to wool show supervisors.

1. The premium book (catalog): The rules of the show must be listed in detail. The wool show section of the premium book should include:

a. Purpose of the show
b. Eligibility of exhibitors
c. Eligibility of fleeces
d. Qualifications (or disqualifications) of fleeces
e. Preparation of fleeces
f. Number of entries allowed per exhibitor
g. Procedures of entry: address, fees, entry and delivery deadlines, and the like
h. Procedures for disposition of fleeces following the show
i. List of classes offered
j. List of premiums and awards
k. Any other items of special concern

Wool show sections of premium books usually reflect the breeds and commercial classes that most nearly represent the types of wool grown in the area served by the show. These sections may be very extensive and detailed, as is the case for the National California Wool Show, or relatively brief and straightforward, as indicated in many county and state fair premium books. A typical wool show section of a premium book (Example 22-1) is shown as a guide for those who may want to initiate a wool show or revise procedures for an existing show.

2. General concerns
    a. Preshow publicity is extremely important. Most wool shows are held in the fall and winter and include fleeces normally shorn the previous spring. Potential exhibitors must be aware of the show by shearing time.
    b. Arrangements need to be made for exhibiting and judging the fleeces. At most shows, it is difficult to provide adequate space. Figure 22-2 shows two methods of giving all fleeces good visibility with minimal space requirements. Tables are required for handling the fleeces during judging and good lighting is essential.
    c. Arrangements for a judge or judges should be made well in advance of the show. A poorly judged show usually results in a reduction of entries the following year.
    d. If special awards are to be included, donors should be contacted before the premium book is printed so their names can be included.
    e. Postshow publicity is often forgotten. The ongoing success of a show is highly dependant on "news" listings (preferably with pictures) of winning fleeces and exhibitors. Award donors are often included in these pictures.
    f. The wool show superintendent must be on hand during the show to ensure that all fleeces are properly entered and judged and the placings recorded.

Selecting Show Fleeces

Selection of show fleeces begins with a well-managed breeding program designed to produce large quantities of wool per sheep. This is followed by a year-long manage-

National Western Wool Show

Frank Craddock .... Superintendent
Leslie Mergelman .... Co-Superintendent
Mike Monell .... Asst. Superintendent

Judges: Mel Lynch, Gillette, WY and James Sachse, Las Cruces, NM

1. Purpose. The purpose of this show is to encourage the production of better wool; to promote proper handling, grading and preparation for market; to create an awareness of the many qualities of wool and its uses.

2. Eligibility of Contestants. Any growers of wool may enter fleeces, but they must be entered in the name of the grower.

3. Entry Fees. An entry fee of $1.00 per fleece in Open Show and 50 cents per fleece Junior Show will be charged on all fleeces entered for competition. In Open Show any exhibitor entering four or more fleeces will receive one exhibitor pass. Unless this fee has been paid by the closing date of the show the fleece shall become the property of the National Western Stock Show and it will sell said fleece, deduct the entry fee and remit the balance of the proceeds from its sale to the exhibitor.

Wool exhibitors who will not attend the show need not buy exhibitor passes.

4. Number of Entries. Exhibitors will be limited to two entries in each class and no fleece can compete in more than one class except for champion or special prizes.

5. Eligibility of Fleeces. In order to compete for prizes each fleece must:

a. Have been shorn in 1979.

b. Represent not more than 12 months' growth, except in the case of yearling fleeces, when 16 months' growth will be allowed.

c. In case of controversy concerning the number of months' growth of any fleece, the acknowledged rate of monthly growth of wool from different breeds and types of sheep shall be used as a guide.

d. The management reserves the right to untie and inspect any fleece entered to determine its eligibility. Fleeces adjudged over-age and/or found to contain parts of other fleeces will not be allowed to compete and the exhibitor will forfeit entry fees for these ineligibile fleeces.

e. Fleeces entered in classes W-18 through W-29 must have been shorn from registered sheep. The appropriate breed registration numbers must be listed on the exhibitor's original entry card.

6. Qualifications. Any fleece entered for competition in the National Western Wool Show shall be discriminated against if:

a. Tied with other than paper twine.

b. Showing any paint or tar brands (soluble branding fluids permitted).

c. Showing excessive dung locks or tags.

d. Discolored by excessive moisture or cotted.

e. Showing a weak staple or break in the fiber.

f. Containing any excess burrs or other vegetable matter.

7. Preparation of Fleeces. The fleeces shall be tied with paper twine, flesh side out, in a manner to allow their being handled without falling apart. It is recommended that exhibitors tie and pack fleeces rather loosely so as not to impair their character or attractiveness.

Figure 22-1 A typical wool show section of a premium book. (Courtesy of National Western Stock Show[2])

Each fleece must be tagged with this information: Breed, sex, number months' growth, catalog classification.

8. Entries Close December 26, 1978, and fleeces must be delivered to the show on or before January 10, 1979.

9. Fleeces Will Be received by the National Western Stock Show, Denver, Colorado, any time during the year prior to the Show and stored free of charge. All fleeces must be delivered to the Show, transportation charges prepaid, and if they are redeemed by the exhibitor, they will be returned to him following the Show, transportation charges collect, unless arranged otherwise. An exhibitor may instruct the National Western Stock Show to sell his wool and return the proceeds to him.

10. In the Case of any controversy relative to the classification of any fleece offered for exhibition, affidavits from the exhibitors may be required. However, in all such controversies and in any interpretation of these rules and regulations, the decision of a committee consisting of the Wool Show Superintendent, the Wool Show Judge and the National Western Stock Show Manager shall be final.

11. Fleeces Judged to be worthy of champion ribbons, trophies or special awards, will be untied and inspected by Wool Show judges and Superintendent to insure their eligibility for competition.

12. Consignors may request their fleeces be picked up at the close of the show, returned at their own expense or sold. All fleeces designated for sale will be marketed as a single lot by sealed bid on Monday, January 15, 1979.

TERRITORY FLEECES

Fleeces from flocks of sheep run under strictly western range conditions shall be eligible to compete in the territory division. All fleeces entered in classes W-1 to W-6 inclusive must have been shorn from ewes. Fleeces entered in classes W-7 to W-10 inclusive must have been shorn from rams owned by exhibitor.

Class		1	2	3	4	5	6	7	8	9
W- 1	64's, 70's, 80's (Fine) Combing	$12	$8	$7	$5	$4	$3	$2	$2	$2
W- 2	62's (½ Blood) Combing	12	8	7	5	4	3	2	2	2
W- 3	60's (½ Blood) Combing	12	8	7	5	4	3	2	2	2
W- 4	58's (⅜ Blood) Combing	12	8	7	5	4	3	2	2	2
W- 5	56's (⅜ Blood) Combing	12	8	7	5	4	3	2	2	2
W- 6	50's, 54's (¼ Blood) Combing	12	8	7	5	4	3	2	2	2
W- 7	64's, 70's, 80's (Fine) Ram Fleece	12	8	7	5	4				
W- 8	60's, 62's (½ Blood) Ram Fleece	12	8	7	5	4				
W- 9	56's, 58's (⅜ Blood) Ram Fleece	12	8	7	5	4				
W-10	50's, 54's (¼ Blood) Ram Fleece	12	8	7	5	4				
	Champion Territory Fleece									Ribbon
	Reserve Champion Territory Fleece									Ribbon

THE CARL OSBORN TROPHY

To promote and encourage the production of better range sheep in Colorado, the Colorado Wool Growers Association offers the Carl Osborn trophy for the best 12-month ewe fleece run under range conditions and entered under territory fleeces. This fleece may not be from a ewe that is registered or that is eligible for registration. The grower must be a member of the Colorado Wool Growers Association. This trophy must be won three times to become the permanent possession of the exhibitor.

Example 22-1 (*Continued*)

FARM FLOCK FLEECES

Ewe fleeces from flocks of sheep run under farm flock conditions shall be eligible to compete in this division.

Class		1	2	3	4	5	6	7	8	9
W-11	64's, 70's, 80's (Fine) Combing	$8	$6	$4	$3	$2	$1			
W-12	60's, 62's (½ Blood) Combing	8	6	4	3	2	1			
W-13	58's (⅜ Blood) Combing	8	6	4	3	2	1			
W-14	56's (⅜ Blood) Combing	8	6	4	3	2	1			
W-15	54's (¼ Blood) Combing	8	6	4	3	2	1			
W-16	50's (¼ Blood) Combing	8	6	4	3	2	1			
W-17	46's, 48's (Low ¼ Blood) Combing	8	6	4	3	2	1			
	Champion Farm Flock Fleece					Ribbon				
	Reserve Champion Farm Flock Fleece					Ribbon				

BREED FLEECES

Only fleeces from purebred sheep of the following breeds shall be eligible to compete in this division.

Class		1	2	3	4	5	6	7	8	9
W-18	Rambouillet Ram Fleece	$8	$6	$4	$3	$2	$1			
W-19	Rambouillet Ewe Fleece	8	6	4	3	2	1			
W-20	Corriedale Ram Fleece	8	6	4	3	2	1			
W-21	Corriedale Ewe Fleece	8	6	4	3	2	1			
W-22	Columbia Ram Fleece	8	6	4	3	2	1			
W-23	Columbia Ewe Fleece	8	6	4	3	2	1			
W-24	*Hampshire, Suffolk, Southdown Ram Fleece	8	6	4	3	2	1			
W-25	*Hampshire, Suffolk, Southdown Ewe Fleece	8	6	4	3	2	1			
W-26	Targhee Ram Fleece	8	6	4	3	2	1			
W-27	Targhee Ewe Fleece	8	6	4	3	2	1			
W-28	Debouillet Ram Fleece	8	6	4	3	2	1			
W-29	Debouillet Ewe Fleece	8	6	4	3	2	1			
	Champion Breed Fleece					Ribbon				
	Reserve Champion Breed Fleece					Ribbon				

*As one breed.

All fleeces being awarded a first prize in the Open and Junior Show may compete for the Champion, Grand Champion and Reserve Grand Champion awards. The second prize fleece in the class from which the champion fleece is taken shall be eligible to compete for the Reserve Grand Champion ribbon.

Grand Champion Fleece — Trophy.

Reserve Grand Champion Fleece — Ribbon.

The Grand Champion Fleece trophy is sponsored by North Central Wool Marketing Corp., Minneapolis, Minn., and Murphy Wool Co., Inc. Harry J. Murphy, Fieldman, Windsor, Co.

Example 22-1 (*Continued*)

BREEDER EXHIBIT OF FIVE FLEECES

W-30 Exhibit of Five Fleeces. This class shall consist of four female fleeces and one ram fleece. These fleeces must be shorn from sheep bred and raised by exhibitor with the exception of the ram fleece, which must have been owned by the exhibitor. Fleeces in this class will be judged on individual merit and on uniformity of type, staple length and character of the combined exhibit. (Fleeces showing in previous classes are eligible to show in this class.)

    1st — $15   2nd — $12   3rd — $10   4th — $8

AMERICAN SHEEP PRODUCERS COUNCIL TROPHY

The American Sheep Producers Council, Denver, Colorado, will offer a beautiful challenge trophy to the winner of the exhibit of five fleeces. This trophy must be won three times to become the permanent possession of the exhibitor.

JUNIOR WOOL SHOW
ENTRY FEE $.50 PER FLEECE

All fleeces entered in the Junior Show must have been shorn from ewes owned by the exhibitor and comply with the rules of the Open Wool Show except 5e. Fleeces entered in classes J-119 through J-123 must have been shorn from registered ewes of the respective breeds.

Class		1	2	3	4	5	6	7	8	9
J-115	Fine Combing	$5	$4	$3	$2	$2	$2	Rib.	Rib.	
J-116	½ Blood Combing	5	4	3	2	2	2	Rib.	Rib.	
J-117	⅜ Blood Combing	5	4	3	2	2	2	Rib.	Rib.	
J-118	¼ Blood and Low ¼ Blood Combing	5	4	3	2	2	2˙	Rib.	Rib.	
J-119	Rambouillet Ewe Fleece	5	4	3	2	2	2	Rib.	Rib.	
J-120	Targhee Ewe Fleece	5	4	3	2	2	2	Rib.	Rib.	
J-121	Columbia Ewe Fleece	5	4	3	2	2	2	Rib.	Rib.	
J-122	Corriedale Ewe Fleece	5	4	3	2	2	2	Rib.	Rib.	
J-123	Downs breed Ewe Fleece	5	4	3	2	2	2	Rib.	Rib.	
	Champion Junior Fleece							Ribbon		
	Reserve Champion Junior Fleece							Ribbon		

Murphy Wool Co., Windsor Co. presents a Trophy to the exhibitor of the Grand Champion Junior Fleece.

Example 22–1 (*Continued*)

Figure 22-2 Fleeces on display at the National Western Stock Show (A) and Wyoming State Fair (B). (University of Wyoming photos)

ment program that minimizes contamination of all the fleeces in a flock. These procedures are followed in many flocks, and a valid assumption is that most of the best fleeces grown in an area are not shown.

The best time to make the initial selection of potential show fleeces is just prior to shearing. Identify those sheep with the longest, densest, most uniform fleeces of the desired grade (Figure 22-3). Pick several "extras" at this point because some

Figure 22-3 Checking a fleece for length and style. (University of Wyoming photo)

that are attractive on the sheep do not weigh as much or tie up as well as anticipated. Be familiar with the classes available in the show to be entered. In many shows, each exhibitor is allowed to enter up to two fleeces per class in several classes. For example, purebred breeders can enter fleeces in both the ewe and ram classes for the breed or breeds they raise. Also, because most purebred flocks are managed under farm flock conditions, fleeces grown by these sheep are usually eligible for entry by appropriate grade in commercial farm flock classes (Example 22-1).

Sheep selected as carrying potential show fleeces should be sheared separately from the bulk of the flock. Extra care should be taken at shearing to minimize second cuts and breaking of the fleece. It is easiest to tie the fleece immediately after it is shorn. However, proper tieing for show requires extra time, care, and patience, and other duties may not permit this at shearing. Well-shorn fleeces can be placed carefully in large, plastic trash bags and tied later. For purebred fleeces, an identification tag with the sheep's number should be included in the sack.

Preparing a Show Fleece

Preparing a show fleece in an attractive manner is a relatively simple matter. Spread the fleece on a large table or clean floor with the skin side down. Be careful not to pull or open its natural continuity as it is handled. Next, compact the fleece to make it appear as if it were a pelt with the wool still attached to the skin. Remove dung locks, urine stains, burrs, and other obvious foreign matter. The belly and neck pieces, which often contain excess vegetable matter, are placed aside for including at the center of the fleece as it is rolled.

To roll the fleece, begin by folding about 8 to 10 in. from each side toward the center (back). Then fold the head-and-neck end over once and place the belly on top. Fold the balance of the fleece toward the shoulder (Figure 22-4). This leaves the britch and head parts on the inside with the more attractive shoulder, side, and back wool on the outside.

Tie loosely with paper twine to give an impression of bulkiness and high yield.

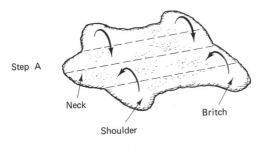

Step A

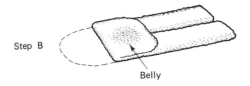

Step B

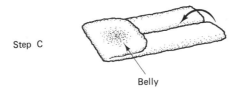

Step C

Figure 22-4 Schematic diagram of steps in folding a show fleece.

The first tie should be made to leave "quarters" of the fleece exposed for examination. The middle of this first string is slipped under the center of the fleece and tied over the top. The free ends are then wrapped around the remainder of the fleece and tied at a 90-degree angle to the first tie (Figure 22-5). One string may be adequate for the fine wools, but the coarser grades usually require a second string either once around the perimeter or placed to divide the exposed quarters into eighths (Figure 22-6). The fleece can be made to lie flat by sliding the wool under the strings toward the perimeter. Identify as to appropriate class and store in a tied plastic bag so the fleece will not dry out. The tendency for fleeces to change from a creamy white to various shades of yellow while stored can be reduced by keeping them as cool as possible during this period.

WOOL JUDGING

The purpose of wool judging is to train people in the art of subjective evaluation of raw wool. Competitive events (contests) provide motivation for practice.

Wool Judging

Figure 22-5 Tying a show fleece with one paper string. (A) Step 1. (B) Step 2. (University of Wyoming photos)

Youth Programs

Wool-judging programs for youth are supported in several states by the State Agricultural Extension Service, County Extension Agents, 4-H Leaders, and parents. 4-H wool-judging programs usually include two parts: grading fleeces and placing classes.

Grading fleeces. This part requires estimating the three most important value-determining factors: grade, length, and yield. At the 4-H level, the blood system is used for describing *grade*. Although no longer used in wool marketing, this system specifies only six grade categories and it is thought more achievable by younger people than the 16 categories included in the spinning count system, and therefore is not so challenging that grading becomes discouraging. Learning to classify wools of different fiber diameters into the relatively broad blood grades is considered an excellent first step in developing skill in grading wool visually.

A useful aid to teachers working with 4-H youth is the *grade set* developed at

Figure 22-6 Two methods for tying a show fleece using two paper strings. (A) Second string at perimeter. (B) Second string tied to permit one-eighths accessible for evaluation.

the University of Wyoming (Figure 22-7). These sets are prepared by selecting and mounting on black paper the locks of wool that show typical crimp formation for each grade.

Another useful teaching aid is a USDA grade set, which can be obtained by contacting the U.S. Department of Agriculture, 711 "0" Street, Greeley, Colorado 80631. The samples in this set (Figure 22-8) have been measured in microns and are provided in spinning count categories.

A good first step in teaching visual wool grading is to note the general relationship between crimp frequency and grade. Finer grades tend to have more crimps per unit of length than the coarser grades. Professional wool graders know that all grades may have nontypical crimp frequencies, but grading by crimp is a good way for beginners to start. As their skill in crimp grading progresses, they can be taught to open locks and evaluate relative fiber thickness to increase their accuracy.

A readily apparent problem that arises in visual grading of individual fleeces is that they vary in grade. All fleeces tend to be finer on the shoulder than the britch,

Wool Judging

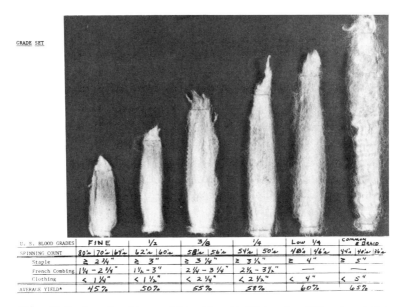

Figure 22-7 A grade set is a useful teaching aid. (University of Wyoming photo)

Figure 22-8 A USDA grade set. (University of Wyoming photo)

and side wool is usually intermediate. The grade of the side wool in a fleece normally represents the average grade of the fleece. However, when removed from the sheep and tied in a bundle, it is difficult to distinguish side wool from the other parts of the fleece. Back wool is similar in grade to side wool and can be identified by the "dirt line." The penetration of soils and small bits of vegetable matter is normally deeper and heavier in wool grown on the backs of sheep than from any other body area. When fleeces appear variable in grade, wool judgers are urged to locate a lock of back wool and estimate the average grade from it.

The second value-determining character of grease wool that is estimated is *staple length*. The limits for lengths, which are described as staple (longest), French combing (intermediate), and clothing (shortest), vary slightly by states (note length discussion, Chapter 18). Teachers should consult their Agricultural Extension Service for length specifications to be used in their county and state contests.

Although length classifications are quite specific, contestants are not permitted to use rulers. Wool judgers should be taught to measure length by placing locks (unstretched) along a finger or thumb that they have measured previously (Figure 22-9). Lock strength is an important consideration when describing staple length. If a fleece contains a *break* (see Chapter 18), the longest end after the break is used to classify length.

The third factor included in fleece grading at the 4-H level is *yield*. In some state contests, participants are required to estimate an exact yield. In other states and the National 4-H Wool Judging Contest, an estimate of high, medium, or low yield is required. Again, as for length, check the rules before beginning a training program.

Regardless of system of description, yield is difficult to estimate because of varying amounts of nonwool components (Chapter 18). As a general rule, coarse fleeces tend to yield more than fine fleeces (Table 22-1) and long-stapled fleeces tend to yield more than shorter fleeces of the same grade.

Yield is a weight-volume relationship (Figure 22-10). A good method is to lift a fleece gently with both hands. If a fleece feels heavier than its size (volume) would suggest, it will probably yield less than average. If it feels lighter than expected, it will probably yield more than average. With some practice, relative accuracy in estimating yield can be achieved.

TABLE 22-1 AVERAGE AND EXPECTED RANGE OF YIELDS FOR FLEECES OF DIFFERENT GRADES

Grade	Average Yield (%)	Range (%)[a]
Fine (64's and finer)	45	30–60
1/2 Blood (60's–62's)	50	35–65
3/8 Blood (56's–58's)	55	40–70
1/4 Blood (50's–54's)	58	43–73
Low 1/4 Blood (46's–48's)	60	45–75

[a] A range is ±15% in yield due to relative amounts of nonwool components is typical.

Wool Judging

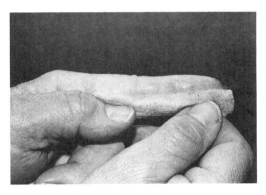

Figure 22-9 Measure the length of a finger (A) for use as a handy ruler (B) for estimating length of unstretched locks of wool. (Photo A from Pexton, Johnson and Craddock[3] with permission. Photo B University of Wyoming photo)

Placing classes. A second part of a wool-judging program (or contest) requires ranking classes of four fleeces. The two types of classes are "breed" and "commercial." Some states and the National 4-H Wool Judging Contest do not include breed classes.

Breed fleeces are ranked with genetic qualities such as uniformity of grade and breed type in mind. In this case, environmental factors such as breaks and vegetable matter are ignored. Grades of wool that are appropriate for fleeces usually used for breed classes are listed in Table 22-2.

Commercial fleeces are compared in terms of their relative processing value to

TABLE 22-2 ACCEPTABLE AND PREFERRED GRADES OF BREED FLEECES NORMALLY USED IN WOOL JUDGING

Breed	Acceptable Grades	Preferred Grades
Rambouillet	60's–80's	64's and finer
Targhee	58's–64's	60's–62's
Columbia	50's–60's	56's–58's
Corriedale	50's–60's	56's–58's

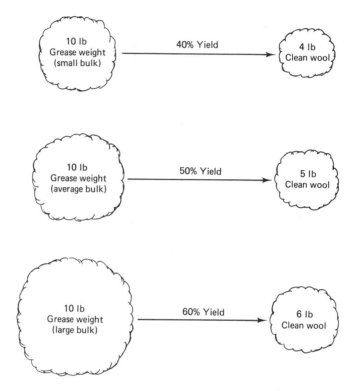

Figure 22-10 The clean content of fleeces of similar weight but different bulkiness can vary greatly.

both producers and manufacturers, so breaks (which result in wastiness), vegetable matter content, and color are considered important.

For both types of classes, *quantity* is the major factor. Participants should be encouraged to always think in terms of the "clean wool content" of fleeces as it is directly related to value. Length and appropriate grade are also extremely important considerations when placing both types of classes.

At the 4-H level, most states and the national contest use a reasons score sheet (Figure 22-11) rather than oral reasons. This procedure requires comparing fleeces in each class in terms of several specific descriptive factors.

University Programs

At present, there are only two university-level wool-judging contests held annually in conjunction with the National Western Stock Show in Denver, Colorado, and the Houston Livestock Show in Houston, Texas. They differ from 4-H level contests only in degree of relative difficulty. The fleece-grading aspect requires grade estimates in spinning counts rather than blood grades, along with estimates of length,

WOOL PLACING AND REASONS SHEET

CONTESTANT NAME OR NO.: _____

COUNTY: _____

CLASS NO.: _____

CLASS NAME: _____

INSTRUCTIONS FOR REASONS:
List the number of the fleece which matches the fleece description.

Fleece Description	Number
LONGEST STAPLE FLEECE	
SHORTEST STAPLE FLEECE	
FLEECE MOST UNIFORM IN STAPLE LENGTH	
FLEECE MOST UNIFORM IN FINENESS (FIBER DIAMETER)	
COARSEST FLEECE IN CLASS	
HEAVIEST GREASE FLEECE WEIGHT	
LIGHTEST GREASE FLEECE WEIGHT	
FLEECE WITH MOST POUNDS OF CLEAN WOOL	
FLEECE WITH LEAST POUNDS OF CLEAN WOOL	
HIGHEST YIELDING FLEECE	
LOWEST YIELDING FLEECE	
FLEECE WITH THE MOST CHARACTER (COLOR, CRIMP, & CONDITION)	
FLEECE WITH THE MOST VEGETABLE MATTER	
FLEECE WITH THE MOST STAINED WOOL	
FLEECE WITH THE LEAST FIBER STRENGTH	
FLEECE WITH NO COLORED FIBERS	

Reasons scoring: 50 total points possible
deduct 3 points for each incorrect answer

REASONS SCORE _____

	Circle Your Placing	Official Score
A	1 2 3 4	
B	1 2 4 3	
C	1 3 2 4	
D	1 3 4 2	
E	1 4 2 3	
F	1 4 3 2	
G	2 1 3 4	
H	2 1 4 3	
I	2 3 1 4	
J	2 3 4 1	
K	2 4 1 3	
L	2 4 3 1	
M	3 1 2 4	
N	3 1 4 2	
O	3 2 1 4	
P	3 2 4 1	
Q	3 4 1 2	
R	3 4 2 1	
S	4 1 2 3	
T	4 1 3 2	
U	4 2 1 3	
V	4 2 3 1	
W	4 3 1 2	
X	4 3 2 1	

Figure 22-11 A typical placings and reasons sheet used in 4-H wool-judging contests.

yield, uniformity of grade, purity, and character. Placing classes in university-level contests includes both breed and commercial classes and oral reasons are required.

Oral reasons in wool-judging contests are presented in the same general format used for defending placings in livestock contests, but using terms unique to wool. As an aid for beginners, a sample set of reasons for placing a commercial class of fleeces is presented as follows:

> I placed this class of 64's and finer commercial fleeces 1-2-3-4. In placing this class, I found two large, bulky, long-stapled fleeces in 1 and 2 for my top pair. I placed 3, a fleece which lacked the volume of my top two, third, leaving 4, the lightest fleece with the least fiber strength at the bottom.
>
> I placed 1 over 2 in a close placing. They were similar in weight, staple length, and strength, and I grant that 2 was a brighter, softer-handling fleece making it more attractive. However, I placed 1 over 2 because it was more uniform in fiber diameter than 2 and because of its higher yield. It contained more clean wool, making it a more valuable fleece for both the producer and manufacturer.
>
> In my middle pair, I placed 2 easily over 3, granting that 3 was at least one spinning count finer and more uniform in its grade than 2. However, I placed 2 over 3 because it was a brighter fleece with a longer staple than 3. It was also a heavier, higher-yielding fleece with distinctly more clean wool in it than 3.
>
> I placed 3 over 4 in my bottom pair, granting that 4 was longer in its original staple and contained less vegetable matter than 3. It was also a brighter fleece with a more uniform distinct crimp formation, giving it more eye appeal. However, I placed 3 over 4 because it was definitely stronger in its staple, and although similar to 4 in its yield, 3 was heavier, indicating that it did contain more clean wool.
>
> I placed 4 at the bottom of this class. It was the fleece with the most character, particularly from the standpoint of its crimp formation, and it carried the least vegetable matter of any fleece in the class. It was also the smallest fleece and because of its definite break, which will result in excess noilage during combing, was the most wastey fleece in the class.
>
> For these reasons, I placed this class of 64's and finer commercial fleeces 1-2-3-4.

Conducting Wool-judging Contests

A great deal of detailed work is involved by the person(s) responsible for conducting wool-judging contests, regardless of level. The following outline is provided as a reminder of some of the concerns normally considered the responsibility of contest supervisors.

1. *Precontest*
   a. Publicity, including date, time, location, eligibility, and entry procedures
   b. Site and physical arrangements
   c. Fleeces for grading and classes
   d. Judges
   e. Awards and their sponsors

2. *Contest*
   a. Contestants' cards and forms
   b. Procedure for handling groups
   c. Group leaders
   d. Timers
   e. Scoring and tabulation
3. *Postcontest*
   a. Presentations of official placings
   b. Presentation of awards
   c. Photographic record of winners and/or award donors
   d. News releases

REFERENCES CITED

1. American Sheep Producers Council, 200 Clayton Street, Denver, Colo. 80206.
2. National Western Stock Show, 1325 E. 46th Avenue, Denver, Colo. 80216.
3. Pexton, Ed., LeRoy Johnson, and Frank Craddock. 1980. 4-H Wool Judging. University of Wyoming Agricultural Extension Service, Laramie, No. 12500h-80.

SUGGESTED READING

4-H Wool Judging, 12500h-80. University of Wyoming Agricultural Extension Service, University of Wyoming, Laramie.

Judging Wool and Mohair, A35-6-058. Texas Agricultural Extension Service, Texas A & M University System, College Station.

Judging and Classification of Wool. Circular 285, Cooperative Extension Service, Montana State University, Bozeman.

Glossary of Wool Terms

Apparel wool: Wool suitable for manufacture into apparel fabrics.

Baby combing wool: Intermediate lengths of 58/56's or 54/50's are sometimes referred to as baby combing rather than French combing.

Bale: A highly compressed package of either grease or scoured wool. A method of increasing weight and reducing volume to reduce shipping costs.

Belly wool: Wool which grows on the belly of a sheep. It is usually shorter, coarser, weaker, and less dense than the rest of the fleece and characterized by a typical prominent crimp.

Belly wool fault: A defect, commonly referred to as "belly wool" where a typical belly wool style is found growing on body areas other than the belly (usually the side), resulting in reduced clean wool production.

Black wool: Any wool that is black, brown, or gray. Black-wooled sheep are used in large flocks as "markers." Fibers from these fleeces can contaminate white fleeces at shearing time if not handled properly. See *Colored wool*.

Blending: Mixing together of various grades and/or lengths of wool in either the raw or semiprocessed state to obtain a specific kind of yarn.

Blood grade: Originally used to describe the proportion of Merino (finewool) breeding represented in a fleece. At present, a U.S. grading term describing the relative fineness of wool.

Bradford system: A method of producing top using the English Noble comb.

Break: A distinct weak place along locks in fleeces due to restricted diameter in wool fibers, which is usually due to illness and fever or severe stress.

Bright wool: Grease wool that is nearly white, showing very little yellow color due to excessive yolk.

Britch or breech wool: Wool from the hindquarters of sheep, usually the coarsest wool in the fleece.

Buck fleeces: Fleeces shorn from mature rams, usually longer and coarser than ewe wool from sheep of the same breed and with a characteristic ram odor.

Bulk grade: The majority grade in a lot of original bag wool.

Burry: Wool with burrs entangled in the fibers. Undesirable because of waste and expense involved in removing them.

Carbonizing: A chemical process used to remove excessive vegetable matter (VM), such as burrs, from wool. The VM is converted to a black, brittle hydrocellulose by the action of certain mineral acids.

Carding wool: Short-stapled wool, suitable only for manufacture of woolen yarn. See *Clothing wool.*

Carpet wool: Coarse wool, often from unimproved sheep, used in the manufacture of carpets.

Character: An ambiguous term usually used to describe the overall attractiveness of wool based on the relative desirability of its color, crimp, and handle. It may also be used to indicate clarity of crimp or refer to grade.

Classing: Separating fleeces into various classes based largely on grade and staple length. A term used in wool-producing countries other than the United States.

Clean basis: A marketing term referring to the value of clean fibers present in grease wool.

Clean content: (Clean wool fibers present.) The portion of grease wool that is free of vegetable matter (VM) and contains by weight 12% moisture, 1.5% grease and 0.5% ash (mineral matter).

Clip: One or more fleeces, one season's production of wool from a common source (ranch, county, state, etc.).

Clothing wool: Wool fibers that are too short to comb, used in the manufacture of woolen yarn; frequently referred to as carding wool.

Color defect: Colors in wool that cannot be removed by scouring, such as stains from manure or excessive yolk.

Colored wool: A term used to describe fleeces of sheep bred specifically to produce naturally colored fibers primarily for use by handcrafters.

Combing: A process in which long fibers are laid relatively parallel and short fibers (noil) are removed.

Comeback: In the wool trade, refers to all-white wool grown by sheep produced by mating crossbred types back to the finewool parent. Typical grades are 60's to 62's.

Condition: An ambiguous term; sometimes refers to the relative care in preparation or degree of grease, dirt, and/or moisture in raw wool or degree of moisture in scoured wool.

Coring: A method of sampling bales, bags, or fleeces of wool. The cores (samples) are then processed to estimate the clean wool content (yield), diameter, and other factors, such as type and quantity of vegetable matter.

Cotted fleeces: Fleeces in which the fibers are excessively matted or tangled.

Crimp: The natural waviness of wool fibers.

Crossbred: In the wool trade, refers to all-white wool grown by sheep or breeds resulting from the crossing of finewool and longwool breeds. Typical grades are in the medium (50's to 60's) range.

Crutchings: Wool removed from the area around the dock and/or udder of sheep.

Defective: Wool with a fault that reduces its value, such as damage by fire, water, or moths. Burry wool is often called defective.

Density: The number of fibers grown on a given area of the skin of sheep: the more fibers, the greater the density.

Domestic wool: All wools grown in the United States, as opposed to foreign wools. Also, wools grown in the farm areas, generally east of the Missouri River; more correctly called *fleece wool*.

Down wool: Medium wool from the black-faced (meat) breeds of British origin: generally harsher than wool from white-faced breeds of similar grades.

Felting: The property of wool fibers to interlock when agitated in warm, moist conditions. This property results in wool fabrics that *shrink* when washed in machines if not treated to resist felting.

Fine: An American grade of wool originally applied to fleeces of pure Merino breeding; presently used to describe fleeces of 64's or finer spinning counts.

Fleece: All the wool shorn from one sheep at one time.

Fleece wool: A term typically used to describe wool grown east of the Missouri River. These wools are frequently incorrectly described as domestic wools.

French combing: An intermediate-length grade describing wool that is long enough to comb on the French comb but too short to comb on the Noble comb.

Frowsy wool: A dry, harsh, lifeless appearing wool, usually caused by excessive weathering or exposure to highly alkaline soils.

Grade: An American term used to describe the relative fineness of wool.

Grading: Separating whole fleeces according to fineness and length of staple.

Grease wool: Wool as it is removed from live sheep, prior to scouring.

Half-blood: An American grade, originally applied to wool from sheep of half finewool breeding; presently used to describe wool of 60's to 62's spinning counts.

Glossary of Wool Terms

Harsh: Wool that lacks softness, feels wiry. Wool from the British meat (Down) breeds is typical.

Handle: Refers to the degree of softness when judged by feel.

Heavy wool: Wool with a high proportion of impurities, especially sand and dirt. A low yielding wool.

In the grease: Grease wool.

Kemp: Opaque, highly medulated wool fibers that shed periodically, considered a serious defect.

Keratin: The type of protein found in hair, wool, hooves, feathers, and horns.

Lamb's wool: Wool shorn from lambs; usually finer, shorter, and softer than wool from the same breeds of mature sheep.

Lanolin: Purified wool grease used extensively as an additive in cosmetics and skin medications.

Lock: A bundle of wool fibers that have been or are growing side by side. See *Staple*.

Lofty wool: See *Harsh*.

Longwool: Long, coarse, lustrous wool shorn from breeds such as the Lincoln, Cotswold, or Leicester.

Lot: A given quantity of wool that has some commonality, such as source, grade, or length.

Luster: A natural gloss or shine caused by light reflection, typical in mohair or wool from the longwool breeds.

Medium wools: The term includes high quarter-blood (54's), three-eighths blood (56's and 58's) and low half-blood (60's) wools. When used in reference to breed type, implies primarily wool from Down breeds. See *Crossbred* and *Comeback*.

Medulated fiber: Wool fibers that are hollow (have a medulla) rather than solid, as in true wool.

Micron: A trade term meaning micrometer (10^{-3} millimeters) used to describe the fineness of wool fibers.

Noil: Short fibers separated from long fibers in the combing process.

Off sorts: The less desirable parts of the fleece removed in skirting and sorting, such as the britch and belly.

Openwool: Fleeces that lack density when growing on the sheep. Coarse fleeces tend to be less dense or more open than fine fleeces.

Original bag: Wools that tend to be less uniform in grade and length than graded lots, but considered uniform enough to require minimal or no sorting before processing.

Pelt: The skin of sheep, including the wool.

Plain: Wool with little crimp.

Pulled wool: Wool pulled from the skin of slaughtered sheep.

Quality: An ambiguous term, which may include characteristics such as handle, color, or crimp; often refers to the relative fineness of wool.

Quarter-blood: An American grade of wool, originally applied to wool from sheep of one-fourth finewool breeding; presently refers to relatively coarse wool of 50's to 54's spinning counts.

Range wool: Wool shorn from sheep raised primarily on native ranges, generally in flocks of 600 or more ewes, often called Territory wool.

Raw wool: See *Grease wool*.

Recycled wool: A term used when labeling fabrics or garments. Either reprocessed or reused wool or a combination of them is included; adjustment in terminology required because of inability to differentiate origin.

Rejects: Fleeces that are rejected for normal processing due to severe faults such as cotted, tender, black, kempy, or excessively stained fleeces.

Reprocessed wool: Wool that has been woven, knitted, or felted into a wool fabric but has not been utilized by consumers. It is then reprocessed into a fibrous state and subsequent fabric. Generally, wool from remnants of the garment industry. See *Recycled wool*.

Reused wool: Similar to reprocessed wool except that the wool products have been used by consumers (rags). See *Recycled wool*.

Run-out: Fleeces with excessive variation in grade; usually much coarser in the britch than the rest of the fleece.

Scoured wool: Clean wool; wool that has been washed to remove grease, soil, and suint.

Scouring: The process of removing grease, soil, and suint, usually by washing in hot water and detergent. A mild alkali is added in some situations.

Second cuts: Short fibers resulting from careless shearing; double cutting.

Seedy: Wools containing seeds, usually from grasses and/or weeds, that require carbonizing for removal.

Shearing: Removing wool from sheep.

Shrink: The nonwool portion of grease wool, including grease, vegetable matter, dirt, suint, paints, insects, etc. See *Yield*.

Skirting: A procedure where the less valuable parts of fleeces, such as bellies, stained pieces, sweat locks, and neck wool, are removed from fleeces.

Sliver: A continuous strand of loosely assembled, untwisted fibers formed by a mechanical card or comb.

Sorting: Breaking a fleece into part or sorts based primarily on fineness and length.

Other criteria such as stain, strength, and vegetable matter may be considered; currently used interchangeably with extensive or maximum skirting.

Sound wool: Strong wool, free of breaks or tenderness.

Spinning count: See *Quality*. A term of British origin used to describe the fineness of wool fibers. The count (such as 64's, 62's, etc.) refers to the number of *hanks* (560-yard lengths) of yarn that can be spun to a minimum thickness from one pound of top.

Stained wool: Wool permanently discolored from urine, manure, some anthelmintics, etc.

Staple: A term used two ways in wool production and marketing: 1. A synonym for lock. 2. A length description used in American wool marketing to describe the longest of the length classifications for each grade. Originally, wool long enough to be combed on British Noble combs.

Strong: Wool free of breaks or tenderness; also used to indicate relatively coarse grades.

Suint: The water-soluble component (largely potassium and organic salts) of grease wool produced by the sudoriferous (sweat) glands in the skin of sheep.

Tags: Dungy locks of wool, described as "dags" in international wool marketing.

Tender fleeces: Fibers in these fleeces lack tensile strength. Weak areas may be found distributed along locks in a general rather than distinct section, as in breaks.

Territory wool: Wool grown in the range areas of Washington, Oregon, the intermountain states including Arizona, New Mexico, and parts of the Dakotas, Nebraska, Kansas, and Oklahoma. Generally, the area west of the Missouri River, excluding Texas and California.

Three-eighths Blood: An American grade originally applied to wool from sheep of three-eighths finewool breeding; presently used to describe wool of 56's and 58's spinning counts.

Top: A continuous untwisted strand of fibers from which the short fibers (noil) have been removed by combing.

Virgin wool: Wool fibers that have been taken directly from sheep (shorn or pulled) and processed into yarn and/or fabric. It has not been reclaimed from any woven, knitted, or felted product. See *Recycled wool*.

Waste: Wool fibers, usually short, lost during carding and combing. See *Noil*.

Wool: The outer covering of sheep.

Wool grease: A complex mixture of esters secreted by the sebaceous (wax) glands in the skin of sheep.

Woolen yarn: Yarns spun from fibers that have been carded but not combed. The fibers are randomly arranged, resulting in a relatively rough yarn.

Worsted yarn: Yarns spun from fibers that have been carded and combed. The fibers are relatively parallel giving a smooth yarn.

Yield: The proportion of clean wool (at standard conditions) present in a given amount of grease wool.

Yolk: A combination of sweat salts and wool grease (waxes) deposited on wool from the sweat and wax glands in the skin of sheep.

Index

A

Abnormalities, death of lambs from, 210, 213
Accelerated lambing programs, 195–97
Acetate, 299
Additive gene action, 112, 114
Adoption, 223
Aerobic packaging, 318
Aftermath, 18–19
Age:
    carcass composition and, 291
    indications of, 8–9
    lamb production and, 100–103
    palatability of meat and, 308
    wool production and, 100–101, 398, 400–402, 414
Age distribution, 185
Agricultural Statistics, 322
Airflow instruments, 388, 389
Alfalfa, 156, 158, 165, 170, 173, 241, 299, 312
All-in-one instrument, 235
Alternating backcross system, 136
American Sheep Producers Council (ASPC), 55, 58–59, 340, 344, 350, 420
American Society for Testing Materials (ASTM), 377, 379, 385
Amino acids, 369, 413
Anaerobic packaging, 318
Animal unit mouths (AUMs), 55–57
Antibiotics, 234
Arrector muscle, 359
Artificial insemination, 95, 193–94
Artificial rearing, 28, 223, 232–33
Ash content, 291, 292
Auction selling, 259–62
Australian Wool Corporation, 340

B

Back wool, 432
Barley, 160–65, 174, 175
Barns, 39–41
Basal cells, 355
Bay wool, 329, 337
Beans, 165
Beets, 165, 175
Belly wool, 122
Bermuda grass hay, 165
Birth, difficult, 214–21
Birth type, 105–6, 130
Blood by-product, 267, 282
Blood grade system, 324–26
Bluegrass, 165, 166, 170
Bluestem, 166, 170
Body surface, 297, 405
Body temperature, 7
Body weight, 8, 147
    at birth, 7
    carcass composition and, 291–95
    palatability of meat and, 308–9
Bones, 5–7
Bradford grade system, 325
Branding, 224
Branding fluids, 335
Break joints, 6, 7
Breaks, 332
Breed adaptability, 32–33
Breed fleeces, 433
Breeding:
    for improved wool production, 396–412
    lambing time, 206–25
    management during gestation, 199–205
    management during lactation, 227–36
    postweaning management, 239–43
    reproductive physiology, 85–106
    weaning, 237–39
    year-round management, 180–97
    (*see also* Inheritance)
Breeding plans, 407–12
Breeding systems, 133–40
Breeding time, 97–99, 181–84
Breeds:
    carcass composition and, 300–305

Breeds (*cont.*)
    crossbreeding and, 134, 137
    inheritance in wool production and, 400, 402
    palatability of meat and, 316–17
    reproductive rates and, 104–5
    types of, 62–83
    (*see also names of breeds*)
Brome, 166, 170
Browse, 19, 187
Budgets, 35, 36
Buildings, 39–41
Bulbourethral glands, 91, 93
Bureau of Land Management (BLM), 54–57
By-products:
    crop, 173
    sheep, 281–83

C

Calcium, 146–60, 165–70
Calories, 286
Capital, 31
Carbonizing, 335, 371
Carcass composition, 289–305
Carcass shrinkage, 293
Carcass traits, 115, 117, 123, 130–31
Carcass value, 266–9
Carotene, 160
Carpet wool, 327
Carrot roots, 166
Castration, 112, 234–35
Centralized cutting, 278–79
Cereal grains, 19
    (*see also names of cereal grains*)
Cervix, 86, 88
Character, 333
Chemical properties of wool, 368–73
Cheviot sheep, 78–79, 104
Chilling:
    of carcasses, 309–10
    death of lambs from, 221
Chlorine, 146, 155
Cholesterol, 286
Chromosomes, 108, 109
Citrus pulp, 166
Clean fleece weight, 397, 399, 403
Clean price, 323, 329, 337
Clean wool, 335, 378–84, 395–96, 403
Climate, 21
Clitoris, 86, 88
Clover hay, 166
Club lambs, 28, 30
Coarse-wooled breeds, 81–83, 137

Cobalt, 146, 155
Colored breeds, 83
Color inheritance, 110–12
Color of wool, 121, 332–33, 392
Columbia sheep, 66–68, 137, 302, 303, 327, 357
Comeback, 327
Commercial fleeces, 433–34
Computerized exchanges, 260–61
Concentrates, 31, 32, 174–76, 241
Conformation, 304–5
Consignment, 339
Contaminants, 334–35, 350
Copper, 146, 155
Corn, 156, 158, 166, 174, 175, 241, 243, 299, 312, 313
Corpus luteum, 86, 87, 94, 95
Corrals (*see* Pens)
Corriedale sheep, 68–69, 137, 301–3, 327, 357, 361
Cortex, 367–68
Cortisol, 97
Cotswold sheep, 82, 327
Cotton and Wool, Outlook and Situation, 322
Cottonseed, 166, 167, 175, 357
Coyotes, 222
Creep feeding, 233–34
Crimp, 333, 363–64, 389
Crisscross system, 136
Crop aftermath, 173
Crossbred wool, 327
Crossbreeding, 134–40
Crossbreeding plans, 407–12
Cryptorchidism, 120
Cuts of meat, 276–84, 290, 293, 303
Cystine, 369

D

Dealers, 251, 259
Death of lambs, causes of, 210–24
Debouillet sheep, 66, 357
Defects, 117–22
Deformities, death of lambs from, 210, 231
Density, fleece, 385–86
Dermis, 355
Development, 7–9, 95–97
Diet (*see* Nutrition)
Digestive system, 3
Direct sales, 250–51, 255
Diseases, 213–14, 415
Disulfide bonds, 369, 370
Docking, 234–36
Dogs, sheep, 33–34
Domestication, 3, 354–57
Dominance, 114

Doppler device, intrarectal, 203
Dorset sheep, 75–76, 104, 112, 137, 139, 197, 301–3, 361
Dressing percentage, 255–56
Drop bands, 206–8
Duties, 340, 341
Dystocia, 214–21

E

Ectoderm, 355
Elastrators, 235
Electrical stimulation, 309
Electric fences, 38–39
Electronic marketing, 260–62
Emasculators, 235
Emasculatone, 93
Embryonic development, 95–97
Endoderm, 355
End-to-end variation, 363
Energy, 145–46, 148–54, 165–69, 413
Enterotoxemia, 213–14, 236
Entropion, 121
Environmental effects on wool production, 412–16
Epidermis, 355
Epididymis, 90–93
Epiphyseal cartilages, 7
Epistasis, 114
Equipment, 37–51
Estrogen, 86, 87, 97
Estrus, 94–95, 188–91
Ether extract content, 291, 292
Evolution, 354–57
Ewes:
    desertion of lambs by, 222–23
    estrus in, 94–95, 188–91
    feeding during lactation, 224–25, 230–31
    gestation (*see* Gestation)
    injury to lambs by, 222
    ova transfer in, 95
    purchasing, 34
    replacement, 263–64
    reproductive system of, 86–89
    slaughter, 262–63
    (*see also* Sheep; Sheep production)

F

Face cover, 405–6
Farm flocks, 24–25, 181
Fat content of lambs:
    amount of, 289–305
    human nutrition and, 286
    palatability of meat and, 312–16
Fat in lamb diet, 299

Fat-rumped sheep, 3
Fat-tailed sheep, 3
Fat thickness, 272
Fatty acids, 314–15
Feeder lambs, 34, 40–44, 240–43, 247–51
Feedlot operations, 26–28, 250–51
Feedlots, 240
Feeds, 31–32
  composition of, 164–71
  efficiency of, 294
  for lambs, 240–43
  providing, 171–78
  sources of, 18–19
Female reproductive system, 85–89
Fencing, 37–39
Fertility, 97–106, 114
Fertilization, 94, 95
Fetal growth and development, 96, 97
Fiber density, 397, 405
Fiber diameter, 330, 386–90, 397, 403–4
Fiber strength, 331–32, 391
Fiber structure and properties, 363–73
Fiber uniformity, 404–7
Fine wool, 323, 324, 326, 333
Fine-wooled sheep, 1, 3, 7, 32, 63–66, 316–17, 327
Finishing lambs, 240–43
Finnish Landrace sheep, 11, 13, 79–80, 103–5, 129, 136–40, 185, 197, 301
Fish and Wildlife Service, 222
Fixed supplement method of balancing rations, 157–58
Fleece traits, 115–17, 130
Fleece weight, 403
Fleece wools, 329, 337
Flock size, 181, 184
Fluorine, 146, 155
Flushing, 187–88
Follicle development, 355, 358–62
Follicle plug, 358
Follicle-stimulating hormone, 94
Food and Drug Administration (FDA), 269
Food Safety and Inspection Service, 269
Forage, 31, 32, 165–60, 174, 200–201, 241
Forbes, 19
Foreign influences, 59
Forest Service, 55–58
Forward contracting, 251
Freezing, 309–10, 317–18

French combing, 328
Funneling corrals, 44

G

Gelatin, 282–83
Generation length, average, 126–28
Genetics (*see* Inheritance)
Germinal cells, 355
Gestation, 95–97
  nutrition during, 147, 148, 153, 155, 157, 199–205
  pregnancy diagnosis, 201–4
  shearing during, 205
Gonadotropic hormones, 94
Government influences, 53–61, 340–44
Graafian follicles, 85–87
Graded wools, 329, 337
Grades:
  meat, 270–75
  wool, 323–27, 351–52, 429–34
Grade set, 429–31
Grama, 167, 170
Grasses, 165–73
  (*see also names of grasses*)
Grazed feeds, 31, 32
Grease, 336
Grease content, 405
Grease price, 329, 337
Grower's organizations, 61
Growth, 7–9, 255–59, 289–90
Growth stimulants, 243
Growth traits, 115, 116, 130

H

Hairiness, 122
Hampshire sheep, 73–74, 104, 110, 137, 173, 211, 302, 303
Handle, 333, 391–92
Handling facilities and equipment, 40–51
Harvested feeds, 31, 32
Harvesting wool, 344–52
Hay, 156, 158–69, 174, 241, 243
Head shape, 5
Health, 106, 185–86, 191, 192
Heart, 267, 282
Heating, 47–48
Hernia, 121
Heterosis, 134–40
Hidden shrink, 336
Hobby flocks, 29
Holding pens, 50
Hominy, 175
Horns, 112, 113, 121
Hospital pens, 48
Hot-carcass-weight basis, 256–59

Hothouse lambs, 28, 275
Housing, 39–41
Hydraulic sackers, 50–51
Hydrogen bonds, 369–70, 372
Hygroscopic property, 373

I

Identification of lambs, 224
Ilium, 5
Immunization, 214
Inbreeding, 133
Incentive programs, 341–44
Incisor teeth, 3, 8, 9
Inclement weather, death of lambs from, 221
Information sources, 35–37
Inheritance, 108–40
  breeding systems, 133–40
  correlations among traits, 122–23
  mechanism of, 108–9
  of qualititative traits, 110–14
  selection, improvement by, 123–33
  of traits influencing productivity, 114–22
  in wool production, 396–412
Inspection, 269–70
Intermuscular fat, 303
International Wool Textile Organization (IWTO) Specifications of Test Methods, 377
Intramuscular fat, 289, 291
Intrarectal Doppler technique, 203
Investment return rate, 20
Iodine, 146, 155
Iron, 146, 155, 286

J

Jaw defects, 118, 119
Jugs, 46–49, 206, 208, 209, 213, 227

K

Kemp fibers, 366, 367
Keratin, 368, 371
Kidney fat, 289–91, 301–2

L

Labor, 22, 33–34, 181
Lactation, 223–25, 227–36
Lamb crops, 13, 17, 18
Lamb feeding operations, 26–28
Lambing attendants, 208–9
Lambing facilities, 46–49

Lambing jugs, 46–49, 206, 208, 209, 213, 227
Lambing pens, 46–49, 227–28
Lambing percent, 105–6, 114, 124, 180
Lambing rate, 129–30, 137, 194
Lambing time, 206–25
Lamb production, 11–15
 income from, 322, 323
 (see also Sheep production)
Lambs:
 birthweight of, 7
 club, 28, 30
 feeder, 34, 40–44, 240–43, 247–51
 finishing, 240–43
 growth of, 7–9, 255–59, 289–90
 hothouse, 28, 275
 meat from (see Meat)
 replacement, 239, 263–64
 slaughter, 252–60
 (see also Sheep; Sheep production)
Lamb scours, 213
Land, 32, 53–58
Leanness, 302, 304
 (see also Fat content of lambs)
Leg bones, 6–7
Legumes, 173, 312–13
 (see also names of legumes)
Leicester sheep, 82, 327, 356, 357, 361
Lespedeza, 167, 170
Lethals, 118
Limestone, 167
Lincoln sheep, 81–82, 327, 357, 361
Linebreeding, 134
Linseed meal, 167, 175
Liver, 267, 282
Long-wooled sheep, 81–83, 327
Lots, 44–45
Luteinizing hormone, 94

M

Macrostructure, 363–64
Magnesium, 146, 155
Male reproductive system, 89–93
Mammary system, 88–89
Management calendar, 182–84
Management systems, 20, 22–30
Manganese, 146, 155
Marketing, 32–33, 245–65, 338–40
Marking harnesses, 192, 193, 202
Marrow fat, 289
Mastitis, 213
Meadow fescue hay, 167
Meat, 266–88
 carcass composition, 289–305
 palatability of, 307–19
Meat breeds, 104
Medium wool, 326
Medium-wooled breeds, 1, 66–81
Medulation, 392
Medulla, 366–67
Merino sheep, 9, 63, 104, 112, 197, 324, 326, 327, 357, 361
Mesoderm, 355
Microminerals, 146, 155
Micron grade system, 325
Microstructure, 364–70
Milk:
 cow's, 167, 312–13
 sheep's, 167
Milk replacers, 232
Milo, 174, 175
Mineral matter in wool, 385
Minerals, 146, 149–55, 177–78, 413
Mixing pens, 228
Moisture content of wool, 379–81, 385
Molasses, 168, 175, 176
Molybdenum, 146, 155
Montadale sheep, 71
Mouflon sheep, 1, 2
Multiple birth, 112, 129–30
Muscle development, 5
Mutton (see Meat)

N

National Research Council method of balancing rations, 160
National Wool Act of 1954, 341
National Wool Growers Association, 61
Native hay, 167
Navajo sheep, 137
Neck, 5
Needle and thread grass, 167
Nutrients, 144–47
Nutrition, 144–78
 balancing rations, 156–60
 at breeding time, 186–88
 carcass composition and, 298–300
 diet formulation, 160–64
 feeds (see Feeds)
 finishing lambs, 240–43
 during gestation, 199–201
 during lactation, 225, 230–31, 233–34
 at lambing time, 224–25
 nutrients, 144–47
 palatability of meat and, 312–16
 for replacements, 239
 requirements, 147–55
 weaning and, 237–39
 wool traits and, 412–15
Nutritional benefits of lamb and sheep meat, 284–88

O

Oats, 160–64, 167, 312
Objective measures, 398
Orchard grass, 167
Order buyers, 251, 259
Ova, 95, 108, 109
Ovaries, 85–87
Ova transfer, 95
Overall reproduction, 114, 115
Oviducts, 86, 87, 95
Ovis aries, 3
Ovis musimon, 1
Ovis vignei, 1
Ovulation, 87, 94
Ovulation rate, 99, 100, 114, 188
Oxford sheep, 76–77, 104

P

Packaging, 317–19
Packers, 254–56, 263
Panama sheep, 69–70
Papilla, 358–59
Parasites, 415
Pastime sheep farming, 28–29
Pasture lambing, 223
Pastures, 171–73
Pearson square method of balancing rations, 158–60
Pelleted feed, 234, 241, 243
Pelt removal, 254
Pelts, 267, 281–82
Pelvic fat, 289, 290, 301
Penis, 91, 93
Pens, 44–50, 227–28
Peptide bonds, 369
Percentage method of balancing rations, 156
Performance testing, 131–33
Phosphate, 168
Phosphorus, 146–60, 165–70, 187
Physiological state, wool traits and, 413–15
Pink eye, 214
Placenta, 97
Pneumonia, 213, 262
Political considerations, 53–61
Polled sheep, 3
Polwarth sheep, 361
Polyethylene twine, 335

Index

Polypay sheep, 80–81
Portable feeders, 41–43
Potassium, 146, 155
Prairie hay, 159, 167
Predation, 21, 222
Predator control, 60–61
Pregnancy (*see* Gestation)
Premium books, 420–25
Prepuce, 91, 93
Presale testing, 339
Preweaning mortality, 211
Pricing:
  meat, 266–69
  raw wool, 336–37
Primary follicles, 359–60
Private treaty, 338
Processing meat, 267, 276, 278–79, 281, 285
Producer-to-consumer transactions, 259–60
Progeny testing, 131–33
Progesterone, 87, 94–95, 97
Projected images of fibers, 387–88
Prolactin, 94, 97
Prolificacy, 114, 115
Prostate gland, 91, 93
Protein in lamb meat, 286, 291, 292
Protein in sheep nutrition, 145, 187
  common feeds, 165–70
  concentrates, 174–76
  daily requirements, 149–54
  fat content of carcass and, 298–300
  ration balancing and, 156–60
  wool traits and, 413
Puberty, 102–3
Public autions, 34
Public lands, 53–58
Public Rangelands Improvement Act of 1978, 55
Purebred flocks, 25–26, 264–65
Purity, 334

Q

Qualitative traits, 110–14
Quality grades, 270–71
Quality number, 325
Quantitative inheritance, 112, 114

R

Rambouillet sheep, 63–66, 104, 112, 113, 136, 137, 140 197, 211, 301–3, 324, 357
Ram flocks, 410

Rams:
  carcass composition of lambs, 295–98
  care of, 191–93
  palatability of lamb meat, 310–12
  purchasing, 35
  reproductive system of, 89–93
  slaughter, 262–63
  (*see also* Sheep; Sheep production)
Range flocks, 23–24, 181
Rape, 312
Rations, balancing, 156–60
Record keeping, 35, 36, 197, 206
Rectal-abdominal palpation, 202
Rectal prolapse, 121
Red clover, 166
Redtop, 168
Referee method, 387
Relative purity, 334
Rendering items, 267
Replacements, 239, 263–64
Reproductive physiology, 85–106
  (*see also* Gestation)
Reproductive rate, 20, 114–16, 123, 127, 180–97
Ribs, 5
Romney sheep, 82–83, 327, 356, 367
Root crops, 174
Rotational crosses, 136
Rye, 168, 174
Ryegrass, 168

S

Sacrum, 5
Sagebrush, 168, 170
Salt-bush, 168
Salt linkages, 369, 371
Sampling, 378–84
Sanitation, 213
Sarcosporidiosis, 262–63
Scouring, 371
Scrotal circumference, 191
Scrotum, 90–91
Scurs, 121
Sealed bids, 339
Sebaceous glands, 4, 359
Secondary follicles, 360–61
Secondary/primary (S/P) ratio, 361–62
Second cuts, 345
Selection, improvement by, 122–33
Selection differential, 125–30

Selenium, 146, 155
Self-feeders, 43–44
Sex:
  carcass composition and, 295–300
  inheritance of wool traits and, 398, 399
  palatability of meat and, 310–12
Shearing, 344–50
Shearing facilities, 49–50
Shearing floor, 50
Shearing trailers, 49–50
Sheep:
  age of (*see* Age)
  breeding (*see* Breeding)
  breeds of (*see* Breeds)
  characteristics of, 1–3
  classification of, 2
  compared with other ruminants, 3–4
  distribution in U.S., 9, 12
  evolution of, 354–57
  external parts of, 6
  form of, 5–7
  growth and develoment in, 7–9
  nutrition in (*see* Nutrition)
  origin of, 1
  purchasing, 34–35
  reproductive physiology of, 85–106
  skeletal structure of, 5–7
  (*see also* Meat; Rams; Sheep production)
Sheep Housing and Equipment Handbook, 37
Sheep Industry Development Program, 251
Sheepman's Production Handbook, 37
Sheep production:
  changes in, 9–15
  opportunities in, 17–22
  management systems, 20, 22–30
  marketing, 32–33, 245–65, 338–40
  meat (*see* Meat)
  political considerations, 53–61
  projected future of, 13
  requirements for, 31–51
  wool (*see* Wool production)
  (*see also* Sheep)
Sheep projects, 29
Sheep publications, 83
Shepherds, 208–9
Shrink, 335
Shropshire sheep, 74–75, 104, 211

Shrubs, 173
Side wool, 432
Silages, 174
Skeletal charts, lamb, 282
Skeletal structure, 5-7
Skin, 4, 355
Skin folds or wrinkles, 120, 406
Skirting, 351
Slaughter ewes and rams, 262-63
Slaughter lambs, 252-60
　(see also Meat)
Sodium, 146, 155, 177-78
Sodium carbonate, 371
Soil type, 415
Sore mouth virus, 213
Sorghum, 168
Sorting, 202, 206-7, 237
Soundness, 185-86, 191, 192
Sound wool, 331
Southdown sheep, 77-78, 104, 211
Soybean meal, 156, 159-64, 168, 175, 299
Specialized sheep production management systems, 28-30
Spermatic cord, 90, 91
Sperm cells, 90, 93, 95, 108, 109
Spinal processes, 5
Spinning count grade system, 325, 326
Spring lambs, 181, 252
Standard condition, 335
Staple, 328
Staple length, 327-30, 390-91, 397, 405, 432, 433
Starvation, 222-23
State:
　number of public land AUMs by, 56-57
　sheep production by, 12
Static three-breed cross, 136
Sterilization, 93
Stillbirth, 223
Storage of meat, 317-18
Stratification, 131, 379, 380
Subcutaneous fat, 289-92, 303
Sudoriferous glands, 359
Suffolk sheep, 71-73, 104, 110-12, 137, 138, 140, 173, 302, 303, 356, 357, 361
Sulfur, 155, 316, 413
Sunflower meal, 169
Support prices, 55, 58-59
Survival rate, 114
Sweet clover hay, 169

T

Tails, 3-6, 234-36
Tallow, 267, 282
Targhee sheep, 70-71, 137, 138, 140, 301-3, 357
Tariffs, 340-41
Taxes, 35
Teaser rams, 194
Teeth, 3, 8-9
Telephone auctions, 260
Temperature, fertility and, 99-100
Tender wool, 331
Terminal markets, 259
Territory wool, 329, 336
Testes, 89-91, 129, 191, 234-35
Testosterone, 90
Texas Delaine sheep, 63
Texas wool, 329, 336-37
Timothy, 169, 170
Tongue, 267, 282
Triplets, 229-30
Turnip roots, 169
Twins, 109, 112, 115-16, 123, 199, 201, 222-23, 229
Type grade system, 326-27

U

Udders, 88-89, 202, 207
Ultrasonic scanning, 203-4
Uniformity of wool, 330
United States Department of Agriculture (USDA), 55, 61, 269, 322
　grade set of, 430, 431
United States Department of the Interior, 54, 55, 60, 61
University of wool-judging programs, 434, 436
Urethra, 91, 93
Urial sheep, 1
Uterus, 86, 88

V

Vaccinations, 234, 236
Vacuum-packaged fresh lamb, 318-19
Vagina, 86, 88
Variability of wool fibers, 378
Vas deferens, 91, 93
Vegetable matter in wool, 334-35, 385, 432
Vertebrae, 5-6
Vesicular glands, 91, 93
Vitamins, 146-47, 149-54, 160, 178, 286, 413
Vulva, 86, 88

W

Wastiness, 334
Water, 145, 176-77
Weaning, 237-39
Weaning age, 7-8
Weaning rates, 20
Weaning time, 237-38
Weaning weight, 238
Weather, 21-22
Weathering, 371
Wensleydale sheep, 303
Wheat, 156, 157, 169, 174, 175, 241
Wheatgrass, 169-71
White clover, 312
Wild sheep, 1
Winterfat, 169
Wool:
　description of, 323-36
　fiber structure and properties, 363-73
　growth of, 354-62
　pricing, 336-37
　protein and, 145
　shearing, 344-50
Wool Act of 1954, 55
Wool blindness, 118-19
Wool breeds, 24
Wool Bureau, Inc., 322
Wool crops, 17, 18
Woolen system, 327-28
Wool Growers Auxiliaries, 420
Wool judging, 428-36
Wool Market News, 340
Wool measurement, 376-92
　choice of method, 376-77
　objective, 377-78, 398
　qualitative, 386-92
　quantitative, 378-86
　subjective, 377
Wool pools, 338-39
Wool production, 321-416
　age of sheep and, 100-101, 398, 400-402, 414
　breeding for improvements, 396-412
　environmental effects on, 412-16
　fleece traits, 115-17
　general observations about, 321-23
　government influences on, 340-44
　harvesting, 344-52
　language of, 323, 438-44
　marketing, 338-40
　measurement (*see* Wool measurement)

specialized, 28
wool judging, 428–36
wool shows, 419–30
(*see also* Sheep production; Wool)
Wool shows, 419–30

Worsted system, 327–28
Wrangling alleys, 50

Y

Yards, 44–45
Yield, 335–36, 378–84, 432, 434

Yield grades, 256–57, 271–75
Yolk, 336

Z

Zinc, 146, 155, 286